Board Review Series

CELL BIOLOGY
AND HISTOLOGY
4th edition

Board Review Series

CELL BIOLOGY AND HISTOLOGY

4th edition

Leslie P. Gartner, PhD
Associate Professor of Anatomy
Department of Oral and Craniofacial Biological Sciences
University of Maryland Dental School
Baltimore, Maryland

James L. Hiatt, PhD
Associate Professor of Anatomy (Retired)
Department of Oral and Craniofacial Biological Sciences
University of Maryland Dental School
Baltimore, Maryland

Judy M. Strum, PhD
Professor
Department of Anatomy and Neurobiology
University of Maryland School of Medicine
Baltimore, Maryland

LIPPINCOTT WILLIAMS & WILKINS
A **Wolters Kluwer** Company
Philadelphia • Baltimore • New York • London
Buenos Aires • Hong Kong • Sydney • Tokyo

Editor: Neil Marquardt
Managing Editor: Bridget Hilferty
Marketing Manager: Scott Lavine
Production Editor: Jennifer D. Weir
Compositor: LWW
Printer: DRC

Printed in the United States of America
First Edition, 1988
Second Edition, 1993
Third Edition, 1998

Library of Congress Cataloging-in-Publication Data
Gartner, Leslie P., 1943-
  Cell biology and histology / Leslie P. Gartner, James L. Hiatt, Judy M. Strum.--4th ed.
    p. ; cm. -- (Board review series)
  Includes index.
  ISBN 0-7817-3310-3
    1. Histology--Outlines, syllabi, etc. 2. Cytology--Outlines, syllabi, etc. I. Hiatt, James
L., 1934- II. Strum, Judy M. (Judy May) III. Title. IV. Series.
    [DNLM: 1. Histology--Outlines. 2. Cytology--Outlines. QS 18.2 G244c 2002]
  QM533 .G37 2002
  611'.018'076--dc21

Preface

We were very pleased with the reception of the third edition of this book, as well as with the many favorable comments we received from students who used it in preparation for the USMLE Step 1 or as an outline and study guide for their histology and/or cell biology courses in professional schools or undergraduate colleges.

Although all of the chapters have been revised and updated to incorporate current information, judicious pruning of material permitted us to curtail the size of the book. We have attempted to refine the content of the text to present Board-driven material as succinctly as possible, but still retain the emphasis on the relationship between cell structure and function through the vehicle of cell and molecular biology. The addition of new tables has allowed us to present detailed information in an easily accessible and student-friendly manner, thus reducing the amount of time a student will spend searching for information.

A tremendous amount of material has been compressed into a concise but highly comprehensive presentation using some new and revised illustrations. The relevancy of cell biology and histology to clinical practice is illustrated by the presence of clinical considerations at the end of each chapter.

The greatest change that occurred in the evolution of this book from its previous edition is the inclusion of a CD-ROM. The major impetus for the inclusion of the electronic medium was to present the student with color photomicrographs and with a number of additional USMLE Step 1-type questions (without an exorbitant increase in the cost). We believe that the included CD-ROM serves a very important purpose in preparing the student not only for the USMLE Step 1, but also for didactic and practical examinations in the histology and/or cell biology course.

As always, we welcome comments, suggestions, and constructive criticism of this book. These may be addressed to our editors at Lippincott Williams & Wilkins or directly to us by e-mail to lgartner@umaryland.edu.

Leslie P. Gartner, PhD
James L. Hiatt, PhD
Judy M. Strum, PhD

Preface

Acknowledgments

We would like to thank the following individuals for their help and support during the preparation of this book. Dr. Jennifer P. Bassiur, who, while a dental student, gave her time generously critiquing the previous edition and made valuable suggestions that improved the present one. Neil Marquardt, our acquisition editor; Katherine Bressler, for her editorial assistance; Rebecca Roberson, who managed the CD project; and especially Bridget Hilferty, our development editor, who helped us weave all the loose ends into a seamless whole.

Contents

1

Plasma Membrane

I. Overview—The Plasma Membrane (plasmalemma; cell membrane)

A. Structure. The plasma membrane is about 7.5 nanometers (nm) thick and consists of a **lipid bilayer** and associated **proteins.**

1. The **inner leaflet** of the plasma membrane faces the cytoplasm, and the **outer leaflet** faces the extracellular environment.

2. The plasma membrane displays a trilaminar **(unit membrane)** structure when examined by transmission electron microscopy (TEM).

B. Function

1. The plasma membrane envelops the cell and maintains its structural and functional integrity.

2. It acts as a **semipermeable** membrane between the cytoplasm and the external environment.

3. It permits the cell to recognize (and be recognized by) other cells and macromolecules.

4. It transduces extracellular signals into intracellular events.

II. Fluid Mosaic Model of the Plasma Membrane

A. The **lipid bilayer** (Figure 1.1) is freely permeable to small, lipid-soluble, nonpolar molecules but is impermeable to charged ions.

1. **Molecular structure.** The lipid bilayer is composed of phospholipids, glycolipids, and cholesterol.

   a. Phospholipids are **amphipathic,** consisting of one polar **(hydrophilic)** head and two nonpolar **(hydrophobic)** fatty acyl tails.

   b. The two leaflets are not identical to one another; instead the distribution of the various types of phospholipids is asymmetrical.

   (1) The **polar head** of each molecule faces the membrane surface, whereas the **tails** project into the interior of the membrane.

   (2) The **tails** of the two leaflets face each other, forming weak bonds that attach the two leaflets to each other.

   c. Glycolipids are restricted to the extracellular aspect of the outer leaflet. **Polar carbohydrate residues** of glycolipids extend from the outer leaflet into the extracellular space and form part of the **glycocalyx.**

   d. Cholesterol constitutes 2% of plasmalemma lipids, is present in both leaflets, and helps maintain the structural integrity of the membrane.

1

e. Cholesterol as well as phospholipids can form microdomains, known as rafts, that can affect the movement of integral proteins of the plasmalemma.

2. **Fluidity** of the lipid bilayer is crucial to **exocytosis, endocytosis, membrane trafficking,** and **membrane biogenesis.**

  a. Fluidity **increases** with increased temperature and with decreased saturation of the fatty acyl tails.

  b. Fluidity **decreases** with an increase in the membrane's cholesterol content.

B. **Membrane proteins** (see Figure 1.1) include integral proteins and peripheral proteins. They constitute approximately 50% of the plasma membrane composition.

1. **Integral proteins** are dissolved in the lipid bilayer.

  a. **Transmembrane proteins** span the entire plasma membrane and function as membrane **receptors** and **transport proteins.**

    (1) Most transmembrane proteins are **glycoproteins.**

    (2) Transmembrane proteins are **amphipathic** and contain **hydrophilic** and **hydrophobic** amino acids, some of which interact with the hydrocarbon tails of the membrane phospholipids.

    (3) Most transmembrane proteins are folded so that they pass back and forth across the plasmalemma; therefore, they are also known as **multipass proteins.**

  b. Integral proteins may also be anchored to the inner (or occasionally outer) leaflet via fatty acyl or prenyl groups.

  c. In freeze-fracture preparations, integral proteins remain preferentially attached to the **P-face**, the external surface of the inner leaflet, rather than the **E-face** (Figure 1.2).

2. **Peripheral proteins** do not extend into the lipid bilayer.

  a. These proteins are located on the cytoplasmic aspect of the inner leaflet.

  b. The outer leaflets of some cells possess covalently linked glycolipids to which peripheral proteins are anchored; these peripheral proteins thus project into the **extracellular space.**

  c. Peripheral proteins bond to the phospholipid polar groups or integral proteins of the membrane via noncovalent interactions.

  d. They usually function as part of the **cytoskeleton** or as part of an **intracellular second messenger system.**

  e. They include a group of anionic, calcium-dependent, lipid-binding proteins known as **annexins,** which act to modify the relationships of other peripheral proteins with the lipid bilayer.

3. **Functional characteristics of membrane proteins**

  a. **The lipid-to-protein ratio** (by weight) in plasma membranes ranges from 1:1 in most cells to 4:1 in myelin.

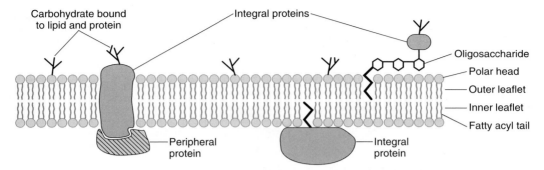

Figure 1.1. Diagrammatic representation of the plasma membrane showing the outer *(top)* and inner *(bottom)* leaflets of the unit membrane. The hydrophobic fatty acyl tails and the polar heads of the phospholipids constitute the lipid bilayer. Integral proteins are embedded in the lipid bilayer. Peripheral proteins are located primarily on the cytoplasmic aspect of the inner leaflet and are attached by noncovalent interactions to integral proteins.

    b. Some membrane proteins **diffuse laterally** in the lipid bilayer; others are **immobile** and are held in place by cytoskeletal components.

C. Glycocalyx (cell coat) is the **sugar coat** located on the outer surface of the outer leaflet of the plasmalemma. When examined by TEM, it varies in appearance (fuzziness) and thickness (up to 50 nm).

    1. Composition. The glycocalyx consists of polar oligosaccharide side chains linked covalently to most proteins and some lipids (glycolipids) of the plasmalemma. It also contains **proteoglycans (glycosaminoglycans** bound to integral proteins).

    2. Function

        a. The glycocalyx aids in **attachment of** some cells (e.g., fibroblasts but not epithelial cells) to extracellular matrix components.

        b. It **binds** antigens and enzymes to the cell surface.

        c. It facilitates **cell-cell recognition** and **interaction.**

        d. It **protects cells** from injury by preventing contact with inappropriate substances.

        e. It assists T cells and antigen-presenting cells in **aligning** with each other in the proper fashion and aids in preventing inappropriate enzymatic cleavage of receptors and ligands.

III. Plasma Membrane Transport Processes. These processes include transport of a single molecule **(uniport)** or cotransport of two different molecules in the same **(symport)** or opposite **(antiport)** direction.

    A. Passive transport (Figure 1.3) includes **simple** and **facilitated diffusion.** Neither of these processes requires energy because molecules move across the plasma membrane **down a concentration or electrochemical gradient.**

    1. Simple diffusion transports small nonpolar molecules (e.g., O_2 and N_2) and small, uncharged, polar molecules (e.g., H_2O, CO_2, and glycerol). It exhibits little specificity, and the diffusion rate is proportional to the concentration gradient of the diffusing molecule.

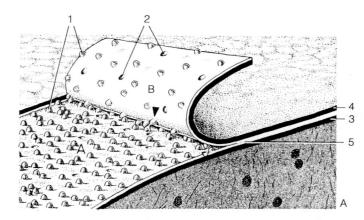

Figure 1.2. Freeze-fracturing cleaves the plasma membrane *(5)*. The impressions *(2)* of the transmembrane proteins are evident on the E-face, between the inner *(3)* and outer leaflets *(4)*. The integral proteins *(1)* remain preferentially attached to the P-face *(A)*, the external surface of the inner leaflet; fewer proteins remain associated with the E-face *(B)*, the internal surface of the outer leaflet. The arrowhead indicates a transmembrane protein attached to both E-face and P-face. (Reprinted with permission from Krstic RV: *Ultrastruktur der Saugertierzelle.* Berlin, Springer Verlag, 1976, p 177.)

2. **Facilitated diffusion** occurs via **ion channel** and/or **carrier proteins,** structures that exhibit **specificity** for the transported molecules. It is faster than simple diffusion; ions and large polar molecules are thus capable of traversing membranes that would otherwise be impermeable to them.

    a. **Ion channel proteins** are multipass transmembrane proteins that form small aqueous pores across membranes through which specific small water-soluble molecules and ions pass down an electrochemical gradient **(passive transport).**

    b. **Carrier proteins** are multipass transmembrane proteins that undergo reversible conformational changes to transport specific molecules across the membrane; these proteins function in both passive transport and **active transport.**

B. **Active transport** is an **energy-requiring process** that transports a molecule **against** an electrochemical gradient via carrier proteins.

    1. **Na^+-K^+ pump**

        a. **Mechanism.** The Na^+-K^+ pump involves the **antiport** transport of Na^+ and K^+ ions mediated by the carrier protein, **Na^+-K^+ adenosine triphosphatase (ATPase).**

            (1) Three Na^+ ions are pumped out of the cell and two K^+ ions are pumped **into** the cell.

            (2) The hydrolysis of a single ATP molecule by the Na^+-K^+ ATPase is required to transport five ions.

        b. **Function**

            (1) The primary function is to **maintain constant cell volume** by decreasing the intracellular ion concentration (and thus the osmotic pressure) and increasing the extracellular ion concentration, thus decreasing the flow of water into the cell.

            (2) The Na^+-K^+ pump also plays a minor role in the maintenance of a **potential difference** across the plasma membrane.

    2. **Glucose transport** involves the **symport** movement of glucose across an epithelium **(transepithelial transport).** Transport is frequently powered by an electrochemical Na^+ gradient, which drives carrier proteins located at specific regions of the cell surface.

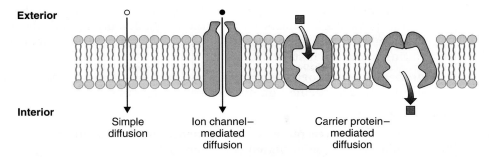

Figure 1.3. Passive transport of molecules across plasma membranes may occur by simple diffusion **(left)** and by either of the two types of facilitated diffusion mediated by ion channel proteins **(center)** and carrier proteins **(right)**.

C. Facilitated diffusion of ions can occur via ion channel proteins or ionophores.

   1. Selective ion channel proteins permit only certain ions to traverse them.

   a. K+ leak channels are the most common ion channels. These channels are ungated and leak K+, the ions most responsible for establishing a potential difference across the plasmalemma.

   b. Gated ion channels open only transiently in response to various stimuli. They include the following types:

   (1) Voltage-gated channels open when the potential difference across the membrane changes (e.g., voltage-gated **Na+** channels, which function in the generation of action potentials; see Chapter 9 VIII B 1 e).

   (2) Mechanically gated channels open in response to a mechanical stimulus (e.g., the tactile response of the hair cells in the inner ear).

   (3) Ligand-gated channels open in response to the binding of a **signaling molecule** or **ion**. These channels include neurotransmitter-gated channels, nucleotide-gated channels, and G-protein-gated K+ channels of cardiac muscle cells.

   2. Ionophores are molecules that form a complex with ions and insert into the lipid bilayer to transport those ions across the membrane.

IV. Cell-to-Cell Communication

   A. Signaling molecules are secreted by cells and function in cell-to-cell communication. Examples include neurotransmitters, which are released into the synaptic cleft (see Chapter 8 IV A 1 b; Chapter 9 IV B 5); endocrine hormones, which are carried in the bloodstream and act on distant target cells; and hormones released into the intercellular space, which act on nearby cells **(paracrine hormones)** or on the releasing cell **(autocrine hormones).**

   1. Lipid-soluble signaling molecules penetrate the plasma membrane and bind to **receptors in the cytoplasm** or **nucleus,** activating intracellular messengers. Examples include hormones that influence gene transcription.

   2. Hydrophilic signaling molecules bind to and activate **cell-surface**

receptors (as do some lipid-soluble signaling molecules) and have diverse physiologic effects (see Chapter 13). Examples include neurotransmitters and numerous hormones (e.g., serotonin, thyroid-stimulating hormone, insulin).

B. Membrane receptors are primarily glycoproteins. They are located on the cell surface, and specific signaling molecules bind to them.

1. **Function**

   a. Membrane receptors **control plasmalemma permeability** by regulating the conformation of ion channel proteins.

   b. They **regulate the entry of molecules** into the cell (e.g., the delivery of cholesterol via low-density lipoprotein receptors).

   c. They **bind extracellular matrix molecules** to the cytoskeleton via **integrins,** which are essential for cell-cell contact and cell-matrix interactions.

   d. They **act as transducers** to transfer extracellular events intracellularly to second messenger systems.

   e. They permit pathogens that mimic normal ligands to enter cells.

2. **Types of membrane receptors**

   a. **Channel-linked receptors** bind a signaling molecule that temporarily opens or closes the gate, permitting or inhibiting the movement of ions across the cell membrane. Examples include **nicotinic acetylcholine receptors** on the muscle-cell sarcolemma at the myoneural junction (see Chapter 8 IV A).

   b. **Catalytic receptors** are single-pass transmembrane proteins.

      (1) Their extracellular moiety is a receptor, and their **cytoplasmic aspect** is a protein kinase.

      (2) Some catalytic receptors lack an extracytoplasmic moiety and as a result are continuously activated; such defective receptors are coded for by some **oncogenes.**

      (3) Examples of catalytic receptors include:

         (a) **Insulin,** which binds to its receptor, which **autophosphorylates.** The cell then takes up the insulin-receptor complex by **endocytosis,** enabling the complex to function within the cell.

         (b) **Growth factors** (e.g., epidermal growth factor, platelet-derived growth factor), which bind to some catalytic receptors and induce mitosis.

   c. **G-protein-linked receptors are** transmembrane proteins **associated with an ion channel or with an enzyme** bound to the cytoplasmic surface of the cell membrane.

      (1) These receptors interact with **heterotrimeric G-protein** [guanosine triphosphate (GTP)-binding regulatory protein] after binding of a signaling molecule. This interaction results in the activation of **intracellular second messengers,** the most common of which are cyclic adenosine monophosphate (cAMP) and **Ca^{2+}.**

(2) Examples include:

(a) **Heterotrimeric G proteins** (Table 1.1), which include stimulatory G protein (**G_s**) [Figure 1.4], inhibitory G protein (**G_i**), phospholipase C activator G protein (**G_p**), olfactory-specific G protein (**G_{olf}**), and transducin (**G_t**)

(b) **Monomeric G proteins (low-molecular-weight G proteins),** which are small single-chain proteins that also function in signal transduction

(i) Various subtypes resemble Ras, Rho, Rab, and ARF proteins.

(ii) These proteins are involved in pathways that regulate cell proliferation and differentiation, protein synthesis, attachment of cells to the extracellular matrix, exocytosis, and vesicular traffic.

V. **Plasmalemma-Cytoskeleton Association.** The plasmalemma and cytoskeleton associate through **integrins.** The extracellular domain of integrins binds to extracellular matrix components, and the intracellular domain binds to cytoskeletal components. Integrins stabilize the plasmalemma and determine and maintain cell shape.

Table 1.1. Functions and Examples of Heterotrimeric G Proteins

Type	Function	Result	Examples
G_s	Activates adenylate cyclase leading to formation of cAMP	Activation of protein kinases	Binding of epinephrine to β-adrenergic receptors increases cAMP levels in cytosol
G_i	Inhibits adenylate cyclase preventing formation of cAMP	Protein kinases remain inactive	Binding of epinephrine to α_2-adrenergic receptors decreases cAMP levels in cytosol
G_p	Activates phospholipase C leading to formation of inositol triphosphate and diacylglycerol	Influx of Ca^{2+} into cytosol and activation of protein kinase C	Binding of antigen to membrane-bound IgE causes the release of histamine by mast cells
G_{olf}	Activates adenylate cyclase in olfactory neurons	Opens cAMP-gated Na^+ ion channels	Binding of odorant to G-protein-linked receptors initiates generation of nerve impulse
G_t	Activates cGMP phosphodiesterase in rod cell membranes leading to hydrolysis of cGMP	Hyperpolarization of rod cell membrane	Photon activation of rhodopsin causes rod cells to fire

cAMP = cyclic adenosine monophosphate; cGMP = cyclic guanosine monophosphate; IgE = immunoglobulin E.

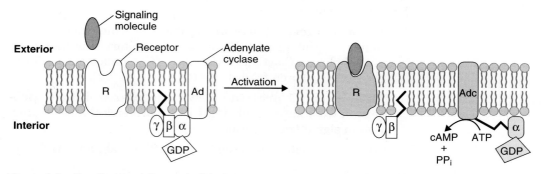

Figure 1.4. Functioning of G_s-protein-linked receptors. The signaling molecule binds to the receptor, which causes the alpha subunit of the G_s protein to bind guanosine triphosphate (GTP) as well as dissociate from the beta and gamma subunits. Activation of adenylate cyclase by the GTP-α-subunit complex stimulates synthesis of cyclic adenosine monophosphate (cAMP), one of the most common intracellular messengers.

A. Red blood cells (Figure 1.5A) have integrins, called **band 3 proteins,** which are located in the plasmalemma. The cytoskeleton of a red blood cell consists mainly of spectrin, actin, band 4.1 protein, and ankyrin.

    1. Spectrin is a long, flexible protein (about 110 nm long), composed of an **α-chain** and a **β-chain,** that forms **tetramers** and provides a scaffold for structural reinforcement.

    2. Actin attaches to binding sites on the spectrin tetramers and holds them together, thus aiding in the formation of a spectrin latticework.

    3. Band 4.1 protein binds to and stabilizes spectrin-actin complexes.

    4. Ankyrin is linked to both band 3 proteins and spectrin tetramers, thus attaching the spectrin-actin complex to transmembrane proteins.

B. The cytoskeleton of **nonerythroid cells** (Figure 1.5B) consists of the following major components:

    1. Actin (and perhaps **fodrin**), which serves as a nonerythroid spectrin

    2. α-Actinin, which cross-links actin filaments to form a meshwork

    3. Vinculin, which binds to α-actinin and to another protein, called **talin,** which in turn attaches to the integrin in the plasma membrane

VI. Clinical Considerations

A. Cystinuria is a hereditary condition caused by abnormal carrier proteins that are unable to remove cystine from the urine, resulting in the formation of kidney stones.

B. Cholera toxin is an exotoxin produced by the bacterium *Vibrio cholerae.*

    1. The cholera toxin **alters G_s** protein so that it is unable to hydrolyze its GTP molecule.

    2. As a result, cAMP levels increase in the surface absorptive cells of the intestine, leading to excessive loss of electrolytes and water and severe diarrhea.

C. Venoms, such as those of some poisonous snakes, **inactivate acetylcholine receptors** of skeletal muscle sarcolemma at neuromuscular junctions.

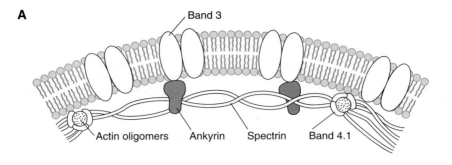

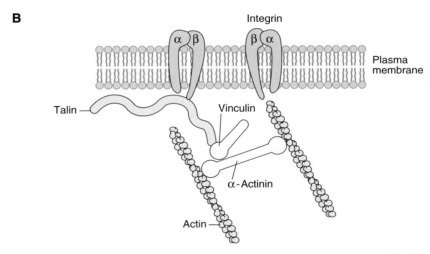

Figure 1.5. Plasmalemma-cytoskeleton association in red blood cells *(A)* and nonerythroid cells *(B)*. (Adapted with permission from Widnell CC, Pfenninger KH: *Essential Cell Biology.* Baltimore, Williams & Wilkins, 1990, p 82.)

D. Autoimmune diseases may lead to the production of antibodies that specifically **bind to and activate certain plasma membrane receptors.** An example is **Graves disease** (hyperthyroidism) [see Chapter 13 VIII B].

E. Genetic defects

1. **Defective G_s proteins** may lead to mental retardation, diminished growth and sexual development, and decreased responses to certain hormones.

2. **Hereditary spherocytosis** results from a **defective spectrin** that has a decreased ability to bind to band 4.1 protein. The disease is characterized by fragile, misshapen red blood cells, or spherocytes; destruction of these **spherocytes** in the spleen leads to anemia.

Review Test

Directions: Each of the numbered items or incomplete statements in this section is followed by answers or by completions of the statement. Select the ONE lettered answer or completion that is BEST in each case.

1. A herpetologist is bitten by a poisonous snake and is brought into the emergency room with progressive muscle paralysis. The venom is probably incapacitating his

(A) Na^+ channels
(B) Ca^{2+} channels
(C) phospholipids
(D) acetylcholine receptors
(E) spectrin

2. Cholesterol functions in the plasmalemma to

(A) increase fluidity of the lipid bilayer
(B) decrease fluidity of the lipid bilayer
(C) facilitate the diffusion of ions through the lipid bilayer
(D) assist in the transport of hormones across the lipid bilayer
(E) bind extracellular matrix molecules

3. The cell membrane consists of various components, including integral proteins. These integral proteins

(A) are not attached to the outer leaflet
(B) are not attached to the inner leaflet
(C) include transmembrane proteins
(D) are preferentially attached to the E-face
(E) function in the transport of cholesterol-based hormones

4. Which one of the following transport processes requires energy?

(A) Facilitated diffusion
(B) Passive transport
(C) Active transport
(D) Simple diffusion

5. Which one of the following substances is unable to traverse the plasma membrane by simple diffusion?

(A) O_2
(B) N_2
(C) Na^+
(D) Glycerol
(E) CO_2

6. Symport refers to the process of transporting

(A) a molecule into the cell
(B) a molecule out of the cell

(C) two different molecules in the opposite direction
(D) two different molecules in the same direction
(E) a molecule between the cytoplasm and the nucleus

7. One of the ways that cells communicate with each other is by secretion of various molecules. The secreted molecule is known as

(A) a receptor molecule
(B) a signaling molecule
(C) a spectrin tetramer
(D) an integrin
(E) an anticodon

8. Adrenocorticotropic hormone (ACTH) travels through the bloodstream, enters connective tissue spaces, and attaches to specific sites on target-cell membranes. These sites are

(A) peripheral proteins
(B) signaling molecules
(C) G proteins
(D) G-protein-linked receptors
(E) ribophorins

9. Examination of the blood smear of a young patient reveals misshapen red blood cells, and the pathology report indicates hereditary spherocytosis. Defects in which one of the following proteins causes this condition?

(A) Signaling molecules
(B) G proteins
(C) Spectrin
(D) Hemoglobin
(E) Ankyrin

10. Which of the following statements concerning plasma membrane components is true?

(A) All G proteins are composed of three subunits.
(B) The glycocalyx is usually composed of phospholipids.
(C) Ion channel proteins are energy-dependent (require adenosine triphosphate).
(D) Gated channels are always open.
(E) Ankyrin binds to band 3 of the red blood cell plasma membrane.

Answers and Explanations

1-D. Snake venom usually blocks acetylcholine receptors, thus preventing depolarization of the muscle cell. The Na^+ and Ca^{2+} channels are not incapacitated by snake venoms.

2-B. The fluidity of the lipid bilayer is decreased in three ways: (1) by lowering the temperature, (2) by increasing the saturation of the fatty acyl tails of the phospholipid molecules, and (3) by increasing the membrane's cholesterol content.

3-C. Integral proteins are not only closely associated with the lipid bilayer but also tightly bound to the cell membrane. These proteins frequently span the entire thickness of the plasmalemma, and are thus termed transmembrane proteins.

4-C. Active transport requires energy. Facilitated diffusion, which is mediated by membrane proteins, and simple diffusion, which involves passage of material directly across the lipid bilayer, are types of passive transport.

5-C. Na^+ and other ions require channel (carrier) proteins for their transport across the plasma membrane. The other substances are small, nonpolar molecules and small, uncharged polar molecules. The molecules can traverse the plasma membrane by simple diffusion.

6-D. The coupled transport of two different molecules in the same direction is termed symport.

7-B. Cells can communicate with each other by releasing signaling molecules, which attach to receptor molecules on target cells.

8-D. G-protein-linked receptors are sites where adrenocorticotropic hormone (ACTH) and some other signaling molecules attach. Binding of ACTH to its receptor causes G_s protein to activate adenylate cyclase, setting in motion the specific response elicited by the hormone.

9-C. Hereditary spherocytosis is caused by a defect in spectrin that renders the protein incapable of binding to band 4.1 protein, thus destabilizing the spectrin-actin complex of the cytoskeleton. Although defects in hemoglobin (the respiratory protein of erythrocytes) also cause red blood cell anomalies, hereditary spherocytosis is not one of them.

10-E. Ankyrin is linked both to band 3 proteins and to spectrin tetramer, thus attaching the spectrin-actin complex to transmembrane proteins of the erythrocyte. There are two types of G proteins, trimeric and monomeric; glycocalyx (the sugar coat on the membrane surface) is composed mostly of polar carbohydrate residues; only carrier proteins can be energy requiring; gated channels are open only transiently.

2

Nucleus

I. Overview—The Nucleus (Figure 2.1)

A. Structure. The nucleus, the largest organelle of the cell, includes the **nuclear envelope, nucleolus, nucleoplasm,** and **chromatin** and contains the genetic material encoded in the **deoxyribonucleic acid** (DNA) of chromosomes.

B. Function. The nucleus directs protein synthesis in the cytoplasm via **ribosomal ribonucleic acid** (rRNA), **messenger RNA** (mRNA), and **transfer RNA** (tRNA). All forms of RNA are synthesized in the nucleus.

II. Nuclear Envelope (Figure 2.2). The nuclear envelope surrounds the nuclear material and consists of **two parallel membranes** separated from each other by a narrow **perinuclear cisterna.** These membranes fuse at intervals, forming openings in the nuclear envelope called **nuclear pores.**

A. Outer nuclear membrane

1. This membrane is about 6 nanometers (nm) thick.

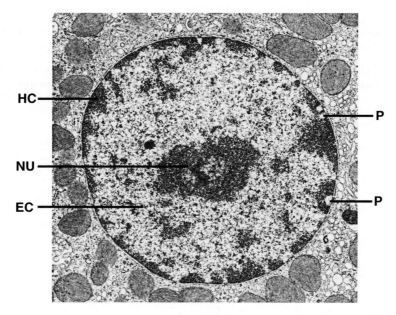

Figure 2.1. Electron micrograph of the cell nucleus. Note that the nuclear envelope is interrupted by nuclear pores *(P)*. The inactive heterochromatin *(HC)* is dense and mostly confined to the periphery of the nucleus. Euchromatin *(EC),* the active form, is less dense and is dispersed throughout. The nucleolus *(NU)* contains fibrillar and granular portions.

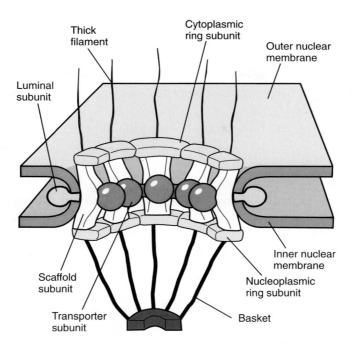

Figure 2.2. Diagram of the nuclear pore complex. (Modified with permission from Alberts B, Bray D, Lewis J, et al: *Molecular Biology of the Cell,* 3rd ed. New York, Garland Publishing, 1994.

2. It faces the cytoplasm and is continuous at certain sites with the rough endoplasmic reticulum (RER).

3. A loosely arranged mesh of intermediate filaments **(vimentin)** surrounds the outer nuclear membrane on its cytoplasmic aspect.

4. **Ribosomes** stud the cytoplasmic surface of the outer nuclear membrane. These ribosomes synthesize proteins that enter the perinuclear cisterna.

B. Inner nuclear membrane

1. The inner nuclear membrane is about 6 nm thick.

2. It faces the nuclear material but is separated from it and supported on its inner surface by the **nuclear lamina,** a fibrous lamina that is 80-300 nm thick and composed primarily of **lamins A, B, and C.** These intermediate filament proteins help organize the nuclear envelope and perinuclear chromatin. Additionally, they are essential during the mitotic events, when they are responsible for the disassembly and reassembly of the nuclear envelope. Phosphorylation of lamins leads to disassembly, and dephosphorylation results in reassembly of the nuclear envelope.

C. Perinuclear cisterna

1. The perinuclear cisterna is located between the inner and outer nuclear membranes and is 20-40 nm wide.

2. It is continuous with the cisterna of the RER.

3. It is perforated by nuclear pores at various locations.

D. Nuclear pores

1. Nuclear pores average 80 nm in diameter and number from dozens to thousands depending upon metabolic activity; they are associated with the nuclear pore complex (NPC).

2. They are formed by fusion of the inner and outer nuclear membranes.

3. They permit passage of certain molecules in either direction between the nucleus and cytoplasm via a 9-nm channel opening.

E. The **NPC** represents protein subunits surrounding the nuclear pore (see Figure 2.2).

1. **Structure.** The NPC is composed of nearly 100 proteins, some of which are arranged in eight-fold symmetry around the margin of the pore. The nucleoplasmic side of the pore exhibits a nuclear basket, whereas the cytoplasmic side displays fibers extending into the cytoplasm. A transporter protein is located in the central core and is believed to be responsible for transporting proteins into and out of the nucleus via **receptor-mediated transport.**

   a. The **cytoplasmic ring** is located around the cytoplasmic margin of the nuclear pore and is composed of eight subunits, each possessing a filamentous fiber extending into the cytoplasm. These fibers may serve as a staging area prior to protein transport.

   b. The **nucleoplasmic ring** is located around the nucleoplasmic margin of the nuclear pore and is composed of eight subunits. Extending from this ring into the nucleoplasm is a basket-like structure, the **nuclear basket.** It is thought to have a function in RNA transport.

   c. The **middle ring** is interposed between the cytoplasmic and nucleoplasmic rings. Eight transmembrane proteins project into the lumen of the nuclear pore, anchoring the complex into the pore rim.

2. **Function.** The NPC permits passive movement across the nuclear envelope via a 9- to 11-nm open channel for simple diffusion. Most proteins, regardless of size, pass in either direction only by **receptor-mediated transport**. These proteins have clusters of certain amino acids known as **nuclear localization segments** (NLS) that act as signals for transport.

3. **Transport mechanisms** involve a group of proteins, **exportins** and **importins.** These are regulated by **Ran,** a group of guanosine triphosphate-binding proteins.

III. Nucleolus

A. Structure.
The nucleolus is a nuclear inclusion that is not surrounded by a membrane. It is present in cells that are actively synthesizing proteins; more than one nucleolus can be present in the nucleus. It is generally detectable only when the cell is in **interphase.** It contains mostly rRNA and protein as well as a modest amount of DNA. It possesses **nucleolar organizer regions (NORs),** portions of those chromosomes (in humans, chromosomes 13, 14, 15, 21, and 22) where rRNA genes are located; these regions are involved in reconstituting the nucleolus during the G_1 phase of the cell cycle. The nucleolus contains four distinct regions.

1. **Fibrillar centers** are composed of **inactive DNA** where DNA is not being transcribed. **NORs** are also located here.

2. **Pars fibrosa** are composed of 5-nm fibrils surrounding the fibrillar centers and contain **transcriptionally active DNA** and the rRNA precursors that are being transcribed.

3. **Pars granulosa** are composed of 15-nm **maturing ribosomal precursor** particles.

4. **Nucleolar matrix** is a fiber network participating in the organization of the nucleolus.

B. **Function.** The nucleolus is involved in the synthesis of **rRNA** and its assembly into ribosome precursors. It has been determined that the nucleolus also sequesters certain nucleolar proteins that function as cell-cycle checkpoint signaling proteins. Three such cell-cycle regulator proteins have been identified within the nucleolus, where they remain sequestered until their release is required for targets in the nucleus and/or cytoplasm.

IV. Nucleoplasm. Nucleoplasm is the protoplasm within the nuclear envelope. It consists of a matrix and various types of particles.

A. **Nuclear matrix** acts as a scaffold that aids in organizing the nucleoplasm.

1. **Structural components** include fibrillar elements, nuclear pore-nuclear lamina complex, residual nucleoli, and a residual ribonucleoprotein (RNP) network.

2. **Functional components** are involved in the transcription and processing of mRNA and rRNA, steroid receptor-binding sites, carcinogen-binding sites, heat-shock proteins, DNA viruses, and viral proteins (T antigen) and perhaps many other functions that are as yet not known.

B. **Nuclear particles**

1. **Interchromatin granules** are clusters of irregularly distributed particles (20-25 nm in diameter) that contain RNP and various enzymes.

2. **Perichromatin granules** (see Figure 2.1) are single dense granules (30-50 nm in diameter) surrounded by a less dense **halo.** They are located at the periphery of heterochromatin and exhibit a substructure of 3-nm packed fibrils.

   a. Perichromatin granules contain **4.7S RNA** and two peptides similar to those found in heterogeneous nuclear RNPs (hnRNPs).

   b. They may represent **messenger RNPs (mRNPs).**

   c. The number of granules increases in liver cells exposed to carcinogens or temperatures above 37°C.

3. The **hnRNP particles** are complexes of **precursor mRNA (pre-mRNA)** and proteins and are involved in processing of pre-mRNA.

4. **Small nuclear RNPs (snRNPs)** are complexes of proteins and **small RNAs** and are involved in hnRNP splicing or in cleavage reactions.

V. Chromatin (see Figure 2.1)

A. **Structure.** Chromatin consists of double-stranded DNA complexed with **histones** and **acidic proteins.** It resides within the nucleus as heterochromatin and euchromatin. The euchromatin-heterochromatin ratio is higher in malignant cells than in normal cells.

1. **Heterochromatin,** condensed inactive chromatin, is concentrated at the periphery of the nucleus and around the nucleolus, as well as scattered throughout the nucleoplasm.

   a. When examined under the light microscope (LM), it appears as basophilic clumps of nucleoprotein.

   b. Although it is **transcriptionally inactive,** recent evidence indicates that heterochromatin plays a role in interchromosomal interactions and chromosomal segregation during meiosis.

   c. Heterochromatin corresponds to **one of two X chromosomes** and is therefore present in nearly all somatic cells of female mammals. During interphase, the inactive X chromosome is visible as a dark-staining body within the nucleus. This structure is called the **Barr body,** or **sex chromatin.**

2. **Euchromatin** is the **transcriptionally active** form of chromatin that appears in the LM as a lightly stained region of the nucleus. It appears in transmission electron microscope (TEM) as electron-lucent regions among heterochromatin, and is composed of 30-nm strings of nucleosomes (see VI) and the DNA double helix.

B. **Function.** Chromatin is responsible for **RNA synthesis.**

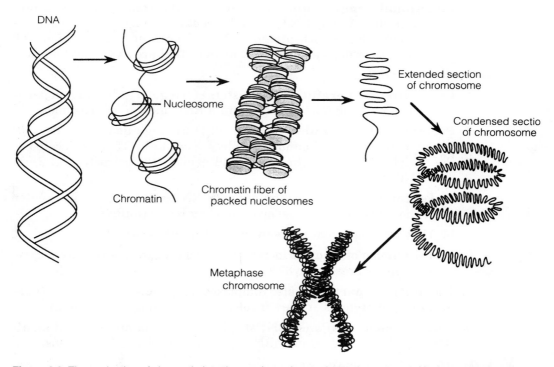

Figure 2.3. The packaging of chromatin into the condensed metaphase chromosome. Nucleosomes contain two copies of histones H2A, H2B, H3, and H4 in extended chromatin. An additional histone, H1, is present in condensed chromatin. DNA = deoxyribonucleic acid. (Adapted with permission from Widnell CC, Pfenninger KH: *Essential Cell Biology.* Baltimore, Williams & Wilkins, 1990, p 47.)

VI. Chromosomes

A. Structure. Chromosomes consist of chromatin extensively folded into loops; this conformation is maintained by DNA-binding proteins (Figure 2.3). Each chromosome contains a single DNA molecule and associated proteins, assembled into **nucleosomes,** the structural unit of chromatin packaging. Chromosomes are visible with the LM only during mitosis and meiosis when their chromatin condenses.

1. **Extended chromatin** forms the nucleosome core, around which the DNA double helix is wrapped two full turns.

    a. The nucleosome core consists of two copies each of **histones H2A, H2B, H3,** and **H4.** Nucleosomes are spaced at intervals of 200 base pairs.

    b. When viewed with TEM, extended chromatin resembles "beads on a string;" the beads represent **nucleosomes** and the string represents **linker DNA.**

2. **Condensed chromatin** contains an additional histone, **H1,** which wraps around groups of nucleosomes forming 30-nm-diameter fibers, the structural unit of the chromosome.

B. G-banding is observed in chromosomes during mitosis after staining with Giemsa, which is specific for DNA sequences rich in **adenine** (A) and **thymine** (T). Banding is thought to represent highly folded DNA loops. G-banding is characteristic for each species and is used to identify chromosomal anomalies.

C. Karyotype refers to the **number and morphology of chromosomes** and is characteristic for each species.

1. **Haploid number** (n) is the number of chromosomes in germ cells (23 in humans).

2. **Diploid number** (2n) is the number of chromosomes in somatic cells (46 in humans).

D. Genome is the **total genetic complement** of an individual, and is stored in its chromosomes. In humans, the genome consists of 22 pairs of **autosomes** and 1 pair of **sex chromosomes** (either **XX** or **XY**), totaling 46 chromosomes.

VII. DNA.
DNA is a long, double-stranded, linear molecule composed of multiple nucleotide sequences. It acts as a **template for the synthesis of RNA.**

A. Nucleotides are composed of a base (purine or pyrimidine), a deoxyribose sugar, and a phosphate group.

1. The **purines** are **adenine** (A) and **guanine** (G).

2. The **pyrimidines** are **cytosine** (C) and **thymine** (T).

B. The **DNA double helix** consists of **two complementary DNA strands** held together by hydrogen bonds between the base pairs A-T and G-C.

C. Exons are regions of the DNA molecule that **code** for specific RNAs.

D. Introns are regions of the DNA molecule, between exons, that **do not code** for RNAs.

    E. A **codon** is a sequence of **three bases** in the DNA molecule that codes for a **single amino acid.**

    F. A **gene** is a segment of the DNA molecule that is responsible for the formation of a single RNA molecule.

VIII. RNA. RNA is a linear molecule similar to DNA, however, it is single stranded and contains **ribose** instead of **deoxyribose sugar** and **uracil** (U) instead of **thymine** (T). RNA is synthesized by **transcription** of DNA. Transcription is catalyzed by three **RNA polymerases:** I for rRNA, II for mRNA, and III for tRNA.

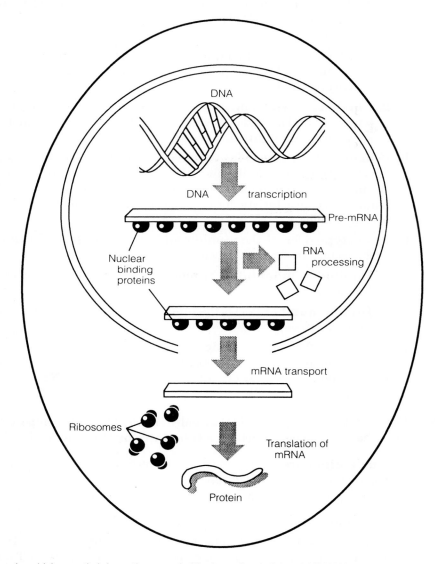

Figure 2.4. Steps by which genetic information encoded in deoxyribonucleic acid (DNA) is transcribed into messenger ribonucleic acid (mRNA) and ultimately converted into proteins in the cytoplasm. (Adapted with permission from Alberts B, Bray D, Lewis J, et al: *Molecular Biology of the Cell*, 2nd ed. New York, Garland Publishing, 1989, p 482.)

A. mRNA carries the genetic code to the cytoplasm to direct **protein synthesis** (Figure 2.4).

    1. This single-stranded molecule consists of hundreds to thousands of nucleotides.

    2. mRNA contains codons that are **complementary** to the DNA codons from which it was transcribed, including one **start codon (AUG)** for **initiating** protein synthesis and one of three **stop codons (UAA, UAG,** or **UGA)** for **terminating** protein synthesis.

    3. mRNA is synthesized in the following series of steps.

        a. **RNA polymerase II** recognizes a **promoter** on a single strand of the DNA molecule and binds tightly to it.

        b. The DNA helix unwinds about two turns, separating the DNA strands and exposing the **codons** that act as the template for synthesis of the complementary RNA molecule.

        c. RNA polymerase II moves along the DNA strand and promotes base pairing between DNA and complementary RNA nucleotides.

        d. When the RNA polymerase II recognizes a **chain terminator** (stop codons—**UAA, UAG,** or **UGA**) on the DNA molecule, it terminates its association with the DNA and is released to repeat the process of transcription.

        e. The primary transcript, **pre-mRNA** after the introns are removed, associates with proteins to form **hnRNP.**

        f. Exons are spliced through several steps, involving **spliceosomes** producing an **mRNP.**

        g. Proteins are removed as the mRNP enters the cytoplasm, resulting in a **functional mRNA.**

B. tRNA is folded into a cloverleaf shape and contains about 80 nucleotides, terminating in adenylic acid (where amino acids attach).

    1. Each tRNA combines with a specific amino acid that has been activated by an enzyme.

    2. One end of the tRNA molecule possesses an **anticodon,** a triplet of nucleotides that recognizes the complementary codon in mRNA. If recognition occurs, the anticodon ensures that the tRNA transfers its activated amino acid molecule in the proper sequence to the growing polypeptide chain.

C. Ribosomal RNA associates with many different proteins (including enzymes) to form **ribosomes.**

    1. rRNA associates with mRNA and tRNA during protein synthesis.

    2. rRNA synthesis takes place in the nucleolus and is catalyzed by RNA polymerase I. A single **45S precursor rRNA (pre-rRNA)** is formed, which is then **processed to form ribosomes** as follows (Figure 2.5).

        a. Pre-rRNA associates with ribosomal proteins and is cleaved into the three sizes (28S, 18S, and 5.8S) of rRNAs present in ribosomes.

        b. The **RNP** containing 28S and 5.8S rRNA then combines with 5S

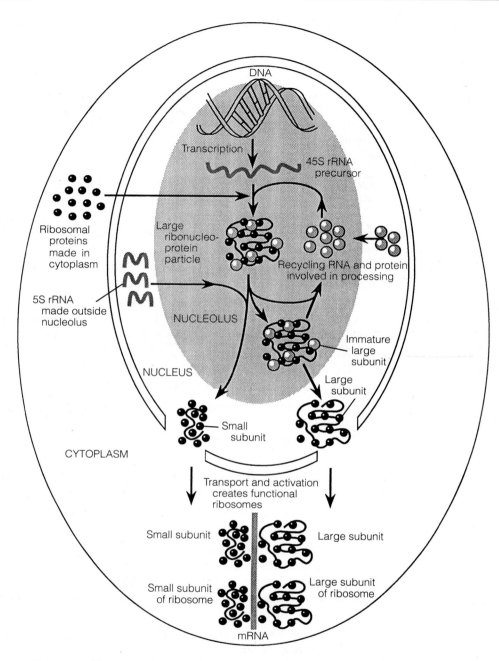

Figure 2.5. Formation of ribosomal ribonucleic acid (rRNA) and its processing into ribosomal subunits, which occurs in the nucleolus. DNA = deoxyribonucleic acid; messenger ribonucleic acid = mRNA. (Adapted with permission from Alberts B, Bray D, Lewis J, et al: *Molecular Biology of the Cell*, 2nd ed. New York, Garland Publishing, 1989, p 542.)

   rRNA, which is synthesized outside of the nucleolus, to form the **large subunit** of the ribosome.

c. The RNP containing 18S rRNA forms the **small subunit** of the ribosome.

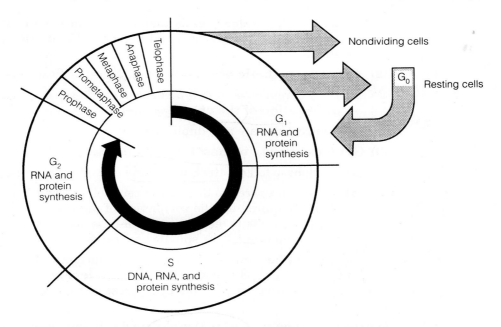

Figure 2.6. Stages of the cell cycle in dividing cells. Differentiated cells that no longer divide have left the cycle, whereas resting cells in the G_0 state may reenter the cycle and begin dividing again. DNA = deoxyribonucleic acid; RNA = ribonucleic acid. (Adapted with permission from Widnell CC, Pfenninger KH: *Essential Cell Biology*. Baltimore, Williams & Wilkins, 1990, p 58.)

IX. Cell Cycle (Figure 2.6)

A. The **cell cycle** varies in length in different types of cells, but is repeated each time a cell divides. It is composed of a series of events that prepare the cell to divide into two daughter cells.

    1. It is **temporarily suspended** in nondividing resting cells (e.g., peripheral lymphocytes), which are in the G_0 state. Such cells may reenter the cycle and begin to divide again.

    2. It is **permanently interrupted** in differentiated cells that do not divide (e.g., cardiac muscle cells and neurons).

B. Two major periods, **interphase** (interval between cell divisions) and **mitosis** (**M phase,** the period of cell division) compose the cell cycle.

    1. Interphase is considerably longer than the M phase and is the period during which **the cell doubles in size and DNA content.**

        a. Interphase is divided into three separate phases **(G_1, S, and G_2),** during which specific cellular functions occur.

            (1) G_1 phase (gap one phase) lasts from hours to several days.

                (a) Occurring after mitosis, it is the period during which the cell grows and proteins are synthesized, thus restoring the daughter cells to normal volume and size.

                (b) Certain "trigger proteins" are synthesized; these proteins enable the cell to reach a threshold **(restriction point)** and proceed to the S phase. Cells that fail to reach the restriction point become resting cells and enter the G_0 (outside phase) state.

(2) S phase (synthetic phase) lasts 8-12 hours in most cells.

    (a) DNA is replicated and proteins are synthesized, resulting in **duplication of the chromosomes.**

    (b) Centrosomes are also duplicated.

(3) G_2 phase (gap two phase) lasts 2-4 hours.

    (a) This phase follows the S phase and extends to mitosis.

    (b) The cell prepares to divide: the centrioles grow to maturity; energy required for the completion of mitosis is stored; and RNA and proteins necessary for mitosis are synthesized, including tubulin for the spindle apparatus.

b. Several **control factors** have been identified, including a category of proteins known as cyclins as well as **cyclin-dependent kinases (CDKs)**, which initiate and/or induce progression through the cell cycle.

    (1) During G_1 phase, **cyclins D and E** bind to their respective CDKs; these complexes enable the cell to enter and advance through the **S phase.**

    (2) Cyclin A binds to its CDKs, thus enabling the cell to leave the **S phase** and enter the **G_2 phase**, and also to manufacture **cyclin B**.

    (3) Cyclin B binds to its CDK, inducing the cell to leave the **G_2 phase** and enter the **M phase**.

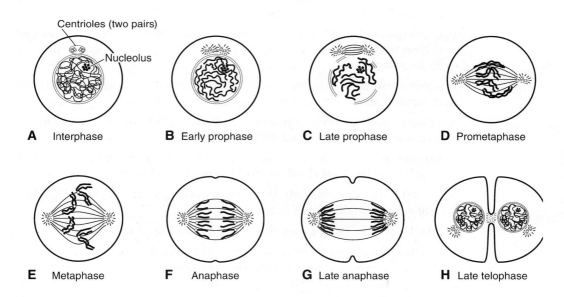

Figure 2.7. Events in various phases of mitosis. (Redrawn with permission from Kelly DE, Wood RL, Enders AC: *Bailey's Textbook of Microscopic Anatomy*, 18th ed. Baltimore, Williams & Wilkins, 1984, p 89.)

Table 2.1. Stages of Mitosis

Stage	DNA Content	Identifying Characteristics
Prophase (early) Prophase (late)	DNA content was doubled in the S phase of interphase (4n); also centrioles were replicated	Nuclear envelope and nucleolus begin to disappear Chromosomes condense; they consist of two sister chromatids attached at centromere Centrioles migrate to opposite poles and give rise to spindle fibers and astral rays
Prometaphase	Double complement of DNA (4n)	Nuclear envelope disappears Kinetochores develop at centromeres and kinetochore microtubules form
Metaphase	Double complement of DNA (4n)	Maximally condensed chromosomes align at the equatorial plate of the mitotic spindle
Anaphase Anaphase (late)	Double complement of DNA (4n)	Daughter chromatids separate at centromere Each chromatid migrates to an opposite pole of the cell along the microtubule (karyokinesis) In late anaphase a cleavage furrow begins to form
Telophase	Each new daughter cell contains a single complement of DNA (2n)	Deepening of furrow (midbody) now forms between newly formed daughter cells (cytokinesis) Nuclear envelope reforms, nucleoli reappear, chromosomes disperse forming new interphase nucleus

DNA = deoxyribonucleic acid.

2. **Mitosis** (Figure 2.7, Table 2.1) lasts 1-3 hours. It follows the G_2 phase and completes the cell cycle. Division of the nucleus **(karyokinesis)** and cytoplasm **(cytokinesis)** results in the production of two **identical** daughter cells. It consists of five major stages.

a. **Prophase** begins when the chromosomes condense; during prophase, the nucleolus and nuclear envelope begin to disappear.

(1) The **centrosome** contains **centrioles** and a pericentriolar cloud of material containing α-tubulin rings, and is the principal **microtubule-organizing center (MTOC)** of the cell. Centrosomes migrate to opposite poles of the cell, and from them **spindle fibers** and **astral rays** of the mitotic spindle polymerize.

(2) Chromosomes consist of two parallel **sister chromatids** (future daughter chromosomes) attached at the **centromere,** a constriction along the chromosome. **Kinetochores** develop at the centromere region and function as MTOCs.

b. **Prometaphase** begins when the nuclear envelope disappears, resulting in the chromosomes being dispersed apparently randomly in the cytoplasm.

(1) The kinetochores complete development and attach to specific spindle microtubules, forming **kinetochore microtubules.**

(2) Spindle microtubules that do not attach to kinetochores are called **polar microtubules.**

c. Metaphase is the phase during which the duplicated condensed chromosomes align at the equatorial plate of the mitotic spindle and become attached to spindle microtubules at their kinetochore.

d. Anaphase begins as the chromatids separate (at the centromere) and daughter chromosomes move to opposite poles of the cell.

(1) The spindle elongates.

(2) In the later stages of anaphase, a **cleavage furrow** begins to form around the cell due to contraction of a band of actin filaments called the **contractile ring.**

e. Telophase is characterized by each set of chromosomes reaching the pole, a deepening of the cleavage furrow; the **midbody** (containing overlapping polar microtubules) is now located between the newly forming daughter cells.

(1) Microtubules in the midbody are depolymerized, facilitating **cytokinesis** and formation of two identical daughter cells.

(2) The **nuclear envelope** is reestablished around the condensed chromosomes in the daughter cells, and **nucleoli reappear.** Nucleoli arise from the specific **NORs** (called secondary constriction sites), which are carried on five separate chromosomes in humans.

(3) The daughter nuclei gradually enlarge, and the condensed chromosomes disperse to form the typical interphase nucleus with heterochromatin and euchromatin.

(4) It appears that at the end of cytokinesis the "mother centriole" of the duplicated pair moves from the newly forming nuclear pole to the intercellular bridge. This event is necessary to initiate disassembly of the midbody microtubules and the complete separation of the daughter cells. If this event fails, DNA replication is arrested at one of the G_1 checkpoints during the next interphase.

X. Apoptosis (Programmed Cell Death). Apoptosis is the method whereby cells are removed from tissues in an orderly fashion, as a part of normal maintenance or during development.

A. There are several **morphological features** of cells that undergo programmed cell death.

  1. They include chromatin condensation, breaking up of the nucleus, and blebbing of the plasma membrane.

  2. Shrinkage of the cell occurs, and it is fragmented into membrane-enclosed fragments called **apoptotic bodies**.

B. Apoptotic cells do not pose a threat to surrounding cells because changes in their plasma membranes make them subject to rapid phagocytosis by macrophages and by neighboring cells.

C. The signals that induce apoptosis may occur through several mechanisms.

1. Genes that code for enzymes, called **caspases**, play an important role in the process.

2. Certain cytokines, such as **tumor necrosis factor** (**TNF**), may also activate caspases that degrade regulatory and structural proteins in the nucleus and cytoplasm, leading to the morphological changes characteristic of apoptosis.

D. **Defects in the process** of programmed cell death contribute to many major diseases.

1. Too much apoptosis causes extensive nerve cell loss in Alzheimer disease and stroke.

2. Not enough apoptosis has been linked to cancer and other autoimmune diseases.

XI. **Meiosis** (Figure 2.8)

A. **Meiosis** is a special form of cell division in germ cells (oogonia and sper-

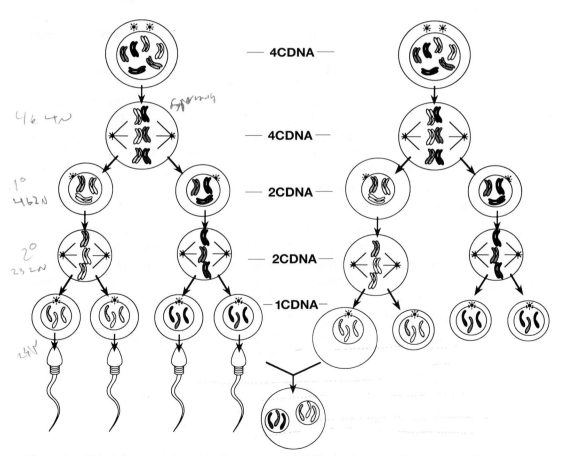

Figure 2.8. Meiosis in men and women. Spermatogenesis in the male gives rise to sperm, each containing the haploid number of chromosomes. Oogenesis in the female gives rise to an ovum with the haploid number of chromosomes. Fertilization reconstitutes the diploid number of chromosomes in the resulting zygote. DNA = deoxyribonucleic acid. (Adapted with permission from Widnell CC, Pfenninger KH: *Essential Cell Biology*. Baltimore, Williams & Wilkins, 1990, p 69.)

matozoa) in which the **chromosome number** is reduced from **diploid (2n)** to **haploid (n).**

1. It occurs in developing germ cells in preparation for sexual reproduction. Subsequent fertilization results in **diploid zygotes.**

2. The DNA content of the original diploid cell is doubled (4n) in the S phase preparatory to meiosis.

   a. This phase is followed by two successive cell divisions that give rise to **four haploid cells.**

   b. In addition, **recombination** of maternal and paternal genes occurs by **crossing over** and **random assortment,** yielding the unique haploid genome of the gamete.

B. **The stages of meiosis** are meiosis I (reductional division) and meiosis II (equatorial division).

   1. **Reductional division (meiosis I)** occurs after interphase when the 46 chromosomes are duplicated, giving the cell a **4CDNA content** (considered to be the total DNA content of the cell).

      a. **Prophase I** is divided into five stages (leptotene, zygotene, pachytene, diplotene, and diakinesis), which accomplish the following events.

         (1) Chromatin condenses into the visible chromosomes, each containing two chromatids joined at the centromere.

         (2) Homologous maternal and paternal chromosomes pair via the **synaptonemal complex,** forming a **tetrad. Crossing over** (random exchanging of genes between segments of homologous chromosomes) occurs at the **chiasmata,** thus **increasing genetic diversity.**

         (3) The nucleolus and nuclear envelope disappear.

      b. **Metaphase I**

         (1) Homologous pairs of chromosomes align on the equatorial plate of the spindle in a random arrangement, facilitating genetic mixing.

         (2) Spindle fibers from either pole attach to the **kinetochore** of any one of the chromosome pairs, thus ensuring that genetic mixing occurs.

      c. **Anaphase I**

         (1) This phase is similar to anaphase in mitosis except that each chromosome consists of **two chromatids that remain held together.**

         (2) Chromosomes migrate to the poles.

      d. **Telophase I** is similar to telophase in mitosis, in that the nuclear envelope is reestablished and two daughter cells are formed via cytokinesis.

         (1) Each daughter cell now contains 23 chromosomes (n) number, but has a **2CDNA content** (the diploid amount).

         (2) Each chromosome is composed of **two** similar **sister chromatids** (not genetically identical).

2. Equatorial division (meiosis II) begins soon after the completion of meiosis I, following a brief interphase **without DNA replication.**

   a. The sister chromatids are portioned out among the two daughter cells formed in meiosis I. The two daughter cells then divide, resulting in the distribution of chromosomes into four daughter cells, each containing its own **unique recombined genetic material (1CDNA;*n*).**

   b. The stages of meiosis II are similar to those of mitosis; thus the stages are named similarly (prophase II, metaphase II, anaphase II, and telophase II).

   c. Meiosis II occurs more rapidly than mitosis.

XII. Clinical Considerations

A. Aneuploidy is defined as an abnormal number of chromosomes and can be detected by karyotyping. Examples include **trisomy** (the presence of a third chromosome of one type) and **monosomy** (the absence of one member of a chromosome pair).

   1. Down syndrome (trisomy 21) is characterized by mental retardation, short stature, stubby appendages, congenital heart malformations, and other defects.

   2. Klinefelter syndrome (XXY) is aneuploidy of the sex chromosomes, characterized by infertility, variable degrees of masculinization, and small testes.

   3. Turner syndrome (XO) is **monosomy** of the sex chromosomes, characterized by short stature, sterility, and various other abnormalities.

B. Transformed cells

   1. Transformed cells have lost their ability to respond to regulatory signals controlling the cell cycle. They may undergo cell division indefinitely, thus becoming cancerous.

   2. Vinca alkaloids may arrest these cells in mitosis; drugs that block purine and pyrimidine synthesis may arrest cells in the S phase of the cell cycle.

C. Oncogenes represent **mutations of certain regulatory genes,** called **proto-oncogenes,** which normally stimulate or inhibit cell proliferation and development.

   1. Genetic accidents or viruses may lead to the formation of oncogenes.

   2. Oncogenes dominate the normal alleles (proto-oncogenes), causing a **deregulation** of cell division, which leads to a cancerous state.

   3. Bladder cancer and acute myelogenous leukemia are caused by oncogenes.

Review Test

Directions: Each of the numbered items or incomplete statements in this section is followed by answers or by completions of the statement. Select the ONE lettered answer or completion that is BEST in each case.

1. The nuclear pore complex

(A) permits free communication between the nucleus and the cytoplasm
(B) is bridged by a unit membrane
(C) is located only at specific nuclear pore sites
(D) permits passage of proteins via receptor-mediated transport

2. Which one of the following nucleotides is present only in ribonucleic acid (RNA)?

(A) Thymine
(B) Adenine
(C) Uracil
(D) Cytosine
(E) Guanine

3. Anticodons are located in

(A) messenger ribonucleic acid (mRNA)
(B) ribosomal RNA (rRNA)
(C) transfer RNA (tRNA)
(D) small nuclear ribonucleoprotein (snRNP)
(E) heterogeneous nuclear ribonucleoprotein (hnRNP)

4. Deoxyribonucleic acid (DNA) is duplicated in the cell cycle during the

(A) G_2 phase
(B) S phase
(C) M phase
(D) G_1 phase
(E) G_0 phase

5. A male child at puberty is determined to have Klinefelter syndrome. Although the parents have been informed of the clinical significance, they have asked for an explanation of what happened. Identify the item that needs to be discussed with the parents.

(A) Trisomy of chromosome 21
(B) Loss of an autosome during mitosis

(C) Loss of the Y chromosome during meiosis
(D) Nondisjunction of the X chromosome

The response options for the next five items are the same. Each item will state the number of options to select. Choose exactly this number.

Questions 6-10

For each description, select the most appropriate structure.

(A) Nuclear pore complex
(B) Nucleolus
(C) Heterochromatin
(D) Outer nuclear membrane
(E) Euchromatin

6. An inclusion not bounded by a membrane that is observable only during interphase (**SELECT 1 STRUCTURE**)

7. Continuous with rough endoplasmic reticulum (**SELECT 1 STRUCTURE**)

8. Controls movement of proteins in and out of the nucleus (**SELECT 1 STRUCTURE**)

9. Site of transcriptional activity (**SELECT 1 STRUCTURE**)

10. Clumps of nucleoprotein concentrated near the periphery of the nucleus (**SELECT 1 STRUCTURE**)

Answers and Explanations

1-D. The nuclear pore complex contains a central aqueous channel, which permits passage of small, water-soluble molecules. However, movement of proteins in and out of the nucleus is selectively controlled by the nuclear pore complex via receptor-mediated transport.

2-C. Deoxyribonucleic acid (DNA) contains the purines, adenine and guanine, and the pyrimidines, cytosine and thymine. In ribonucleic acid (RNA), uracil, a pyrimidine, replaces thymine.

3-C. Each transfer ribonucleic acid (tRNA) possesses a triplet of nucleotides, called an anticodon, which recognizes the complementary codon in messenger RNA (mRNA).

4-B. The S (synthesis) phase of the cell cycle is the period during which deoxyribonucleic acid (DNA) replication and histone synthesis occur, resulting in duplication of the chromosomes. At the end of the S phase, each chromosome consists of two identical chromatids attached to one another at the centromere.

5-D. Klinefelter syndrome occurs only in males. This condition results from nondisjunction of the X chromosome during meiosis, resulting in an extra X chromosome in somatic cells. These cells therefore have a normal complement of autosomal chromosomes (22 pairs), and instead of one pair of sex chromosomes (XY) there is an extra X chromosome. These individuals have an XXY genotype, resulting in 47 total chromosomes rather than the normal complement of 46. This syndrome is an example of trisomy of the sex chromosomes. Down syndrome is an example of an autosomal trisomy, specifically trisomy of chromosome number 21. Both syndromes have profound complications.

6-B. The nucleolus is an inclusion, not bounded by a membrane, located within the nucleus. It is observable during interphase, but disappears during mitosis.

7-D. The outer nuclear membrane is continuous with the rough endoplasmic reticulum.

8-A. The nuclear pore complex selectively controls movements of water-soluble molecules and proteins in and out of the nucleus.

9-E. The pale-staining euchromatin is the transcriptionally active chromatin in the nucleus.

10-C. Heterochromatin is the dark-staining nucleoprotein located near the periphery of the nucleus. It is transcriptionally inactive but may be responsible for proper chromosome segregation during meiosis.

3

Cytoplasm

I. **Overview—The Cytoplasm.** The cytoplasm contains three main structural components: **organelles, inclusions,** and the **cytoskeleton.** The fluid component is called the cytosol. The functional interactions among certain organelles result in the uptake and release of material by the cell, protein synthesis, and intracellular digestion.

II. **Structural Components**

    A. **Organelles** (Figure 3.1) are metabolically active units of cellular matter.

        1. The **plasma membrane** is discussed in Chapter 1.

        2. **Ribosomes**

            a. **Structure.** Ribosomes are 12 nanometers (nm) wide and 25 nm long and consist of a **small** and **large** subunit. The subunits are composed of several types of ribosomal ribonucleic acid (rRNA) and numerous proteins (Table 3.1; see Figure 2.5).

            b. Ribosomes may exist free in the cytosol or bound to membranes of the rough endoplasmic reticulum (RER) or outer nuclear membrane. Whether free or bound, the ribosomes constitute a single interchangeable population.

            c. A **polyribosome (polysome)** is a cluster of ribosomes along a single strand of messenger ribonucleic acid (mRNA); polyribosomes have a spiral configuration.

            d. **Function.** Ribosomes are the sites where **mRNA is translated into protein.**

                (1) The **small ribosomal subunit** binds mRNA and activated transfer ribonucleic acid (tRNAs); the **codons** of the mRNA then **base-pair** with the corresponding **anticodons** of the tRNA.

                (2) Next, an initiator tRNA recognizes the **start codon (AUG)** on the mRNA.

                (3) The **large ribosomal subunit** then binds to the complex. **Peptidyl transferase** in the large subunit catalyzes peptide-bond formation, resulting in addition of amino acids to the growing polypeptide chain.

                (4) A **chain-terminating codon (UAA, UAG,** or **UGA)** causes release of the polypeptide from the ribosome, and the ribosomal subunits dissociate from the mRNA.

        3. **RER** [see Figure 3.1]

Table 3.1. Ribosome Composition

Subunit	rRNA Types	Number of Proteins
Large (60S)	5S	49
	5.8S	
	28S	
Small (40S)	18S	33

rRNA = ribosomal ribonucleic acid.

    a. Structure. RER is a system of sacs, or cavities, bounded by membranes. The outer surface of RER is studded with ribosomes (thus appearing rough). The interior region of RER is called the **cisterna.** The outer nuclear membrane is **continuous** with the RER membrane,

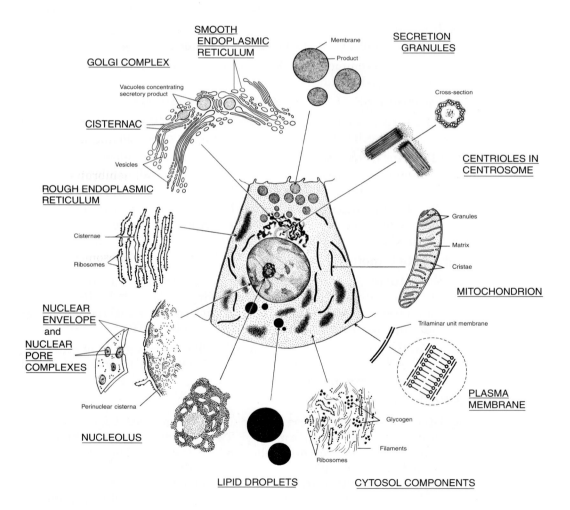

Figure 3.1. Diagram of a eukaryotic cell and its major organelles and inclusions. (Adapted with permission from Fawcett DW: *Bloom and Fawcett's Textbook of Histology*, 12th ed. New York, Chapman & Hall, 1994, p 2.)

thus bringing the perinuclear cisterna into continuity with the cisternae of the RER. The RER membrane also has receptors **(ribophorins)** in its membrane to which the large ribosomal subunit binds.

    b. RER is abundant in cells synthesizing **secretory proteins;** in such cells, the RER is organized into many parallel arrays.

    c. The RER sac closest to the Golgi apparatus gives rise to buds free of ribosomes that form vesicles, and is known as a **transitional element**.

    d. Function. The RER is the site where **membrane-packaged proteins** are synthesized, including secretory, plasma-membrane, and lysosomal proteins. In addition, the RER **monitors** the assembly, retention, and even the degradation of certain proteins.

4. Smooth endoplasmic reticulum (SER)

    a. Structure. SER is an irregular network of membrane-bounded channels that lacks ribosomes on its surface (thus appearing smooth).

    b. It usually appears as branching anastomosing **tubules,** or **vesicles,** whose membranes do **not** contain ribophorins.

    c. SER is less common than RER but is prominent in cells synthesizing steroids, triglycerides, and cholesterol.

    d. Function. SER has different functions in different cell types.

        (1) Steroid hormone synthesis occurs in SER-rich cells such as the Leydig cells of the testis, which make testosterone.

        (2) Drug detoxification occurs in hepatocytes following proliferation of the SER in response to phenobarbital; oxidases that metabolize this drug are present in the SER.

        (3) Muscle contraction and relaxation involves the release and recapture of calcium ions by the sarcoplasmic reticulum (the term for SER in skeletal muscle cells).

5. Annulate lamellae

    a. Structure. Annulate lamellae are parallel stacks of membranes (usually 6 to 10) that resemble the nuclear envelope, including its pore complexes. They are often arranged with their **annuli** (pores) in register and are frequently **continuous** with the RER.

    b. Function. Annulate lamellae are found in rapidly growing cells (e.g., germ cells, embryonic cells, and tumor cells), but their function and significance remain unknown.

6. Mitochondria (see Figure 3.1)

    a. Structure. Mitochondria are rod-shaped organelles [0.2 micrometers (μm) wide and up to 7 μm long]. They possess an outer membrane, which surrounds the organelle, and an inner membrane, which invaginates to form **cristae.** They are subdivided into an **intermembrane compartment,** which is located between the two membranes, and an inner **matrix compartment.** Granules within the matrix bind the divalent cations Mg^{2+} and Ca^{2+}.

    b. Enzymes and genetic apparatus. Mitochondria contain:

        (1) All the enzymes of the **Krebs [tricarboxylic acid (TCA)] cycle**

in the matrix, except for succinate dehydrogenase, which is located on the inner mitochondrial membrane

(2) **Elementary particles** (visible on negatively stained cristae) that contain adenosine triphosphate (**ATP**) **synthase,** a special enzyme involved in **coupling oxidation to phosphorylation** of adenosine diphosphate (ADP) to form ATP

(3) A **genetic apparatus** in the matrix composed of circular deoxyribonucleic acid (DNA), mRNA, tRNA, and rRNA (with a limited coding capacity), although most mitochondrial proteins are encoded by nuclear DNA

c. Origin and proliferation

(1) Mitochondria may have **originated as symbionts** (intracellular parasites). According to this theory, anaerobic eukaryotic cells endocytosed aerobic microorganisms that evolved into mitochondria, which function in oxidative processes.

(2) Mitochondria proliferate by division (fission) of preexisting mitochondria and typically have a 10-day life span.

d. Mitochondrial ATP synthesis

(1) Mitochondria synthesize ATP via the Krebs cycle, which traps chemical energy and produces ATP by **oxidation** of fatty acids, amino acids, and glucose.

(2) ATP is also synthesized via a **chemiosmotic coupling mechanism** involving enzyme complexes of the **electron transport chain** and **elementary particles** present in the cristae (Figure 3.2).

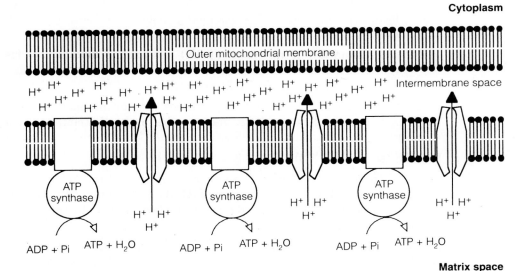

Figure 3.2. Chemiosmotic coupling mechanism for generating adenosine triphosphate (ATP) in mitochondria. During electron transport, H+ ions (protons) are pumped from the inner matrix compartment across the inner mitochondrial membrane into the intermembrane compartment. The electrochemical proton gradient thus created drives ATP synthase to catalyze the conversion of adenosine diphosphate (ADP) and Pi to ATP.

e. **Condensed mitochondria** result from a **conformational change** in the orthodox form (typical morphology). The change occurs in response to an uncoupling of oxidation from phosphorylation.

(1) In condensed mitochondria, the size of the inner compartment is decreased and the matrix density is increased. The intermembrane compartment is enlarged.

(2) Condensed mitochondria are present in **brown fat cells,** which produce **heat** rather than ATP due to a special transport protein in their inner membrane that **uncouples** respiration from ATP synthesis (see Chapter 6 IV B 5 b).

(3) They **swell** in response to calcium, phosphate, and thyroxine, which induce an increase in water uptake and an uncoupling of phosphorylation; ATP reverses the swelling.

7. **Golgi complex (apparatus)** [see Figure 3.1; Figure 3.3]

a. **Structure.** The Golgi complex consists of several membrane-bounded **cisternae (saccules)** arranged in a **stack** and positioned and held in place by microtubules. Cisternae are disk-shaped and slightly curved, with flat centers and dilated rims, but their size and shape may vary. A distinct **polarity** exists across the cisternal stack.

b. **Regions**

(1) The **cis face** of the Golgi complex is now also called the **cis Golgi network (CGN).** It is located at the side of the Golgi stack facing a separate endoplasmic reticulum-Golgi-intermediate compartment called ERGIC (see Figure 3.3).

(2) The **medial compartment** of the Golgi complex is comprised of a few cisternae lying between the cis and trans faces.

(3) The **trans face** of the Golgi complex is composed of the cisternae located at the side of the stack facing vacuoles and secretory granules.

(4) The **trans Golgi network (TGN)** lies apart from the last cisterna at the trans face and is separated from the Golgi stack. It sorts proteins for their final destinations.

c. **Functions.** The Golgi complex processes membrane-packaged proteins synthesized in the RER and recycles and redistributes membranes.

8. **ERGIC** (see Figure 3.3)

a. **Structure.** Lying between the **e**ndoplasmic **r**eticulum and the **G**olgi is an **i**ntermediate **c**ompartment (**ERGIC**). The components of this compartment are also referred to as **v**esicular **t**ubular **c**lusters (**VTC**) based on their morphology.

b. **Function.** The ERGIC has enzymes and proteins that are unique and distinct from both the RER and the Golgi. It appears to be a first way station for the segregation of anterograde versus retrograde transport, and also functions to concentrate some proteins to a limited extent.

9. **Coated vesicles** are characterized by a visible cytoplasmic surface coat.

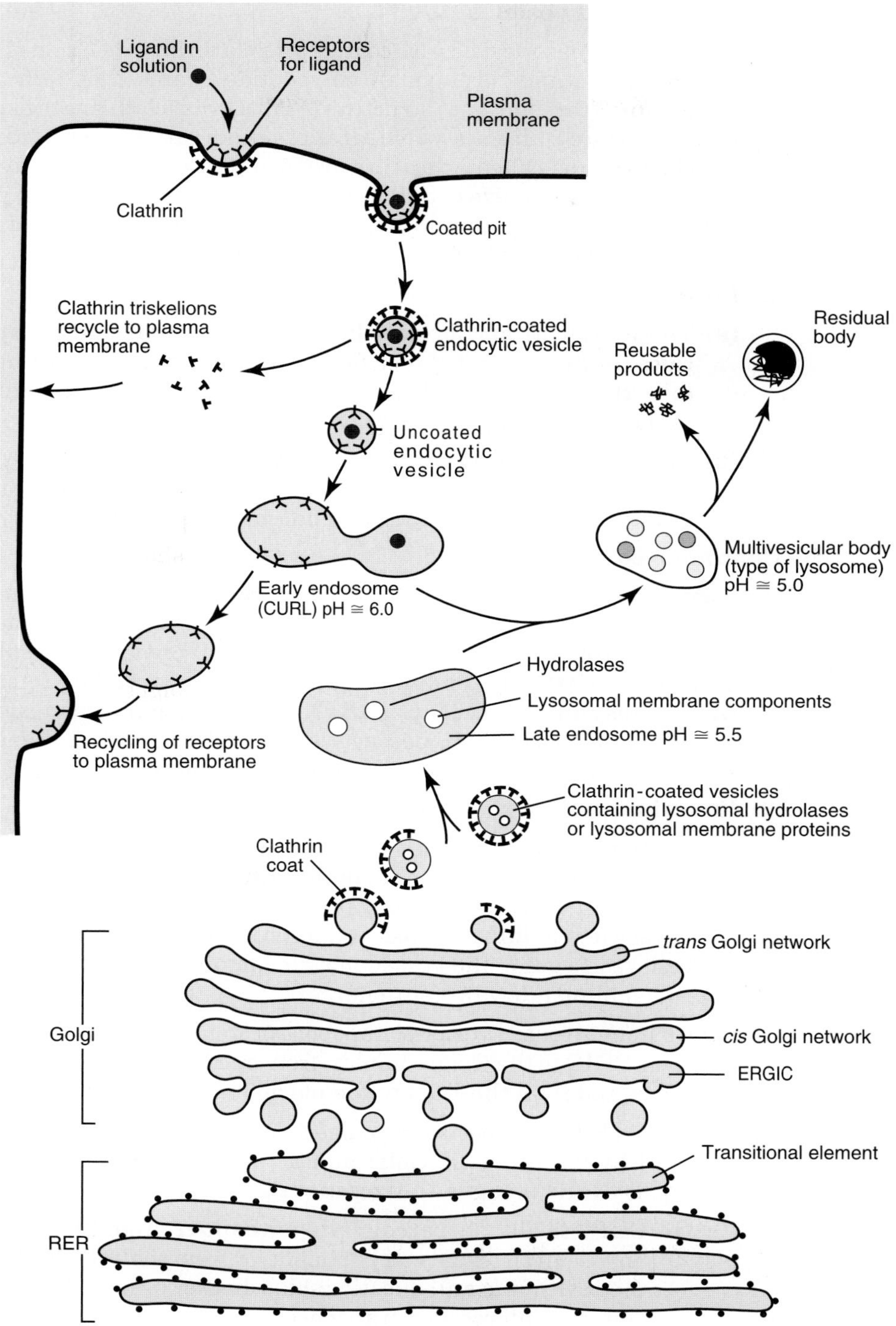

Figure 3.3. Diagram illustrating receptor-mediated endocytosis of a ligand (e.g., low-density lipoproteins) and the lysosomal degradative pathway. Clathrin triskelions quickly recycle back to the plasma membrane. The receptors and ligands then uncouple in the early endosome (**c**ompartment for **u**ncoupling of **r**eceptors and **l**igands, CURL), which is followed by recycling of receptors back to the plasma membrane. The late endosome is the primary intermediate in the formation of lysosomes (e.g., multivesicular bodies). Material that is phagocytosed or organelles that undergo autophagy do *not* use the early endosomal pathway.

a. Clathrin-coated vesicles

(1) **Structure.** These vesicles are coated with clathrin, which consists of three large and three small polypeptide chains that form a **triskelion** (three-legged structure). Thirty-six clathrin triskelions associate to form a polyhedral cage-like lattice around the vesicle. Proteins called **adaptins** are also part of clathrin-coated vesicles. They recognize both the clathrin triskelions and the cargo receptors, thus they help to form the vesicle curvature and also capture the cargo molecules.

(2) **Function**

(a) These vesicles are formed during **receptor-mediated uptake (endocytosis)** of specific molecules by the cell. After uptake, the vesicles quickly lose their coats, and clathrin returns to the plasma membrane for recycling (see Figure 3.3).

(b) They also function in the **signal-directed (regulated) transport** of proteins from the TGN either to the secretory granule pathway or to the late endosome-lysosome pathway.

b. Coatomer-coated vesicles

(1) **Structure.** These vesicles have coats consisting of coatomer, which does not form a cage-like lattice around vesicles. Coatomer is a large protein complex formed by individual **co**at **p**rotein subunits called **COPs.** Assembly of coatomer depends on the protein **ADP-ribosylation factor (ARF),** which binds guanosine triphosphate (GTP), becomes activated, and recruits coatomer subunits. ARF also helps to select the cargo molecules.

(2) **Function**

(a) Coatomer-coated vesicles mediate the continuous **constitutive protein transport** (default pathway; bulk flow) within the cell. Specific GTP binding proteins are present at each step of vesicle budding and fusion, and proteins called **snares** are believed to guide the vesicle movement. Vesicle v-snares bind to complimentary target t-snares.

(b) Coatomer-coated vesicles transport proteins from the RER to the ERGIC to the Golgi complex, from one Golgi cisterna to another, and from the TGN to the plasma membrane.

(i) COP-II transports molecules forward from the RER to the ERGIC to the cis Golgi and across the cisternae to the TGN **(anterograde transport).**

(ii) COP-I facilitates **retrograde transport** (from ERGIC, or any Golgi cisternal compartment, or from the TGN back to the RER). It is still questionable whether or not COP-I facilitates anterograde transport.

c. Caveolin-coated vesicles. These coated vesicles are less common and less understood than those of the previous two categories.

(1) **Structure.** Caveolae are invaginations of the plasma membrane in endothelial cells and smooth muscle cells. They possess a distinct coat formed by the protein caveolin.

(2) Function. Caveolae have been associated with cell signaling and a variety of transport processes, such as transcytosis and endocytosis.

10. **Lysosomes**

   a. **Structure.** Lysosomes are dense, membrane-bound organelles of diverse shape and size that function to degrade material. They may be identified in sections of tissue by cytochemical staining for **acid phosphatase.** Lysosomes possess special membrane proteins and about 50 acid hydrolases, which are synthesized in the RER. ATP-powered proton pumps in the lysosome membrane maintain an **acid pH** (≈ 5).

   b. **Formation.** Lysosomes are formed when sequestered material fuses with a **late endosome** and enzymatic degradation begins. Formation of a lysosome via one lysosomal pathway (see Figure 3.3) involves the following intermediates.

   (1) Early endosomes

   (a) These irregular, peripherally located vesicles form part of the pathway for receptor-mediated endocytosis and contain receptor-ligand complexes.

   (b) They are also known as the compartment for uncoupling of receptors and ligands **(CURL).**

   (c) Their acidic interiors (pH ≈ 6) are maintained by ATP-driven proton pumps. The acidity aids in the uncoupling of receptors and ligands; receptors return to the plasma membrane and ligands move to a late endosome.

   (2) Late endosomes

   (a) Late endosomes play a **key role** in a variety of lysosomal pathways and therefore are sometimes known as the intermediate compartment.

   (b) These irregular vesicles (pH ≈ 5.5) located deep within the cell receive ligands via microtubular transport of vesicles from early endosomes.

   (c) Late endosomes contain **both lysosomal hydrolases and lysosomal membrane proteins;** these are formed in the RER, transported to the Golgi complex for processing, and delivered in separate vesicles to late endosomes.[1]

   (d) Once late endosomes have received a full complement of lysosomal enzymes, they begin to degrade their ligands and are classified as lysosomes.

   c. **Types of lysosomes** (Figure 3.4). Lysosomes are named after the content of recognizable material; otherwise, the general term **lysosome** is used.

   (1) Multivesicular bodies are formed by fusion of an early endosome containing endocytic vesicles with a late endosome.

   (2) Phagolysosomes are formed by fusion of a **phagocytic vacuole** with a late endosome or a lysosome.

---

[1]The terms **primary** and **virgin lysosomes**, formerly used for tiny vesicles believed to be lysosomes that have not yet engaged in digestive activity, are no longer used.

A   HETEROPHAGY

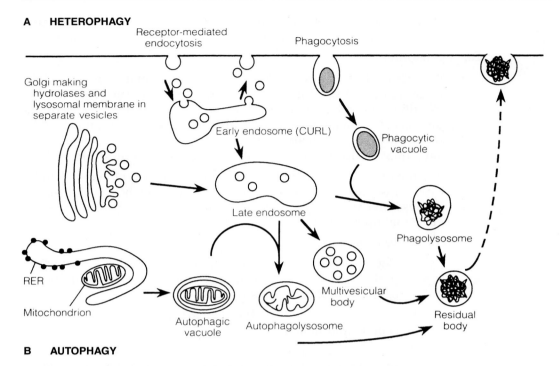

Figure 3.4. Diagram illustrating the different pathways of intracellular lysosomal digestion and the types of lysosomes involved in each.

B   AUTOPHAGY

(3) **Autophagolysosomes** are formed by fusion of an **autophagic vacuole** with a late endosome or lysosome. Autophagic vacuoles are formed when cell components targeted for destruction become enveloped by smooth areas of membranes derived from the RER.

(4) **Residual bodies** are lysosomes of any type that have expended their capacity to degrade material. They contain **undegraded material** (e.g., lipofuscin and hemosiderin) and eventually may be excreted from the cell.

11. Peroxisomes

a. **Structure.** Peroxisomes (also known as **microbodies**) are membrane-bound, spherical, or ovoid organelles that may be identified in cells by a cytochemical reaction for **catalase.** In stained preparations, they appear as small organelles (0.15-0.25 μm in diameter); they may be larger in hepatocytes. Peroxisomes may contain a **nucleoid,** a crystalline core consisting of urate oxidase (uricase); the human peroxisome lacks a nucleoid.

b. They originate from preexisting peroxisomes, which grow by importing specific cytosolic proteins that are recognized by receptor proteins (called **peroxins**) in the peroxisomal membrane. Then the peroxisome divides by fission; it has a life span of approximately 5-6 days.

c. **Function.** Peroxisomes contain a variety of **enzymes** whose functions vary from the oxidation of long chain fatty acids, to the synthesis of cholesterol, to the detoxification of substances such as ethanol.

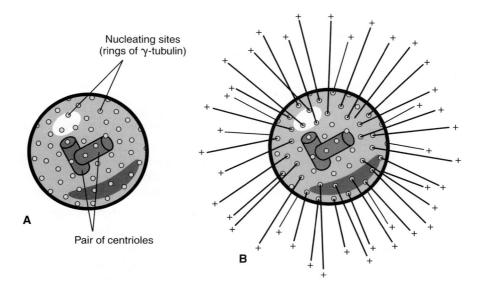

Figure 3.5. Polymerization of tubulin at a centrosome. *(A)* A centrosome consists of an amorphous cloud of material containing γ-tubulin rings that initiate microtubule polymerization. Within the cloud is a pair of centrioles. *(B)* A centrosome with attached microtubules. The minus end of each microtubule is embedded in the centrosome, having grown from a nucleating ring, whereas the plus end of each microtubule is free in the cytoplasm. (Modified from Alberts B, Bray D, Johnson A, et al: *Essential Cell Biology.* New York, Garland Publishing, 1998, p 521.)

B. Inclusions. Inclusions are accumulations of material that is **not metabolically active.** They usually are present in the cytosol only **temporarily.**

1. **Glycogen** appears as small clusters (or in hepatocytes as larger aggregates, known as **rosettes**) of electron-dense, 20- to 30-nm β-particles, which are similar in appearance to, but larger than, ribosomes. Glycogen is not bound by a membrane but frequently lies close to the SER. Glycogen serves as a **stored energy source** that can be degraded to glucose, which enters the bloodstream to elevate blood sugar levels.

2. **Lipid droplets** vary markedly in size and appearance depending on the method of fixation and are not bound by a membrane. Lipid droplets are storage forms of **triglycerides** (an energy source) and **cholesterol** (used in the synthesis of steroids and membranes).

3. **Lipofuscin** appears as membrane-bound, electron-dense granular material varying greatly in size and often containing lipid droplets. Lipofuscin represents a residue of undigested material present in residual bodies. Because the amount of this material increases with age, it is called **age pigment.** It is most common in nondividing cells (e.g., cardiac muscle cells, neurons) but also is found in hepatocytes).

4. **Centrosome** (see Figures 3.1, 2.7, and 3.5)

   a. **Structure.** The centrosome is located near the nucleus and contains two **centrioles** and a cloud of **pericentriolar material.** The centrioles exist as a pair of cylindrical rods (each 0.2 μm wide and 0.5 μm long) oriented at right angles to one another. Each member of the pair is composed of nine triplets of microtubules (9 + 0 axoneme pattern) arranged radially in the shape of a pinwheel.

 b. The centrioles **self-duplicate** in the S phase of the cell cycle, as each parent centriole forms a **procentriole** at right angles to itself.

 c. Centrioles also form **basal bodies,** which appear identical to unpaired centrioles and which give rise to the axonemes of cilia and flagella.

 d. Function

  (1) The centrosome is the major **microtubule organizing center** in the cell.

  (2) The **pericentriolar cloud** of material contains hundreds of ring-shaped structures composed of **γ-tubulin**, and each ring serves as a starting point for the polymerization of one microtubule.

  (3) Centrioles play no role in nucleating microtubules, but help to maintain the organization of the centrosome.

  (4) The centrosome itself is also duplicated during interphase (S phase), then separates to form the poles of the mitotic spindle where microtubules originate and converge.

C. Cytoskeleton. The cytoskeleton is the structural framework within the cytosol. It functions in maintaining cell shape, stabilizing cell attachments, facilitating endocytosis and exocytosis, and promoting cell motility. It includes the following major components:

 1. Microtubules

  a. Structure. Microtubules are straight, hollow tubules 25 nm in diameter and made of **tubulin**. They have a rigid wall composed of 13 protofilaments, each of which consists of a linear arrangement of tubulin dimers; each dimer consists of nonidentical **α and β tubulin subunits.**

  b. Microtubules are **polar,** with polymerization (assembly) and depolymerization (disassembly) occurring preferentially at the **plus end** as **GTP** is bound to tubulin dimers.

  c. Microtubules have **microtubule-associated proteins** (MAPs), which stabilize them and bind them to other cytoskeletal components and organelles; they also are associated with **kinesin and cytoplasmic dynein,** two force-generating proteins, which serve as "motors" for vesicle or organelle movement.

  d. Function. Microtubules maintain cell shape; aid in the transport of macromolecules within the cytosol; and promote the movement of chromosomes, cilia, and flagella.

 2. Microfilaments

  a. Structure. Microfilaments are also known as **F actin** or **actin filaments.** They are 7 nm in diameter and are composed of globular actin monomers **(G actin)** linked into a **double helix.**

  b. They display a **polarity** similar to that of microtubules; that is, their polymerization and depolymerization occurs preferentially at the **plus end** when **ATP** is bound by G actin.

Table 3.2. Major Classes of Intermediate Filaments

Protein	Location	Function
Keratin: 29 distinct isoforms (16 acidic, 13 neutral/basic)	Epithelial cells	Structural support and tension-bearing role; enables cells to withstand the stress caused by stretching; keratin tonofilaments are associated with desmosomes and hemidesmosomes Keratin serves as an immunological marker for tumors arising from epithelia
Vimentin-containing filaments	Fibroblasts, endothelial cells, chondroblasts, and various mesenchymal cells	Forms a cage-like structure around nucleus; structural support for cell Vimentin serves as an immunological marker for tumors arising from connective tissue
Desmin plus vimentin*	Skeletal muscle Cardiac muscle Smooth muscle	Forms a framework linking myofibrils and myofilaments Desmin serves as an immunological marker for tumors arising from muscle
GFAP plus vimentin*	Astrocytes Oligodendrocytes Schwann cells	Provides structural support GFAP serves as immunological marker for tumors arising from glia.
Neurofilaments NF-L (70 kDa) NF-M (140 kDa) NF-H (210 kDa)	Neurons	Provide support for axons and dendrites Neurofilaments serve as immunological markers for tumors of neuronal origin
Lamins A, B, and C	Nuclear lamina of all cells	Form a two-dimensional meshwork lining the inner surface of the inner nuclear membrane; organize peripheral nuclear chromatin

GFAP = glial fibrillary acidic protein.
*Desmin and GFAP are shown with vimentin because they may copolymerize with it; they are sometimes categorized as vimentin-like filaments.

    c. Many **actin-binding proteins** associate with microfilaments and modify their properties.

    d. Microfilaments are abundant at the periphery of the cell, where they are anchored to the plasma membrane via one or more intermediary proteins (e.g., α-actinin, vinculin, talin).

    e. Function. Microfilaments are involved in many **cellular processes,** such as establishing focal contacts between the cell and the extracellular matrix, locomotion of nonmuscle cells, formation of the contractile ring (in dividing cells), and the folding of epithelia into tubes during development.

  3. Intermediate filaments are 8 to 10 nm in diameter. They constitute a population of heterogeneous filaments that includes keratin, vimentin, desmin, glial fibrillary acid protein (GFAP), lamins, and neurofilaments (Table 3.2). [Desmin and GFAP sometimes co-polymerize with vimentin and may be categorized as vimentin-like filaments.] In general, intermediate filaments **provide mechanical strength** to cells. They lack polarity and do not require GTP or ATP for assembly, which occurs along the entire length of the filament.

III. Interactions Among Organelles. Organelles are involved in important cellular processes, such as the uptake and release of material by cells, protein synthesis, and intracellular digestion. These interactions provide the basis for a functional approach to the dynamics of cell biology.

A. Uptake and release of material by cells

1. **Endocytosis** is the uptake **(internalization) of material by cells.** Endocytosis includes pinocytosis, receptor-mediated endocytosis, and phagocytosis.

   a. **Pinocytosis ("cell drinking")** is **the nonspecific (random) uptake** of extracellular fluid and material in solution into pinocytic vesicles.

   b. **Receptor-mediated endocytosis** is the **specific uptake** of a substance [e.g., low-density lipoproteins (LDLs) and protein hormones] by a cell that has a plasma-membrane receptor for that substance (which is termed a **ligand**). It involves the following sequence of events (see Figure 3.3):

      (1) A ligand **binds specifically** to its receptors on the cell surface.

      (2) Ligand-receptor complexes cluster into a **clathrin-coated** pit, which invaginates and gives rise to a clathrin-coated vesicle containing the ligand.

      (3) The cytoplasmic clathrin coat is rapidly **lost,** leaving an **uncoated endocytic vesicle** containing the ligand.

   c. **Phagocytosis ("cell eating")** is the uptake of microorganisms, other cells, and particulate matter (frequently of foreign origin) by a cell. Phagocytosis usually involves cell-surface receptors. It is characteristic of cells—particularly **macrophages**—that degrade proteins and cellular debris and involves the following sequence of events:

      (1) A macrophage binds, via its **Fc receptors,** to an antibody-coated [immunoglobulin G (IgG)-coated] bacterium, or via its C3b receptors to a complement-coated bacterium.

      (2) Binding progresses until the plasma membrane completely envelops the bacterium, forming a phagocytic vacuole.

2. **Exocytosis** is the **release of material** from the cell via fusion of a secretory granule membrane with the plasma membrane. It **requires interaction of receptors in both the granule and plasma membrane,** as well as the **coalescence** (adherence and joining) of the two phospholipid membrane bilayers. Exocytosis takes place in both regulated and constitutive secretion.

   a. **Regulated secretion** (signal directed) is the release, in response to an **extracellular signal,** of proteins and other materials **stored** in the cell.

   b. **Constitutive secretion** (default pathway) is the more-or-less **continuous** release of material (e.g., collagen and serum proteins) without any intermediate storage step. An extracellular signal is **not** required for constitutive secretion.

3. **Membrane recycling** maintains a relatively constant plasma-membrane surface area following exocytosis. In this process, the secretory

granule membrane added to the plasma-membrane surface during exocytosis is **retrieved** during endocytosis via clathrin-coated vesicles. This vesicular membrane is returned to the TGN (via early endosomes) for recycling. Figure 3.6 illustrates endocytic pathways used by cells.

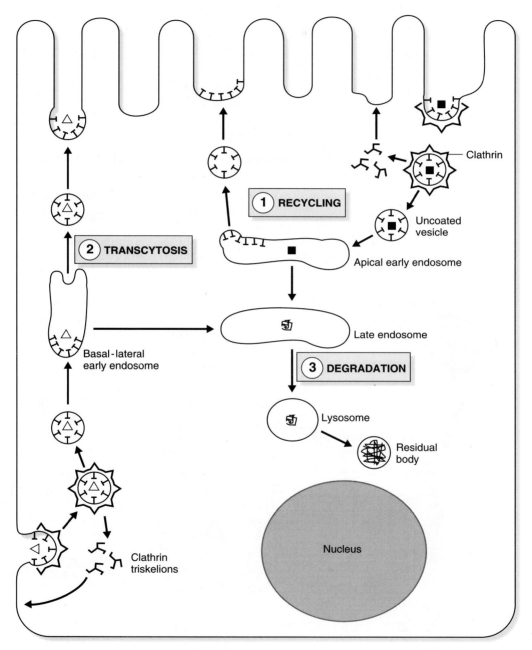

Figure 3.6. Pathways used by receptors and ligands following endocytosis. *(1)* Recycling of receptors to the same plasma membrane surface. *(2)* Transcytosis from one surface (e.g., basal-lateral) to another (e.g., apical). Transcytosis can occur in either direction, but separate early endosome compartments exist near the domain of vesicle entry. *(3)* Degradation. If not retrieved from either early endosome compartment, the ligands move to a common late endosome, which subsequently becomes a lysosome, where degradation is completed.

B. Protein synthesis

1. **Synthesis of membrane-packaged proteins** involves translation of mRNAs encoding **secretory, membrane,** and **lysosomal proteins** on ribosomes at the surface of the RER; transport of the growing polypeptide chain across the RER membrane and into the cisterna; and its processing within the RER.

   a. **Transport of newly formed peptide into the RER cisterna** is thought to occur by a mechanism described by the **signal hypothesis** as follows (Figure 3.7):

      (1) mRNAs for secretory, membrane, and lysosomal proteins contain codons that encode a **signal sequence.**

      (2) When the signal sequence is formed on the ribosome, a **signal recognition particle (SRP)** present in the cytosol binds to it.

      (3) Synthesis of the growing chain stops until the SRP facilitates the relocation of the polysome to **SRP receptors** in the RER membrane.

      (4) The large subunits of the ribosomes interact with ribosome receptor proteins, which "bind" them to the RER membrane. The SRP detaches, and multisubunit protein translocators form a **pore** across the RER membrane. Synthesis resumes, and the newly formed polypeptide is threaded through the pore and into the RER cisterna (lumen).

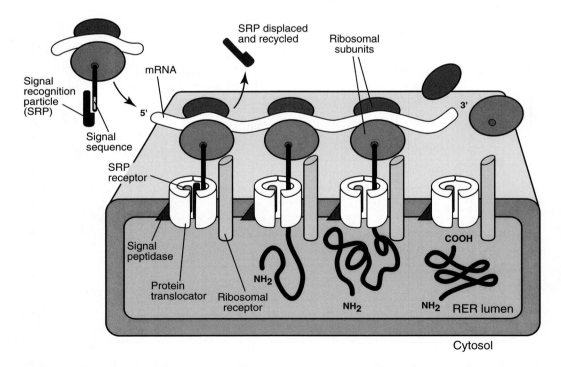

Figure 3.7. Diagram illustrating the signal hypothesis. The signal sequence of a newly-formed secretory polypeptide binds to a signal recognition particle (SRP) that delivers the ribosome-peptide-SRP complex to a receptor on the rough endoplasmic reticulum (RER). The SRP is recycled, and the polypeptide is translocated into the cisterna of the RER, where a signal peptidase cleaves off the signal sequence.

b. Posttranslational modification in the RER

   (1) After the newly formed polypeptide enters the cisterna, a **signal peptidase** cleaves the signal sequence from it.

   (2) The polypeptide is glycosylated.

   (3) Disulfide bonds form, converting the linear polypeptide into a globular form.

c. Protein transport from the RER to the ERGIC (Figure 3.8)

   (1) Transitional elements of the RER give rise to **COP-II-coatomer coated vesicles** containing newly synthesized protein.

   (2) These vesicles move to the **ERGIC,** where they deliver the protein.

   (3) The ERGIC is the **first recycling compartment** in the secretory pathway. Proteins either move forward toward the CGN, or (if they are RER-resident proteins) they are captured by a specific membrane receptor protein and returned in **COP-I-coatomer vesicles** to the RER along a microtubule-guided pathway.

d. Anterograde transport from the ERGIC to the CGN (cis Golgi) is via COP-II-coatomer coated vesicles.

e. Movement of material anterograde among the Golgi subcompartments may occur by cisternal maturation and/or by vesicular transport as follows:

   (1) Cisternae containing proteins may change in biochemical composition as they move intact across the stack.

   (2) COP-II-coated vesicles may bud off of one cisterna and fuse with the dilated rim of another cisterna.

   (3) Although both of the above mechanisms have been observed, the precise way that anterograde transport occurs across the Golgi stack of cisternae is currently unresolved.

   (4) *Retrograde* vesicular transport occurs between Golgi cisternae and between the Golgi and the ERGIC or RER via COP-I-coated vesicles.

f. Protein processing in the Golgi complex (see Figure 3.8) occurs as proteins move from the cis to the trans face of the Golgi complex through **distinct cisternal subcompartments.** Protein processing may include the following events, each of which occurs in a different cisternal subcompartment:

   (1) Proteins targeted for lysosomes are tagged with mannose 6-phosphate in the CGN.

   (2) Mannose residues are removed in cis and medial cisternae.

   (3) Terminal glycosylation of some proteins with sialic acid residues and galactose occurs.

   (4) Sulfation and phosphorylation of amino acid residues takes place.

   (5) A membrane similar in composition and thickness to the plasma membrane is acquired.

g. Sorting of proteins in TGN (see Figure 3.8)

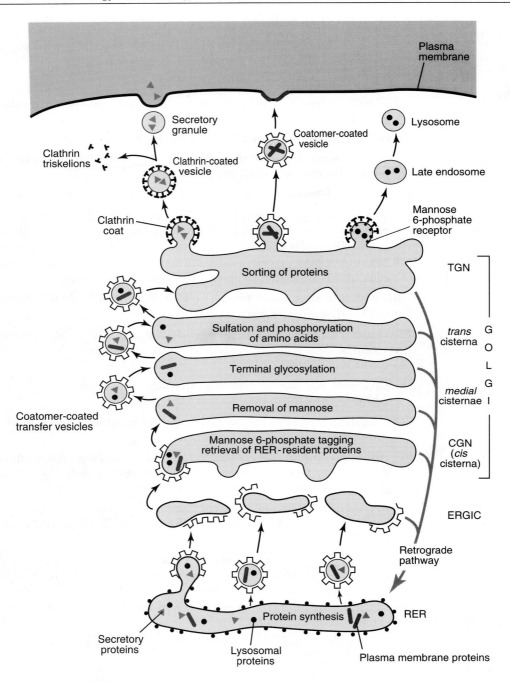

Figure 3.8. The pathways of secretory proteins in separate compartments of the Golgi complex. Proteins synthesized in the RER include secretory (▲), membrane (∕), and lysosomal (●) proteins. These proteins bud off the transitional element of the rough endoplasmic reticulum (RER) via coat protein (COP)-coated vesicles and enter an endoplasmic reticulum-Golgi-intermediate compartment (ERGIC), also called vesicular-tubular clusters (VTC). From here they are transported to the cis Golgi [also called cis Golgi network (CGN)] via COP-coated vesicles. Anterograde passage through the Golgi cisternae is either via COP-coated vesicles or by cisternal maturation. Cisternal maturation coupled with retrograde vesicular transport of Golgi enzymes is currently a favored view. Only COP-I vesicles function in retrograde transport (from Golgi to ERGIC or RER) but both COP-II and COP-I may function in anterograde transport. All proteins do not undergo all of the chemical modifications (e.g., only lysosomal proteins undergo tagging with mannose-6-phosphate). Final sorting occurs in the trans Golgi network (TGN).

(1) **Regulated secretory proteins** are sorted from membrane and lysosomal proteins and delivered via clathrin-coated vesicles to condensing vacuoles, where removal of water, via ionic exchanges, yields **secretory granules.**

(2) **Lysosomal proteins** are sorted into clathrin-coated regions of the TGN that have receptors for mannose 6-phosphate, and are delivered to late endosomes via clathrin-coated vesicles.

(3) **Plasma-membrane proteins** are sorted into coatomer-coated regions of the TGN and delivered to the plasma membrane in COP-II-coatomer-coated vesicles.

2. **Synthesis of cytosolic proteins** takes place on polyribosomes **free** in the cytosol and is directed by mRNAs that **lack** signal codons. The proteins (e.g., protein kinase and hemoglobin) are released directly into the cytosol.

C. **Intracellular digestion**

1. **Nonlysosomal digestion** is the degradation of cytosolic constituents by mechanisms outside of the vacuolar lysosomal pathway. The major site for the degradation of unwanted proteins is the **proteosome,** a cylindrical complex of nonlysosomal proteases. Proteins marked for destruction are enzymatically tagged with **ubiquitin,** which delivers them to the proteosome, where they are broken down to small peptides.

2. **Lysosomal digestion** (see Figure 3.4) is the degradation of material within various types of lysosomes by lysosomal enzymes. Different lysosomal compartments are involved, depending on the origin of the material to be degraded.

   a. **Heterophagy** is the ingestion and degradation of **foreign material** taken into the cell by receptor-mediated endocytosis or phagocytosis.

      (1) Digestion of **endocytosed** ligands occurs in **multivesicular bodies** (see Figure 3.4).

      (2) Digestion of **phagocytosed** microorganisms and foreign particles begins and may be completed in **phagolysosomes.**

   b. **Autophagy** is the segregation of an organelle or other cell constituents within membranes from the RER to form an **autophagic vacuole,** which is subsequently digested in an **autophagolysosome.**

   c. **Crinophagy** is the fusion of **hormone secretory granules** with lysosomes and their subsequent digestion. Crinophagy is used to remove **excess numbers** of secretory granules from the cell.

IV. Clinical Considerations

A. **Lysosomal storage diseases** are hereditary conditions in which the synthesis of specific lysosomal acid hydrolases is impaired. They are characterized by the **inability of lysosomes to degrade certain compounds,** which accumulate and interfere with cell functioning.

1. In **Tay-Sachs disease,** glycolipids accumulate in neurons.

2. In **glycogen storage disease,** glycogen is abundant in the liver and muscle.

3. In **Hurler syndrome,** glycosaminoglycans accumulate in many tissues and organs.

B. **Familial hypercholesterolemia** is associated with a **decreased ability of cells to take in cholesterol,** which normally is ingested by receptor-mediated endocytosis of **LDLs.**

1. This disease is caused by an inherited genetic defect that results in an inability to synthesize LDL receptors, or in the synthesis of defective receptors unable to bind either to LDLs or to clathrin-coated pits.

2. It is characterized by an elevated level of cholesterol in the bloodstream, which facilitates early development of **atherosclerosis,** which may be fatal.

C. **Peroxisomal diseases**

1. **Zellweger syndrome** is a genetic disease in which normal peroxisomes are absent. Infants with this syndrome have profound neurological, liver, and kidney problems and usually die within a few months. Electron micrographs of biopsies from these patients reveal empty peroxisomes, lacking enzymes. Although peroxisomal enzymes may be synthesized, they become mislocated in the cytosol.

2. **Adrenoleukodystrophy** is caused by the inability of peroxisomes to metabolize fatty acids. Therefore, lipids accumulate in the nervous system and adrenal glands, impairing their function.

D. **Tumor diagnosis** is often based on immunocytochemical identification of the intermediate filaments in tumor cells because the type of intermediate filament present identifies the tissue from which the metastatic cancer cells originated.

Review Test

Directions: Each of the numbered items or incomplete statements in this section is followed by answers or by completions of the statement. Select the ONE lettered answer or completion that is BEST in each case.

1. Which of the following organelles divides by fission? *milochondria or peroxsina*

(A) Golgi complex
(B) Rough endoplasmic reticulum
(C) Peroxisome
(D) Smooth endoplasmic reticulum
(E) Centriole

2. A 30-year-old man with very high blood cholesterol levels (290 milligrams per deciliter) has been diagnosed with premature atherosclerosis. His father died of a heart attack at age 45, and his mother, age 44, has coronary artery disease. Which of the following is the most likely explanation of his condition?

(A) He has a lysosomal storage disease and cannot digest cholesterol.
(B) He suffers from a peroxisomal disorder and produces low levels of hydrogen peroxide.
(C) The smooth endoplasmic reticulum (SER) in his hepatocytes has proliferated and produced excessive amounts of cholesterol.
(D) He has a genetic disorder and synthesizes defective low-density lipoprotein (LDL) receptors.
(E) He is unable to manufacture endosomes.

The response options for the next five items are the same. Each item will state the number of options to select. Choose exactly this number.

Questions 3-7

For each process, select the option that fits best.

(A) Clathrin-coated vesicle
(B) Coatomer-coated vesicle
(C) Both
(D) Neither

3. Movement of protein from the rough endoplasmic reticulum (RER) to the endoplasmic reticulum-Golgi-intermediate compartment (ERGIC) (**SELECT 1 OPTION**)

membian recycling after secretion *via*

4. Retrieval of secretion-granule membrane after exocytosis (**SELECT 1 OPTION**)

5. Vesicular movement of protein from trans to cis Golgi cisternae (**SELECT 1 OPTION**)

6. Uncoupling of endocytosed ligands from receptors (**SELECT 1 OPTION**)

7. Movement of acid hydrolases from the trans Golgi network to a late endosome (**SELECT 1 OPTION**)

The response options for the next four items are the same. Each item will state the number of options to select. Choose exactly this number.

Questions 8-11

For each description, select the cell component that it best describes.

(A) Microfilament
(B) Intermediate filament
(C) Microtubule

8. Is associated with kinesin (**SELECT 1 CELL COMPONENT**)

9. Consists of globular actin monomers linked into a double helix (**SELECT 1 CELL COMPONENT**)

10. Has a rigid wall composed of 13 protofilament strands (**SELECT 1 CELL COMPONENT**)

11. Provides structural support to astrocytes (**SELECT 1 CELL COMPONENT**)

Answers and Explanations

1-C. A peroxisome originates from preexisting peroxisomes. It imports specific cytosolic proteins and then undergoes fission. The other organelle that divides by fission is the mitochondrion.

2-D. Cells import cholesterol by the receptor-mediated uptake of low-density lipoproteins (LDLs) in coated vesicles. Certain individuals inherit defective genes and cannot make LDL receptors, or they make defective receptors that cannot bind to clathrin-coated pits. The result is an inability to internalize LDLs, which leads to high levels of LDLs in the bloodstream. High LDL levels predispose a person to premature atherosclerosis and increase the risk of heart attacks.

3-B. Transport of protein from the rough endoplasmic reticulum (RER) to the endoplasmic reticulum-Golgi-intermediate compartment (ERGIC) occurs via (COP-II) coatomer-coated vesicles.

4-A. Membrane recycling after exocytosis of the contents of a secretion granule occurs via clathrin-coated vesicles.

5-B. Transfer of material among the cisternae of the Golgi complex in a retrograde direction takes place via (COP-I) coatomer-coated vesicles.

6-D. The uncoupling of ligands and receptors internalized by receptor-mediated endocytosis occurs in the early endosome.

7-A. Proteins targeted for lysosomes (via late endosomes) leave the trans Golgi network in clathrin-coated vesicles.

8-C. Kinesin is a force-generating protein associated with microtubules. It serves as a molecular motor for the transport of organelles and vesicles.

9-A. Globular actin monomers (G actin) polymerize into a double helix of filamentous actin (F actin), also called a microfilament, in response to the regulatory influence of a number of actin-binding proteins.

10-C. A microtubule consists of α- and β-tubulin dimers polymerized into a spiral around a hollow lumen to form a fairly rigid tubule. When cross-sectioned, the microtubule reveals 13 protofilament strands, which represent the tubulin dimers present in one complete turn of the spiral.

11-B. Glial filaments are a type of intermediate filament, composed of glial fibrillary acidic protein, present in fibrous astrocytes. These filaments are supportive, but they may also play additional roles in both normal and pathologic processes in the central nervous system.

4

Extracellular Matrix

I. Overview—The Extracellular Matrix

A. Structure. The extracellular matrix is an **organized meshwork of macromolecules** surrounding and underlying cells. Although it varies in composition, in general it consists of an amorphous **ground substance** [containing primarily glycosaminoglycans (GAGs), proteoglycans, and glycoproteins] and **fibers** (Figure 4.1).

B. Functions. The extracellular matrix, along with water and other small molecules (e.g., nutrients, ions), constitutes the **extracellular environment.** By affecting the metabolic activities of cells in contact with it, the extracellular matrix may alter the cells as well as influence their shape, migration, division, and differentiation.

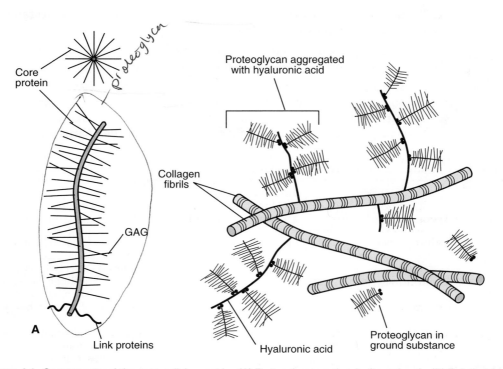

Figure 4.1. Components of the extracellular matrix. *(A)* Proteoglycan molecule (two views). *(B)* Relationships among various extracellular matrix molecules. GAG = glycosaminoglycan. (Adapted with permission from Henrikson RC, Kaye GI, Mazurkiewicz JE: *NMS Histology.* Baltimore, Lippincott Williams and Wilkins, 1997, p 104.

II. Ground Substance

A. GAGs are long, unbranched polysaccharides composed of **repeating disaccharide units.**

1. An **amino sugar** (either *N*-acetylglucosamine or *N*-acetylgalactosamine) is always one of the repeating disaccharides.

2. Because GAGs are commonly **sulfated** and usually possess a **uronic acid sugar** (which has a carboxyl group) in the repeating disaccharide unit, they have a strong **negative charge.**

3. GAGs are generally linked to a **core protein.**

4. The attraction of osmotically active cations (e.g., Na+) to GAGs results in a heavily hydrated matrix that strongly **resists compression.**

5. Their extended random coils occupy large volumes of space because they are unable to fold compactly.

6. GAGs may be classified into four main groups based on their chemical structure (Table 4.1).

   a. **Hyaluronic acid** is a very large, nonsulfated molecule up to 20 micrometers (μm) in length that is not attached to a core protein.

   b. The other three GAG groups are **chondroitin sulfate** and **dermatan sulfate; heparin** and **heparan sulfate;** and **keratan sulfate.**

B. Proteoglycans consist of a **core protein from which many GAGs extend.** These large molecules are shaped like a bottle brush (see Figure 4.1A).

1. Proteoglycans may attach to hyaluronic acid, via their core proteins, to form large, complex aggregates.

2. Their core proteins, their molecular size, and the number and types of GAGs they contain show marked heterogeneity.

3. **Function.** Proteoglycans act as binding sites for **growth factors** (e.g., fibroblast growth factor) **and other signaling molecules** and confer unique attributes to the extracellular matrix in certain locations (e.g., **selective permeability** in the filtration barrier of the glomerulus).

Table 4.1. Classification of Glycosaminoglycans

Group	Glycosaminoglycans	Linked to Core Protein	Sulfated	Major Locations in Body
I	Hyaluronic acid	No	No	Synovial fluid, vitreous humor, cartilage, skin, most connective tissues
II	Chondroitin sulfate	Yes	Yes	Cornea, cartilage, bone, adventitia of arteries
	Dermatan sulfate	Yes	Yes	Skin, blood vessels, heart valves
III	Heparin	Yes	Yes	Lung, skin, liver, mast cells
	Heparan sulfate	Yes	Yes	Basal laminae, lung, arteries, cell surfaces
IV	Keratan sulfate	Yes	Yes	Cornea, cartilage, nucleus pulposus of intervertebral disks

C. Glycoproteins

1. **Fibronectin**

   a. **Types and location**

      (1) **Matrix fibronectin,** an **adhesive glycoprotein,** forms fibrils in the extracellular matrix.

      (2) **Cell-surface fibronectin** is a protein that transiently attaches to the surface of cells.

      (3) **Plasma fibronectin** is a circulating plasma protein that functions in blood clotting, wound healing, and phagocytosis.

   b. **Function.** Fibronectin is a **multifunctional** molecule.

      (1) Fibronectin has domains for **binding collagen, heparin,** various **cell-surface receptors,** and **cell-adhesion molecules (CAMs).**

      (2) It **mediates cell adhesion** to the extracellular matrix by binding to fibronectin receptors on the cell surface.

2. **Laminin** is located in basal laminae, where it is synthesized by adjacent epithelial cells, and in external laminae surrounding muscle cells and Schwann cells.

   a. The "arms" of this large **cross-shaped glycoprotein** have binding sites for cell-surface receptors (integrins), heparan sulfate, type IV collagen, and entactin.

   b. **Function.** Laminin mediates interaction between epithelial cells and the extracellular matrix by anchoring the cell surface to the basal lamina.

3. **Entactin** is a component of all basal (external) laminae.

   a. This sulfated adhesive glycoprotein **binds laminin.**

   b. **Function.** Entactin links laminin with type IV collagen in the lamina densa.

4. **Tenascin** is an adhesive glycoprotein, most abundant in embryonic tissues.

   a. Tenascin is secreted by glial cells in the developing nervous system.

   b. **Function.** Tenascin promotes cell-matrix adhesion and thus plays a role in cell migration.

5. **Chondronectin** is a glycoprotein in cartilage and attaches chondrocytes to type II collagen.

   a. This **multifunctional** molecule has binding sites for collagen, proteoglycans, and cell-surface receptors.

   b. **Function.** By influencing the composition of its extracellular matrix, chondronectin plays a role in the development and maintenance of cartilage.

6. **Osteonectin**

   a. This extracellular matrix glycoprotein is found in bone.

   b. **Function.** Osteonectin links minerals to type I collagen and influences calcification by inhibiting crystal growth.

D. Fibronectin receptors belong to the **integrin family of receptors** and are transmembrane proteins, consisting of two polypeptide chains.

1. Because they enable cells to adhere to the extracellular matrix, they are known as cell-adhesion molecules **(CAMs).**

2. They bind to fibronectin via a specific tripeptide sequence (Arg-Gly-Asp; RGB sequence); other extracellular adhesive proteins also contain this sequence.

3. **Function.** They link fibronectin outside the cell to cytoskeletal components (e.g., actin) inside the cell (Figure 4.2).

III. Fibers

A. **Collagen** is the most abundant structural protein of the extracellular matrix. It exists in at least 20 molecular types, which vary in the amino acid sequence of their three α chains (Table 4.2).

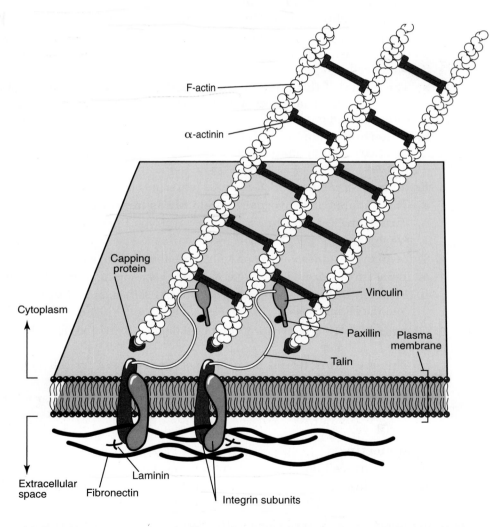

Figure 4.2. Integrin receptors, such as the fibronectin receptor, link molecules outside the cell with components inside the cell. This is common at focal contacts (adhesion plaques), where the integrins serve as transmembrane linkers that mediate reciprocal interactions between the cytoskeleton and the extracellular matrix. (Adapted from Gartner LP, Hiatt JL: *Color Textbook of Histology,* 2nd ed. New York, Saunders, 2001, p 46.)

Table 4.2. Characteristics of Common Collagen Types*

Molecular Type	Cells Synthesizing	Major Locations in Body	Function
I	Fibroblast Osteoblast Odontoblast	Dermis of skin, bone, tendons, ligaments, fibrocartilage	Resists tension
II	Chondroblast	Hyaline cartilage	Resists intermittent pressure
III	Fibroblast Reticular cell Smooth muscle Schwann cell Hepatocyte	Lymphatic system, cardio-vascular system, liver, lung, spleen, intestine, uterus, endoneurium	Forms structural framework in expandable organs
IV	Endothelial cell Epithelial cell	Basal lamina	Provides support and filtration
	Muscle cell Schwann cell	External lamina	Acts as scaffold for cell migration
V	Mesenchymal cell	Placenta	Unknown
VII	Keratinocyte	Dermal-epidermal junction	Forms anchoring fibrils that secure lamina densa to underlying connective tissue

*The six most abundant collagen types are included in this table.

1. Collagen synthesis and assembly into **fibrils** takes place though a series of **intracellular** and **extracellular** events (Figure 4.3).
   a. **Intracellular events in collagen synthesis** occur in the following sequence.
      (1) **Preprocollagen synthesis** occurs at the rough endoplasmic reticulum (RER) and is directed by mRNAs that encode the different types of α chains to be synthesized.
      (2) **Hydroxylation** of specific proline and lysine residues of the forming polypeptide chain occurs within the RER. The reaction is catalyzed by specific **hydroxylases,** which require vitamin C as a co-factor.
      (3) **Attachment of sugars (glycosylation)** to specific hydroxylysine residues also occurs within the RER.
      (4) **Procollagen (triple-helix) formation** takes place in the RER and is precisely regulated by **propeptides** (extra nonhelical amino acid sequences), located at both ends of each α chain. The three α chains align and coil into a triple helix.
      (5) **Addition of carbohydrates** occurs in the Golgi complex to which procollagen is transported via transfer vesicles. With the addition of carbohydrates, the oligosaccharide side chains are completed.

INTRACELLULAR EVENTS

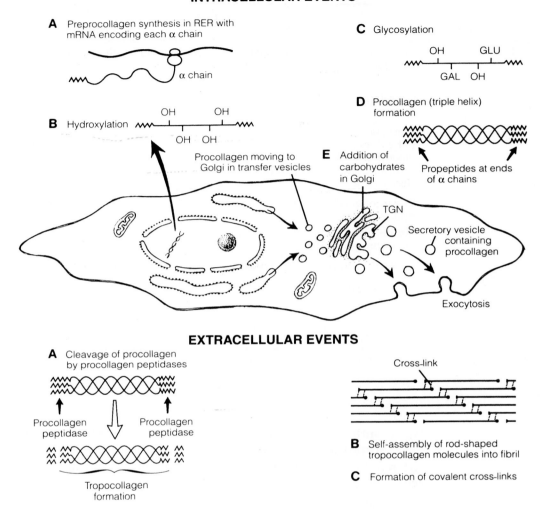

A Preprocollagen synthesis in RER with mRNA encoding each α chain

α chain

B Hydroxylation

OH   OH

OH   OH

C Glycosylation

OH   GLU

GAL   OH

D Procollagen (triple helix) formation

Propeptides at ends of α chains

Procollagen moving to Golgi in transfer vesicles

E Addition of carbohydrates in Golgi

TGN

Secretory vesicle containing procollagen

Exocytosis

EXTRACELLULAR EVENTS

A Cleavage of procollagen by procollagen peptidases

Procollagen peptidase

Procollagen peptidase

Tropocollagen formation

Cross-link

B Self-assembly of rod-shaped tropocollagen molecules into fibril

C Formation of covalent cross-links

Figure 4.3. The intracellular and extracellular steps involved in the synthesis of a collagen fibril. RER = rough endoplasmic reticulum; mRNA = messenger ribonucleic acid; TGN = trans Golgi network. (Adapted with permission from Junqueira LC, Carneiro J, Kelley RO: *Basic Histology*, 9th ed. Norwalk, CT, Appleton & Lange, 1998, p 101.)

    (6) Secretion of procollagen occurs by exocytosis after secretory vesicles from the trans Golgi network are guided to the cell surface along microtubules.

  b. Extracellular events in collagen synthesis occur in the following sequence.

    (1) Cleavage of procollagen is catalyzed by **procollagen peptidases,** which remove most of the propeptide sequences at the ends of each α chain, yielding **tropocollagen** (or simply collagen).

    (2) Self-assembly of tropocollagen occurs as insoluble tropocollagen molecules aggregate near the cell surface.

      (a) Fibrils characteristic of types I, II, III, V, and VII collagen are produced.

(b) These fibrils have a transverse banding periodicity of 67 nanometers (nm) in types I, II, and III collagen (Figure 4.4); the periodicity varies in other types of collagen.

(3) Covalent bond formation (cross-linking) occurs between adjacent tropocollagen molecules and involves formation of lysine- and hydroxylysine-derived aldehydes. This cross-linking imparts great tensile strength to collagen fibrils.

2. Synthesis of **type IV collagen** is unique in that it assembles into a meshwork, rather than fibrils.

   a. Type IV collagen constitutes most of the **lamina densa** of basal laminae (and external laminae).

   b. It differs from other collagen types as follows.

   (1) The propeptide sequences are not removed from the ends of its procollagen molecules.

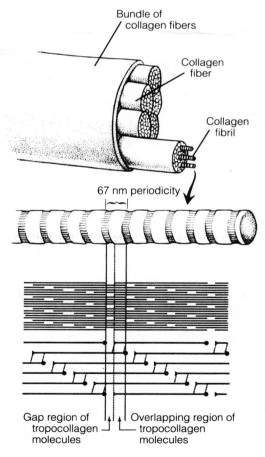

Bundle of collagen fibers

Collagen fiber

Collagen fibril

67 nm periodicity

Gap region of tropocollagen molecules

Overlapping region of tropocollagen molecules

Figure 4.4. The levels of organization in collagen fibers. As seen by light microscopy, collagen fibers consist of collagen fibrils (which typically reveal a 67-nanometer (nm) cross-banding when observed by electron microscopy). The periodicity along the collagen fibril is due to the precise arrangement of tropocollagen molecules, which overlap each other, producing gap regions where electron-dense stains penetrate and produce a transverse banding across the fibril. (Adapted with permission from Junqueira LC, Carneiro J, Kelley RO: *Basic Histology*, 9th ed. Norwalk, CT, Appleton & Lange, 1998, p 99.)

      (2) Its triple-stranded helical structure is interrupted in many regions.

      (3) It forms head-to-head dimers, which interact to form lateral associations, creating a sheet-like meshwork.

B. Elastic fibers

  1. Components

    a. Elastin, an amorphous structural protein, imparts remarkable elasticity to the extracellular matrix. Ninety percent of elastic fibers or elastic sheets are composed of elastin.

      (1) Elastin is unusual in that its lysine molecules form unique linkages with one another.

      (2) Lysine residues of four different chains form covalent bonds (called desmosine cross-links) to create an extensive elastic network.

      (3) Once the tensile force is released, the elastin returns to its original shape after being stretched (similar to a rubber band).

    b. Fibrillin, a glycoprotein, organizes elastin into fibers and is the main component of the **peripheral microfibrils** of elastic fibers.

  2. Synthesis of elastic fibers is carried out by **fibroblasts** (in elastic ligaments), **smooth muscle cells** (in large arteries), and **chondrocytes** and **chondroblasts** (in elastic cartilage).

    a. Synthesis begins with the elaboration of **fibrillin microfibrils,** which appear near the surface of the cell.

    b. Elastin then forms among the bundles of microfibrils.

IV. Clinical Considerations

A. Scurvy is associated with a **deficiency of vitamin C.**

  1. This disease is caused by the synthesis of poorly hydroxylated tropocollagen, which is unable to form either a stable triple helix or collagen fibrils.

  2. Symptoms include bleeding gums and eventual tooth loss.

  3. Administration of vitamin C reverses the disease.

B. Wound healing in adults involves the formation of **fibronectin "tracks"** along which cells migrate to their destinations.

  1. In **connective tissue,** wound healing is often characterized by migration of fibroblasts across blood clots, where they adhere to fibronectin.

  2. In **epithelia,** wound healing involves re-epithelialization, which depends on the **basal lamina serving as a scaffold** for cell migration to cover the denuded area; epithelial cell proliferation and replacement then occurs.

C. Ehlers-Danlos type IV syndrome

  1. This syndrome results from a genetic defect in transcription of deoxyribonucleic acid (DNA) or translation of messenger ribonucleic acid (mRNA) encoding **type III collagen,** the major component of **reticular fibers.**

2. Patients often present with a rupture of the bowel and/or large arteries, where reticular fibers normally ensheath smooth muscle cells.

D. Marfan syndrome results from mutations in the genes encoding **fibrillin,** a critical component of elastic fibers.

1. Patients with this condition have unusually long, slender limbs and long fingers.

2. The lens of the eye often dislocates; cardiovascular problems are common; and the aorta may rupture, causing death.

3. Treatment includes drugs that decrease blood pressure, and in severe cases, surgically replacing the aorta.

Review Test

Directions: Each of the numbered items or incomplete statements in this section is followed by answers or by completions of the statement. Select the ONE lettered answer or completion that is BEST in each case.

1. Which one of the following statements about the fibronectin receptor is true?

(A) It is located exclusively in the basal lamina.
(B) It is a cross-shaped glycoprotein.
(C) It mediates the linkage of molecules outside the cell with cytoskeletal elements inside the cell.
(D) It belongs to the entactin family of receptors.

2. Which one of the following events in collagen synthesis occurs outside of the cell?

(A) Synthesis of preprocollagen
(B) Hydroxylation of lysine residues
(C) Triple-helix formation
(D) Carbohydrate addition to procollagen
(E) Cleavage of procollagen by procollagen peptidases

3. A medical student presents in the emergency room and is diagnosed with a ruptured bowel, the result of a genetic condition called Ehlers-Danlos type IV syndrome. Which one of the following statements about this patient's condition is true?

(A) He has a defect in the synthesis of messenger ribonucleic acid (mRNA) encoding type I collagen.
(B) He has a defect in the genes encoding type IV collagen.
(C) He has defective type II collagen.
(D) He has an increased risk for breaking his bones.
(E) He has a defect in the translation of mRNA for type III collagen.

4. Which one of the following statements about hyaluronic acid is true?

(A) It is a component of elastic fibers.
(B) It is a glycosaminoglycan.
(C) It is a proteoglycan with a shape resembling a bottle brush.
(D) It is sulfated.
(E) It is a small molecule.

5. Which one of the following statements about osteonectin is true?

(A) It is present in the lacunae of bone.
(B) It is a proteoglycan.
(C) It binds to type II collagen.
(D) It influences calcification of bone.

6. Which of the following statements about scurvy is true?

(A) One of its symptoms is "bow legs".
(B) It is a condition caused by excessive glycosylation of tropocollagen.
(C) It is caused by a deficiency of vitamin A.
(D) It is associated with structurally defective elastic fibers.
(E) It is alleviated by eating citrus fruits.

The response options for the next four items are the same. Each item will state the number of options to select. Choose exactly this number.

Questions 7-10

For each description, select the most appropriate protein.

(A) Fibrillin
(B) Fibronectin
(C) Elastin
(D) Entactin
(E) Laminin
(F) Tenascin

7. A glycoprotein across which fibroblasts migrate during wound healing (**SELECT 1 PROTEIN**)

8. An adhesive glycoprotein that links type IV collagen with laminin in the lamina densa (**SELECT 1 PROTEIN**)

9. The main component of peripheral microfibrils in an elastic fiber (**SELECT 1 PROTEIN**)

10. It is present in the basement membrane and is manufactured by connective tissue cells (**SELECT 1 PROTEIN**)

Answers and Explanations

1-C. The fibronectin receptor is a transmembrane protein that enables cells to adhere to the extracellular matrix. Laminin is a cross-shaped glycoprotein located in the basal lamina, where entactin is also present.

2-E. In the extracellular space, peptidases cleave off end sequences of procollagen to yield tropocollagen, which then self-assembles to form collagen fibrils.

3-E. Ehlers-Danlos type IV syndrome is associated with a defect in the synthesis and translation of messenger ribonucleic acid (mRNA) for type III (reticular) collagen.

4-B. Hyaluronic acid is a glycosaminoglycan, not a proteoglycan. The core protein of proteoglycans can attach to hyaluronic acid, forming large aggregates.

5-D. Osteonectin influences the calcification of bone and binds to type I collagen in the bone matrix. Type II collagen is found in cartilage.

6-E. Scurvy is caused by a deficiency of vitamin C, which is a necessary cofactor in the hydroxylation of preprocollagen. Citrus fruits are rich in vitamin C.

7-B. Fibronectin forms "tracks" along which cells migrate. During wound healing in connective tissue, fibroblasts adhere to fibronectin in blood clots, facilitating the healing process.

8-D. Entactin is a sulfated adhesive glycoprotein in basal (external) laminae that binds both type IV collagen and laminin.

9-A. Fibrillin is the major component of the peripheral microfibrils of elastic fibers.

10-B. Fibronectin is synthesized by cells of the connective tissue, usually fibroblasts, and is located in the lamina reticularis near the lamina densa.

5

Epithelia and Glands

I. Overview—Epithelia

A. Structure. Epithelia are **specialized layers** that line the internal and cover the external surfaces of the body. An epithelium consists of a **sheet of cells** lying close together, with little intercellular space. These cells have distinct biochemical, functional, and structural domains that confer **polarity, or sidedness,** to epithelia.

1. The **basement membrane** separates the epithelium from underlying connective tissue and blood vessels.

2. Epithelia are **avascular** and receive nourishment by diffusion of molecules through the **basal lamina.**

B. Classification (Table 5.1). Epithelia are classified into various types based on the **number of cell layers** (one cell layer is **simple;** more than one is **stratified**) and the **shape of the superficial cells** (Figure 5.1). **Pseudostratified** epithelia appear to have multiple cell layers, but all cells are in contact with the basal lamina.

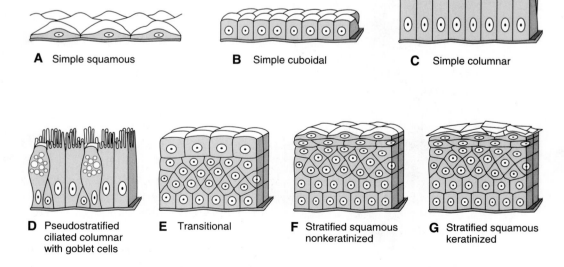

A Simple squamous   **B** Simple cuboidal   **C** Simple columnar

D Pseudostratified ciliated columnar with goblet cells   **E** Transitional   **F** Stratified squamous nonkeratinized   **G** Stratified squamous keratinized

Figure 5.1. Different classifications of epithelia.

Table 5.1. Classification of Epithelia

Type	Shape of Superficial Cell Layer	Typical Locations
One cell layer		
Simple squamous	Flattened	Endothelium (lining of blood vessels), mesothelium (lining of peritoneum and pleura)
Simple cuboidal	Cuboidal	Lining of distal tubule in kidney and ducts in some glands, surface of ovary
Simple columnar	Columnar	Lining of intestine, stomach, and excretory ducts in some glands
Pseudostratified	All cells rest on basal lamina, but not all reach the lumen; thus the epithelium appears falsely stratified	Lining of trachea, primary bronchi, nasal cavity, and excretory ducts in parotid gland
More than one cell layer		
Stratified squamous (nonkeratinized)	Flattened (nucleated)	Lining of esophagus, vagina, mouth, and true vocal cords
Stratified squamous (keratinized)	Flattened (and without nuclei)	Epidermis of skin
Stratified cuboidal	Cuboidal	Lining of ducts in sweat glands
Stratified columnar	Columnar	Lining of large excretory ducts in some glands and cavernous urethra
Transitional	Dome-shaped (when relaxed), flattened (when stretched)	Lining of urinary passages from renal calyces to the urethra

C. Function

1. **Transcellular transport** of molecules from one epithelial surface to another occurs by various processes, including the following:

   a. **Diffusion of oxygen and carbon dioxide** across the epithelial cells of lung alveoli and capillaries

   b. **Carrier protein-mediated transport** of amino acids and glucose across intestinal epithelia

   c. **Vesicle-mediated transport** of immunoglobulin A (IgA) and other molecules

2. **Absorption** occurs via **endocytosis** or **pinocytosis** (see Chapter 3 III A) in various organs (e.g., the proximal convoluted tubule of the kidney).

3. **Secretion** of various molecules (e.g., hormones, mucus, proteins) occurs by **exocytosis.**

4. **Selective permeability** results from the presence of **tight junctions** between epithelial cells and permits fluids with different compositions and concentrations to exist on separate sides of an epithelial layer (e.g., intestinal epithelium).

5. **Protection** from abrasion and injury is provided by the **epidermis,** the epithelial layer of the skin.

II. Lateral Epithelial Surfaces (Figure 5.2). These surfaces contain specialized **junctions** that provide adhesion between cells and restrict movement of materials into and out of lumina.

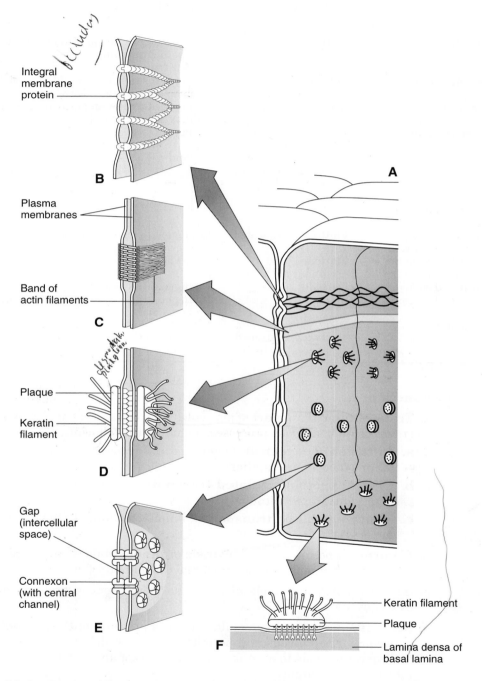

Figure 5.2. Junctional specializations. *A,* apical surface of epithelium; *B,* tight junction; *C,* intermediate junction, *D,* desmosome; *E,* gap junction; *F,* hemidesmosome. (Adapted with permission from Gartner LP, Hiatt JL: *Color Atlas of Histology,* 2nd ed. Baltimore, Williams & Wilkins, 1994.)

A. The **junctional complex** is an intricate arrangement of membrane-associated structures that functions in cell-to-cell attachment of columnar epithelial cells. It corresponds to the "terminal bar" observed in epithelia by light microscopy and consists of three distinct components that are visible by electron microscopy.

  1. The **tight junction (zonula occludens)** is a **zone that surrounds the entire apical perimeter** of adjacent cells and is formed by **fusion of the outer leaflets** of the cells' plasma membranes (see Figure 5.2).

     a. In freeze-fracture preparations of this zone, the tight junction is visible as a branching anastomosing network of intramembrane **strands** on the P-face and **grooves** on the corresponding E-face. The strands consist of transmembrane proteins of each cell **attached directly to one another,** thus sealing off the intercellular space.

     b. The tight junction **prevents movement of substances into the intercellular space** from the lumen. This ability (its "tightness") is directly related to the number and complexity of the intramembrane strands.

     c. The tight junction is analogous to the **fascia occludens,** a ribbon-like area of fusion between transmembrane proteins on adjacent **endothelial cells** lining capillaries.

  2. The **intermediate junction (belt desmosome; zonula adherens) is the zone that surrounds the entire perimeter** of epithelial cells just basal to the tight junction (see Figure 5.2).

     a. It is characterized by a 10- to 20-nanometer (nm) separation between the adjacent plasma membranes, where the extracellular portions of cadherin molecules occupy the intercellular space.

     b. A mat of **actin filaments** is located on each of its cytoplasmic surfaces. The actin filaments are linked, via α-actinin and vinculin, to the transmembrane glycoprotein **E-cadherin.** This protein is markedly dependent on calcium ions for promoting **adhesion** at this structurally supportive junction.

     c. It is analogous to the **fascia adherens,** a ribbon-like adhesion zone present in the **intercalated disks** of cardiac muscle.

  3. A **desmosome (macula adherens)** is a small, discrete, disk-shaped **adhesive site.** It is also commonly found at sites other than the junctional complex, where it joins epithelial cells.

     a. It is characterized by a **dense plaque** of **intracellular** attachment proteins, called **desmoplakins,** on the cytoplasmic surface of each opposing cell.

     b. Intermediate **keratin** filaments **(tonofilaments)** loop into and out of the dense plaque from the cytoplasm.

     c. It also contains transmembrane linker glycoproteins, called desmogleins and desmocollins, that are cadherin molecules (see Figure 5.2).

B. The **gap junction (communicating junction; nexus)** is not part of the junctional complex and is common in certain tissues other than epithelia (e.g., central nervous system, cardiac muscle, and smooth muscle).

  1. Gap junctions **couple adjacent cells metabolically and electrically.**

2. This plaque-like entity is composed of an **ordered array of subunits** called connexons, which extend beyond the cell surface into the **gap** to keep the opposing plasma membranes approximately 2 nm apart (see Figure 5.2).

    a. Connexons consist of six cylindrical subunits (composed of proteins called **connexins**), which are arranged radially around a central channel with a diameter of 1.5 nm.

    b. Precise **alignment of connexons** on adjacent cells produces a junction where **cell-to-cell channels** permit passage of ions and small molecules with a molecular weight of less than 1200 daltons.

    c. Connexins may **alter their conformation** to shut off communication between cells.

C. Lateral interdigitations are irregular, finger-like projections that **interlock** adjacent epithelial cells.

III. Basal Epithelial Surfaces (Figure 5.3)

A. The **basal lamina** is an **extracellular supportive structure** (20-100 nm thick) that is visible only by electron microscopy. It is produced by the epithelium resting upon it and is composed mainly of **type IV collagen, laminin, entactin,** and **proteoglycans** (rich in heparan sulfate).

1. It consists of two zones: the **lamina lucida** (or **lamina rara**), which lies next to the plasma membrane, and the **lamina densa,** a denser meshwork (of type IV collagen, glycoproteins, and glycosaminoglycans), which lies adjacent to the reticular lamina of the deeper connective tissue.

2. The basal lamina plus the underlying **reticular lamina** constitute the **basement membrane,** which is observable by light microscopy.

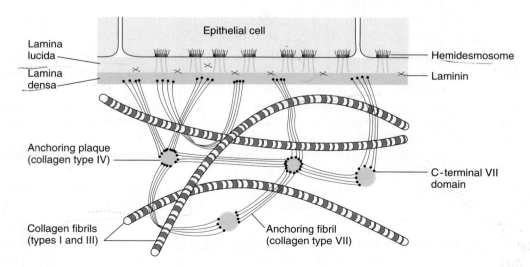

Figure 5.3. Basal specializations beneath an epithelium. (Adapted with permission from Keene DR, Sakai LY, Lunstrum GP, et al. *J Cell Biol* 104:611, 1987.)

B. Hemidesmosomes are specialized junctions that resemble one-half of a desmosome. They mediate **adhesion** of epithelial cells to the underlying extracellular matrix.

1. These junctions are present on the basal surface of **basal cells** in certain epithelia (e.g., tracheal epithelium and stratified squamous epithelium) and on **myoepithelial cells,** where they lie adjacent to the basal lamina.

2. They consist of a **dense cytoplasmic plaque,** which is linked via transmembrane receptor proteins **(integrins)** to laminins in the basal lamina. Anchoring filaments (type VII collagen) from the basal lamina extend deeper into the underlying connective tissue and insert into plaques of type IV collagen.

3. Keratin filaments (tonofilaments) in the cell terminate in the hemidesmosome plaque, thus allowing these junctions to **link the cytoskeleton with the extracellular matrix.**

C. Basal plasma-membrane infoldings are common in **ion-transporting epithelia** (e.g., distal convoluted tubule of the kidney, striated ducts in salivary glands).

1. **They form deep invaginations that compartmentalize** mitochondria.

2. **Function.** They increase the surface area and bring ion pumps [Na^+-K^+ adenosine triphosphatase (ATPase)] in the plasma membrane close to their energy supply (ATP produced in mitochondria).

IV. Apical Epithelial Surfaces. These surfaces may possess specialized structures such as microvilli, stereocilia, and cilia.

A. Microvilli are finger-like **projections of epithelia** [approximately 1 micrometer (μm) in length] that **extend into a lumen** and increase the cell's surface area.

1. A **glycocalyx** (sugar coat) is present on their surfaces (see Chapter 1 II C).

2. A bundle of approximately 30 **actin filaments** runs longitudinally through the core of each microvillus and extends into the **terminal web,** a zone of intersecting filaments in the apical cytoplasm.

3. Microvilli constitute the **brush border** of kidney proximal tubule cells and the **striated border** of intestinal absorptive cells.

B. Stereocilia are very **long microvilli** (not cilia) and are located in the **epididymis** and **vas deferens** of the male reproductive tract.

C. Cilia are **actively motile** processes (5-10 μm in length) extending from certain epithelia (e.g., tracheobronchial and oviduct epithelium) that **propel substances** along their surfaces. They contain a **core of longitudinally arranged microtubules** (the axoneme), which arises from a basal body during ciliogenesis.

1. The **axoneme** (Figure 5.4A) consists of nine doublet microtubules

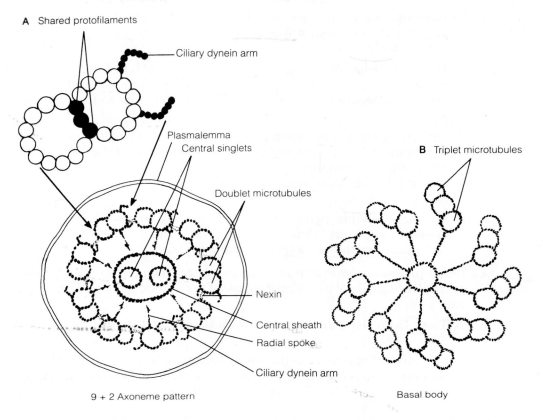

Figure 5.4. Cross-sectional diagrams of a cilium and basal body. (*Part A* adapted with permission from Junqueira LC, Carneiro J, Kelley RO: *Basic Histology*, 9th ed. Norwalk, CT, Appleton & Lange, 1998, p 45. *Part B* adapted with permission from West JB: *Best and Taylor Physiological Basis of Medical Practice*, 12th ed. Baltimore, Williams & Wilkins, 1991, p 12.)

uniformly spaced around two central microtubules **(9 + 2 configuration)** as well as the following components:

   a. **Ciliary dynein arms,** which extend unidirectionally from one member of each doublet microtubule and interact with adjacent doublets, so that they slide past one another. These arms consist of **ciliary dynein,** with a head that is **an ATPase,** which splits ATP to liberate the energy necessary for active movement of a cilium.

   b. **Radial spokes**, which extend from each of the nine outer doublets toward the central sheath

   c. **Central sheath,** which surrounds the two central microtubules; it and the radial spokes regulate the ciliary beat

   d. **Nexin,** an elastic protein that connects adjacent doublet microtubules and helps maintain the shape of the cilium

2. The **basal body** (see Figure 5.4*B*) is a cylindrical structure located at the **base of each cilium** that consists of nine triplet microtubules arranged radially in the shape of a pinwheel **(9 + 0 configuration).** It resembles a centriole (see Figure 3.1) but has a less complex central organization. The inner two triplets of the basal body give rise to the doublet microtubules of the cilium axoneme.

V. Glands. They originate from an epithelium that penetrates the connective tissue and forms secretory units.

A. Structure. A gland consists of a functional portion of secretory and ductal epithelial cells **(parenchyma),** which is separated by a basal lamina from supporting connective tissue elements **(stroma).**

B. Classification. Glands are classified into three types based on the site of secretion: **exocrine glands** secrete into a duct or onto a surface; **endocrine glands** secrete into the bloodstream; and **paracrine glands** secrete into the local extracellular space.

   1. Exocrine glands

      a. Unicellular glands are composed of a single cell (e.g., goblet cells in tracheal epithelium).

      b. Multicellular glands

         (1) Classification is based on two criteria.

            (a) Multicellular glands are classified according to **duct branching** as **simple glands** (duct does not branch) or **compound glands** (duct branches).

            (b) They are further classified according to the **shape of the secretory unit** as **acinar** or **alveolar** (sac or flask-like) or **tubular** (straight, coiled, or branched).

         (2) A connective tissue capsule may surround the gland, or septae of connective tissue may divide the gland into **lobes** and smaller **lobules.**

         (3) Glands may have **ducts** that are located between lobes **(interlobar),** within lobes **(intralobar),** between lobules **(interlobular),** or within lobules **(intralobular)** such as striated and intercalated ducts.

         (4) Multicellular glands secrete various substances.

            (a) Mucus is a viscous material, which usually protects or lubricates cell surfaces.

            (b) Serous secretions are watery and often rich in enzymes.

            (c) Mixed secretions contain both mucus and serous components.

         (5) Mechanisms of secretion vary.

            (a) In **merocrine** glands (e.g., parotid gland), the secretory cells release their contents by exocytosis.

            (b) In **apocrine** glands (e.g., lactating mammary gland), part of the apical cytoplasm of the secretory cell is released along with the contents.

            (c) In **holocrine** glands (e.g., sebaceous gland), the entire secretory cell along with its contents is released.

   2. Endocrine glands may be **unicellular** (e.g., individual endocrine cells in gastrointestinal and respiratory epithelia) or **multicellular** (e.g., adrenal gland), and they **lack a duct system.** In multicellular glands, secretory material is released in the vicinity of fenestrated capillaries, which are abundant just outside the basal lamina of the glandular epithelium.

VI. Clinical Considerations

A. Immotile cilia syndrome results from a genetic defect that causes an abnormal ciliary beat or the absence of a beat.

1. In this syndrome, cilia have axonemes that lack ciliary dynein arms and have other abnormalities.

2. The syndrome is associated with recurrent lower respiratory tract infections, reduced fertility in women, and sterility in men.

B. Epithelial cell tumors occur when cells fail to respond to normal growth regulatory mechanisms.

1. These tumors are **benign** when they remain **localized.**

2. They are **malignant** when they **metastasize** to other parts of the body.
   a. **Carcinomas** are malignant tumors that arise from **surface epithelia.**
   b. **Adenocarcinomas** are malignant tumors that arise from **glands.**

C. Bullous pemphigoid is an **autoimmune disease** in which antibodies against hemidesmosomes are produced.

1. This disease is characterized by chronic, generalized **blisters** in the skin.

2. These blisters cause the epithelium to separate from the underlying substratum.

Review Test

Directions: Each of the numbered items or incomplete statements in this section is followed by answers or by completions of the statement. Select the ONE lettered answer or completion that is BEST in each case. ~~zonula occludens~~ ~~disk-shaped~~.

1. Which one of the following statements about the desmosome is true?

(A) It is sometimes called a nexus.
(B) It permits the passage of large proteins from one cell to an adjacent cell.
(C) It has a plaque made up of many connexons.
(D) It facilitates metabolic coupling between adjacent cells.
(E) It is a disc-shaped adhesion site between epithelial cells. ⌒

2. A medical student who has chronic lower respiratory infections seeks the advice of an ear, nose, and throat specialist. A biopsy of the student's respiratory epithelium reveals alterations in certain epithelial structures. This patient is most likely to have abnormal

(A) microvilli
(B) desmosomes
(C) cilia
(D) hemidesmosomes
(E) basal plasmalemma infoldings

3. Which one of the following statements about the gap junction is true?

(A) It extends as a zone around the apical perimeter of adjacent cells.
(B) It possesses dense plaques composed in part of desmoplakins.
(C) It permits the passage of ions from one cell to an adjacent cell.
(D) Its adhesion is dependent upon calcium ions.
(E) It possesses transmembrane linker glycoproteins.

4. Which one of the following statements about glands is true?

(A) Exocrine glands lack ducts.
(B) Simple glands have ducts that branch.
(C) Endocrine glands secrete into ducts.
(D) Serous secretions are watery.
(E) Holocrine glands release their contents by exocytosis.

5. Which one of the following statements about epithelia is true?

(A) They are polarized.
(B) They are vascular.
(C) They are completely surrounded by a basal lamina.
(D) They contain wide intercellular spaces.
(E) They are not part of the wall of blood vessels.

6. Which one of the following statements about cilia is true?

(A) They possess a 9 + 0 configuration of microtubules.
(B) They do not contain an axoneme.
(C) They contain ciliary dynein arms.
(D) They are nearly identical to centrioles.

7. Which one of the following statements about stratified squamous epithelium is true?

(A) The surface layer of cells is always keratinized.
(B) The cells in its most superficial layer are flattened.
(C) Its basal cells rest on an elastic lamina.
(D) Its cells lack desmosomes.

The response options for the next three items are the same. Each item will state the number of options to select. Choose exactly this number.

Questions 8–10

For each description, select the disease that is related to it.

(A) Immotile cilia syndrome
(B) Carcinoma
(C) Bullous pemphigoid
(D) Adenocarcinoma

8. An autoimmune disease (**SELECT 1 DISEASE**)

9. A hereditary disease that may be associated with infertility (**SELECT 1 DISEASE**)

10. A tumor arising from glandular epithelium (**SELECT 1 DISEASE**)

Answers and Explanations

1-E. Desmosomes are sites of adhesion characterized by dense cytoplasmic plaques and associated keratin filaments. Only gap junctions permit cell-to-cell communication of small molecules via their connexon "channels."

2-C. Individuals with abnormal respiratory cilia often have recurrent respiratory infections if the cilia are unable to clear the respiratory epithelium of microorganisms, debris, and so forth. The student may have immotile cilia syndrome, which is caused by a genetic defect resulting in cilia with axonemes that lack ciliary dynein arms and thus are unable to beat.

3-C. The gap junction "channel" regulates the passage of ions and small molecules from cell to cell, excluding those having a molecular weight greater than 1200 daltons. The tight junction is the zone of adhesion around the apical perimeter of adjacent cells. The other statements are characteristics of desmosomes.

4-D. Serous secretions produced by glands are often rich in enzymes and watery in consistency. Exocrine glands secrete into ducts, and endocrine glands lack ducts. Merocrine glands use exocytosis to release their products.

5-A. Epithelia are polarized, meaning they show a "sidedness" and have apical and basolateral surfaces with specific functions.

6-C. Cilia contain an axoneme with ciliary dynein arms extending unidirectionally from one member of each doublet. Ciliary dynein has ATPase activity, and when it splits adenosine triphosphate (ATP), the adjacent doublets slide past one another and the cilium moves.

7-B. Stratified squamous epithelium is characterized by flattened cells (with or without nuclei) in its superficial layer. It may or may not be keratinized, and it rests on a basal lamina produced by the epithelium.

8-C. Bullous pemphigoid is an autoimmune disease. Affected individuals form antibodies against their own hemidesmosomes.

9-A. The immotile cilia syndrome results from a genetic defect that prevents synthesis of ciliary dynein adenosine triphosphatase (ATPase), resulting in cilia that cannot actively move. Men are sterile because their sperm are not motile (the flagella in their tails lack this enzyme). Women may be infertile because cilia along their oviducts may fail to move oocytes towards the uterus.

10-D. Adenocarcinomas are epithelial tumors that originate in glandular epithelia. Carcinomas originate from surface epithelia.

6

Connective Tissue

I. Overview—Connective Tissue

A. Structure. Connective tissue is formed primarily of **extracellular matrix,** consisting of **ground substance**, extracellular fluid, and **fibers,** in which various connective tissue **cells** are embedded.

B. Function. Connective tissue **supports** organs and cells, acts as a **medium for exchange** of nutrients and wastes between the blood and tissues, **protects** against microorganisms, **repairs** damaged tissues, and **stores fat.**

II. Extracellular Matrix. The extracellular matrix provides a medium for the transfer of nutrients and waste materials between connective tissue cells and the bloodstream.

A. Ground substance is a colorless, transparent, gel-like material in which the cells and fibers of connective tissue are embedded.

1. It is a complex mixture of **glycosaminoglycans, proteoglycans,** and **glycoproteins** (see Chapter 4 for details on these components).

2. Ground substance is a lubricant, helps prevent invasion of tissues by foreign agents, and resists forces of compression.

B. Fibers (collagen, reticular, and elastic) are long, slender protein polymers present in different proportions in different types of connective tissue.

1. **Collagen fibers.** There are at least 19 molecular types of collagen. The most common collagen types in connective tissue proper are **type I** and **type III collagen**, both consisting of many closely packed **tropocollagen** fibrils. The average diameter of type I collagen fibrils is 75 nanometers (nm) [see Figure 4.4].

   a. Collagen fibers are produced in a two-stage process involving intracellular events (within fibroblasts) and extracellular events (see Figure 4.3).

   b. Collagen fibers have great tensile strength, which imparts both flexibility and strength to tissues containing them.

   c. Bone, skin, cartilage, tendon, and many other structures of the body contain collagen fibers.

2. **Reticular fibers** are extremely **thin** [0.5-2.0 micrometers (μm) in diameter] and are composed primarily of **type III collagen;** they have a higher carbohydrate content than other collagen fibers.

   a. Type III collagen fibers constitute the architectural framework of certain organs and glands.

    b. Because of their high carbohydrate content, they stain black with silver salts.

  3. Elastic fibers are coiled, branching fibers 0.2-1.0 μm in diameter that sometimes form loose networks.

    a. These fibers may be stretched up to 150% of their resting length.

    b. They are composed of microfibrils of **elastin** and **fibrillin** embedded in amorphous elastin (see Chapter 4 III B).

    c. Elastic fibers require special staining in order to be observed by light microscopy.

III. Connective Tissue Cells. Connective tissue cells include many types with different functions. Some originate locally and remain in the connective tissue (fixed or **resident cells**), whereas others originate elsewhere and remain only temporarily in connective tissue **(transient cells)** (Figure 6.1). **Resident connective tissue cells** include fibroblasts, pericytes, adipose cells, mast cells, and "fixed" macrophages. **Transient cells include** certain macrophages, lymphocytes, plasma cells, neutrophils, eosinophils, and basophils.

A. Fibroblasts arise from mesenchymal cells, are the **predominant cells** located in connective tissue proper, and often have an oval nucleus with two or more nucleoli. Fibroblasts seldom undergo mitosis except in wound healing. They may differentiate into other cell types (chondrocytes, osteoblasts, adipose cells) under certain conditions.

  1. Active fibroblasts are spindle-shaped (fusiform) and contain well-developed rough endoplasmic reticulum (RER) and many Golgi complexes. **Synthetically active,** they produce procollagen and other components of the extracellular matrix.

  2. Quiescent fibroblasts are small, flattened cells containing little RER. Although synthetically **inactive,** they may revert to the active state if stimulated (e.g., during wound healing).

B. Pericytes are derived from embryonic mesenchymal cells and may retain the **pluripotential role.**

  1. They possess characteristics of smooth muscle cells and endothelial cells.

  2. They are smaller than fibroblasts and are located mostly along capillaries, yet lie within their own basal lamina.

  3. Pericytes may function as contractile cells that modify capillary blood flow.

  4. They may differentiate into smooth muscle cells as well as endothelial cells during blood vessel formation and repair.

C. Adipose cells arise from mesenchymal cells and perhaps from fibroblasts. They do not normally undergo cell division. They are surrounded by a basal lamina and are responsible for the **synthesis, storage, and release of fat.**

  1. Unilocular adipose cells (white adipose tissue) contain a single large fat droplet. To accommodate the droplet, the cytoplasm and nucleus are squeezed into a thin rim around the cell's periphery.

    a. These cells have plasmalemma **receptors** for insulin, growth hormone, norepinephrine, and glucocorticoids to **control the uptake and release of free fatty acids and triglycerides.**

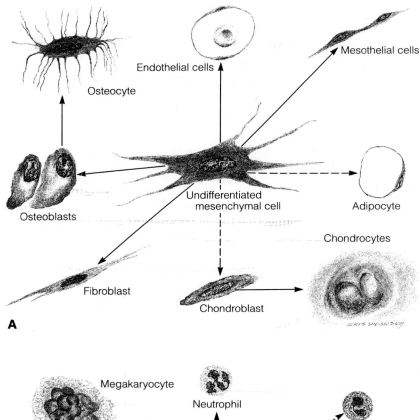

A

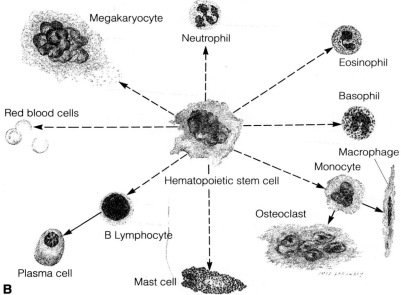

B

Figure 6.1. Origin of connective tissue cells. *(A)* Cells arising from undifferentiated mesenchymal cells are formed in connective tissue and remain there. *(B)* Cells arising from hematopoietic stem cells are formed in the bone marrow and reside transiently in connective tissue. Dotted lines indicate that intermediate cell types occur between those shown.

     b. They generally do not increase in number after a limited postnatal period.

  2. Multilocular adipose cells (brown adipose tissue) contain many small fat droplets, and thus the spherical nucleus is centrally located.

D. Mast cells arise from myeloid stem cells during hemopoiesis and usually reside near small blood vessels. Although they share many of their characteristics with basophils, they develop from different precursors and are not related.

  1. These cells are one of the largest cells of connective tissue proper. They possess a centrally placed, spherical nucleus; their cytoplasm is filled with coarse, deeply stained metachromatic granules; their contents (known as **primary mediators**) are listed in Table 6.1.

  2. Their surfaces are folded, and in electron micrographs they have a well-developed Golgi complex, scant RER, and many dense lamellated granules.

  3. When mast cells become activated during a type I hypersensitivity reaction (see below), phospholipids of their cell membranes can be converted into arachidonic acid by the enzyme phospholipase A_2. Arachidonic acid is in turn converted into **secondary mediators** (listed in Table 6.1).

  4. Mast cells mediate **immediate (type I) hypersensitivity** reactions **(anaphylactic reactions)** as follows:

    a. After the first exposure to an allergen, plasma cells manufacture immunoglobulin E (IgE) antibodies, which bind to **Fc receptors (FcεRI receptors)** on the surface of mast cells (and basophils), causing these cells to become **sensitized.** Common antigens that may evoke this response include plant **pollens, insect venoms,** certain **drugs,** and **foreign serum.**

    b. During the **second exposure** to the same allergen, the membrane-bound IgE binds the allergen. Subsequent cross-linking and clustering of the allergen-IgE complexes trigger **degranulation** of mast cells and the release of primary and secondary mediators (Figure 6.2; Table 6.1).

E. Macrophages are the principal **phagocytosing cells** of connective tissue. They are responsible for removing large particulate matter and assisting in the **immune response.** They also secrete substances that function in wound healing.

  1. Macrophages originate in the bone marrow as **monocytes,** circulate in the bloodstream, then migrate into the connective tissue, where they mature into functional macrophages (see Chapter 10 VI E).

  2. When activated, they display filopodia, an eccentric kidney-shaped nucleus, phagocytic vacuoles, lysosomes, and residual bodies.

  3. When stimulated, they may fuse to form **foreign body giant cells.** These **multinucleated cells** surround and phagocytose large foreign bodies.

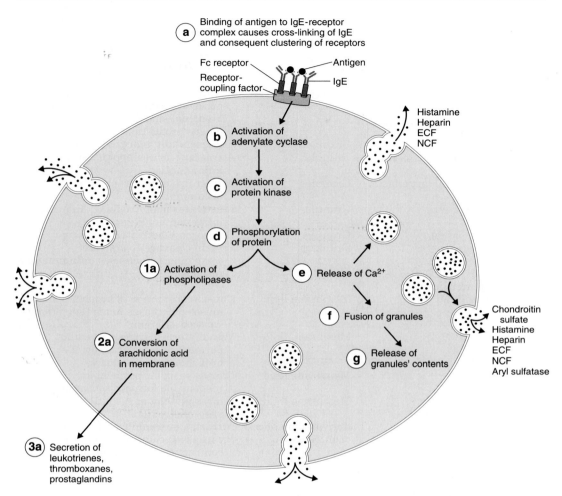

Figure 6.2. Activation and degranulation of the mast cell. ECF = eosinophil chemotactic factor; IgE = immunoglobulin E; NCF = neutrophilic chemotactic factor. (Adapted with permission from Gartner LP, Hiatt JL: *Color Textbook of Histology.* Philadelphia, Saunders, 1997, p 100.)

F. Lymphoid cells arise from lymphoid stem cells during hemopoiesis (see Chapter 10 VI G). They are located throughout the body in the subepithelial connective tissue and accumulate in the respiratory system, gastrointestinal tract, and elsewhere in areas of chronic inflammation. (For more information concerning lymphoid cells see Chapter 12 II, III, and IV.)

1. **T lymphocytes (T cells)** initiate the **cell-mediated immune response.**

2. **B lymphocytes (B cells),** following activation by an antigen, differentiate into plasma cells, which function in the **humoral immune response.**

3. **Natural killer cells (NK cells)** lack the surface determinants characteristic of T and B lymphocytes but may display **cytotoxic activity** against tumor cells.

Table 6.1. Major Mediators Released by Mast Cells

Substance	Intracellular Source	Action
Primary mediators		
Histamine	Granules	Vasodilator; increases vascular permeability; causes contraction of bronchial smooth muscle; increases mucus production
Heparin	Granules	Anticoagulant; inactivates histamine
Eosinophil chemotactic factor (ECF)	Granules	Attractant for eosinophils to site of inflammation
Neutrophil chemotactic factor (NCF)	Granules	Attractant for neutrophils to site of inflammation
Aryl sulfate	Granules	Inactivates leukotriene C_4, limiting inflammatory response
Chondroitin sulfate	Granules	Binds and inactivates histamine
Neutral proteases	Granules	Protein cleavage to activate complement; increases inflammatory response
Secondary mediators		
Prostaglandin D_2	Membrane lipid	Causes contraction of bronchial smooth muscle; increases mucus secretion; vasoconstriction
Leukotrienes C_4, D_4, and E_4	Membrane lipid	Vasodilators; increases vascular permeability; contraction of bronchial smooth muscle
Bradykinins	Membrane lipid	Causes vascular permeability; responsible for pain sensation
Thromboxane A_2	Membrane lipid	Causes platelet aggregation; vasoconstriction
Platelet-activating factor	Activated by phospholipase A_2	Attracts neutrophils and eosinophils; causes vascular permeability; contraction of bronchial smooth muscle

    G. Plasma cells are **antibody-manufacturing cells** that arise from activated B lymphocytes and are responsible for **humoral immunity.**

        1. These ovoid cells contain an eccentrically placed nucleus possessing **clumps of heterochromatin,** which appear to be arranged in a wheel-spoke fashion.

        2. Their cytoplasm is deeply basophilic due to an abundance of RER.

        3. A prominent area adjacent to the nucleus appears pale and contains the Golgi complex (negative Golgi image).

        4. They are most abundant at wound entry sites or in areas of chronic inflammation.

    H. Granulocytes are white blood cells that possess cytoplasmic granules and arise from myeloid stem cells during hemopoiesis. At sites of inflammation, they leave the bloodstream and enter the loose connective tissue, where they perform their specific functions (see Chapter 10 II B 2 A).

        1. Neutrophils phagocytose, kill, and digest bacteria at sites of acute inflammation. **Pus** is an accumulation of dead neutrophils, bacteria, extracellular fluid and additional debris at an inflammatory site.

2. Eosinophils bind to antigen-antibody complexes on the surface of parasites (e.g., helminths) and then release cytotoxins that damage the parasites.

    a. They are most prevalent at sites of chronic or allergic inflammation.

    b. Eosinophils are attracted by eosinophil chemotactic factor (ECF), which is secreted by mast cells and basophils, to sites of allergic inflammation. There, eosinophils release enzymes that cleave histamine and leukotriene C, thus **moderating the allergic reaction.**

    c. These cells also phagocytose antibody-antigen complexes.

3. Basophils are similar to mast cells in that they possess Fc receptors (FcεRI receptors), their granules house the same primary mediators, and the same secondary mediators are manufactured de novo from the phospholipids of their plasmalemma. They differ, however, in that they circulate via the blood stream, whereas mast cells do not (see Table 10.2).

IV. Classification of Connective Tissue. Classification is based on the proportion of cells to fibers as well as on the arrangement and type of fibers (embryonic connective tissue, connective tissue proper, or specialized connective tissue).

A. Embryonic connective tissue

1. Mucous tissue (Wharton jelly) is loose connective tissue that is the main constituent of the umbilical cord. It consists of a jelly-like matrix with some collagen fibers in which large stellate-shaped fibroblasts are embedded.

2. Mesenchymal tissue is found only in **embryos.** It consists of a gel-like amorphous matrix containing only a few scattered reticular fibers in which star-shaped, pale-staining mesenchymal cells are embedded. Mitotic figures are often observed in these cells.

B. Connective tissue proper

1. Loose connective tissue (areolar tissue) possesses relatively fewer fibers but more cells than dense connective tissue.

    a. This tissue is **well vascularized, flexible, and not very resistant to stress.**

    b. It is **more abundant** than dense connective tissue.

2. Dense connective tissue contains more fibers but fewer cells than loose connective tissue. It is classified by the orientation of its fiber bundles into two types:

    a. Dense irregular connective tissue (most common), which contains fiber bundles that have no definite orientation. This tissue is characteristic of the **dermis** and **capsules of many organs.**

    b. Dense regular connective tissue, which contains fiber bundles and attenuated fibroblasts that are arranged in a uniform parallel fashion

        (1) It is present only in **tendons** and **ligaments.**

        (2) This tissue may be collagenous or elastic.

3. **Elastic tissue** is composed of coarse, branching elastic fibers with a sparse network of collagen fibers and some fibroblasts filling the interstitial spaces. It is present in the dermis, lungs, elastic cartilage, and elastic ligaments as well as in large (conducting) blood vessels, where it forms fenestrated sheaths.

4. **Reticular tissue** consists mostly of a network of branched reticular fibers **(type III collagen).**

   a. This tissue invests liver sinusoids, smooth muscle cells, and fat cells and forms the stroma of lymphatic organs, bone marrow, and endocrine glands.

   b. It also forms the reticular lamina of basement membranes.

5. **Adipose tissue** is the primary site for storage of energy (in the form of **triglycerides**) and has a rich neurovascular supply.

   a. **White adipose tissue** is comprised of **unilocular** adipose cells.

      (1) This tissue constitutes nearly all the adult adipose tissue throughout the body.

      (2) It **stores and releases lipids** as follows:

         (a) Adipose cells synthesize the enzyme **lipoprotein lipase,** which is transferred to the luminal aspect of the capillary endothelium.

         (b) Dietary fat is transported to adipose tissue as **very-low-density lipoproteins** (VLDLs) and **chylomicrons.** Lipoprotein lipase then hydrolyzes these substances into fatty acids and glycerol.

         (c) The free fatty acids enter the adipose cells, where they are re-esterified and stored as triglycerides (in fat droplets). Adipose cells also synthesize fatty acids from glucose.

         (d) Lipid storage is stimulated by **insulin,** which increases the rate of synthesis of lipoprotein lipase and the uptake of glucose by adipose cells.

         (e) Release of lipids is affected by neural impulses and/or adrenaline. Stored triglycerides are hydrolyzed by **hormone-sensitive lipase** [which is activated by cyclic adenosine monophosphate (cAMP)]. The free fatty acids are released into the extracellular matrix and then enter the capillary lumen.

   b. **Brown adipose tissue** is composed of **multilocular** adipose cells, which contain many large mitochondria.

      (1) This tissue is capable of **generating heat by uncoupling oxidative phosphorylation. Thermogenin,** a transmembrane protein located in mitochondria, causes the release of protons away from adenosine triphosphate (ATP) synthesis, resulting in heat production.

      (2) This tissue is found in infants (also in hibernating animals) and is much reduced in adults.

C. **Specialized connective tissue**

    1. Cartilage and bone are discussed in Chapter 7.

    2. Blood is discussed in Chapter 10.

V. Clinical Considerations

A. **Edema** is a pathologic process resulting in an **increased volume of tissue fluid.**

    1. Edema may be caused by venous obstruction or decreased venous blood flow (as in congestive heart failure), increased capillary permeability (due to injury), starvation, excessive release of histamine, and obstruction of lymphatic vessels.

    2. Edema that is responsive to localized pressure (i.e., depressions persist after release of pressure) is called **pitting edema.**

B. **Hay fever** and **asthma**

    1. **Hay fever** is characterized by nasal congestion caused by localized edema in the nasal mucosa. This edema results from the increased permeability of small blood vessels due to excessive release of **histamine** from mast cells in the nasal mucosa.

    2. People with **asthma** experience difficulty breathing due to bronchospasms resulting from **leukotrienes** released in the lungs.

C. **Anaphylactic shock** results from the effects of powerful mediators released during an **immediate hypersensitivity reaction** following a second exposure to an allergen.

    1. This reaction can occur within seconds or minutes after contact with an allergen.

    2. Signs and symptoms include shortness of breath, decreasing blood pressure, and other signs and symptoms of shock.

    3. Anaphylactic shock may be life-threatening if untreated.

D. Obesity occurs as either **hypertrophic obesity,** characterized by an increase in adipose cell **size** resulting from increased fat storage **(adult onset),** or **hyperplastic obesity,** characterized by an increase in the **number** of adipose cells, which begins in childhood and is usually lifelong.

Review Test

Directions: Each of the numbered items or incomplete statements in this section is followed by answers or by completions of the statement. Select the ONE lettered answer or completion that is BEST in each case.

1. Which one of the following statements regarding collagen is true?

(A) It is composed of tropocollagen.
(B) Reticular fibers are composed of type II collagen.
(C) It is synthesized mostly by mast cells.
(D) Elastic fibers are composed of type IV collagen.

2. Dense regular connective tissue is present in

(A) capsules of organs
(B) basement membrane
(C) tendons
(D) skin

3. Of the following cell types found in connective tissue, which is most often present along capillaries and resembles fibroblasts?

(A) Plasma cell
(B) Lymphocyte
(C) Macrophage
(D) Pericyte

4. Synovial fluids of normal joints are usually devoid of collagen. It has been demonstrated that patients suffering from rheumatoid diseases will have different types of collagen in their synovial fluid, depending on the tissue being damaged. If a patient has type II collagen in his synovial joint, which of the following tissues is being eroded?

(A) Vascular endothelium
(B) Compact bone
(C) Vascular smooth muscle
(D) Articular cartilage
(E) Synovial membrane

5. Which one of the following cells arises from monocytes?

(A) Plasma cells
(B) Fibroblasts
(C) Lymphocytes
(D) Macrophages

6. Foreign-body giant cells are formed by the coalescence of

(A) macrophages
(B) lymphocytes
(C) fibroblasts
(D) adipose cells

7. Which one of the following cells located in the connective tissue arises from myeloid stem cells?

(A) Pericytes
(B) Eosinophils
(C) Fibroblasts
(D) Osteoblasts
(E) Adipocytes

8. Which of the following cells is responsible for anaphylactic shock?

(A) Fibroblasts
(B) Eosinophils
(C) Pericytes
(D) Mast cells
(E) Macrophages

9. Which one of the following statements regarding proteoglycans is true?

(A) They consist of a core of fibrous protein covalently bound to glycoproteins.
(B) They are attached to ribonucleic acid.
(C) They are binding sites for deoxyribonucleic acid (DNA).
(D) They are composed of a protein core to which glycosaminoglycans are attached.

10. Which one of the following statements concerning loose connective tissue is true?

(A) It is less abundant than dense connective tissue.
(B) It has a lower proportion of cells to fibers than does dense connective tissue.
(C) It acts as a medium for exchange of nutrients and wastes between the blood and tissues.
(D) It provides structural support for organs.
(E) It consists of fibers in which various types of cells are embedded.

Answers and Explanations

1-A. Collagen is composed of closely packed tropocollagen molecules. Reticular fibers are composed of type III collagen, whereas elastic fibers are composed of elastin microfibrils rather than collagen. Fibrocytes are inactive nonsecreting fibroblasts that synthesize the procollagen molecules.

2-C. Tendons are composed of dense regular connective tissue containing collagen fibers arranged in a uniform parallel fashion.

3-D. Pericytes are pluripotential cells that resemble fibroblasts (although they are smaller) and are located adjacent to capillaries.

4-D. Type II collagen is present only in hyaline and elastic cartilages; therefore, finding type II collagen in the synovial fluid of a joint is indicative of erosion of the articular cartilage.

5-D. Monocytes leave the bloodstream and migrate into the connective tissue, where they mature into functional macrophages.

6-A. Foreign-body giant cells result when macrophages coalesce.

7-B. Eosinophils arise from myeloid stem cells during hemopoiesis and migrate to sites of inflammation within the connective tissue. Pericytes, fibroblasts, osteoblasts, and adipocytes arise from undifferentiated mesenchymal cells.

8-D. Mast cells. After first exposure to an allergen, plasma cells make immunoglobulin E (IgE) antibodies that bind to Fc (FcεRI) receptors on mast cells (and basophils), thus sensitizing them. At the second exposure, the allergen binds to IgE, initiating degranulation of mast cells, thus releasing several mediators that give rise to type I hypersensitivity reaction.

9-D. Proteoglycans consist of a protein core to which glycosaminoglycans are attached.

10-C. Both loose and dense connective tissue are composed of three elements: an amorphous ground substance, fibers, and various types of cells. The amorphous ground substance of loose connective tissue is the medium of exchange between the connective tissue cells and the bloodstream.

7

Cartilage and Bone

I. Overview—Cartilage. Cartilage is an **avascular** specialized **fibrous connective tissue.** It has a **firm extracellular matrix** that is less pliable than that of connective tissue proper, and it contains **chondrocytes** embedded in the matrix. Cartilage **functions** primarily to support soft tissues and in the development and growth of long bones. The three types of cartilage—**hyaline cartilage, elastic cartilage,** and **fibrocartilage**—vary in certain matrix components (Table 7.1).

A. Hyaline cartilage (Figure 7.1; see Table 7.1) is the most abundant cartilage and serves as a temporary skeleton in the fetus until it is replaced with bone.

1. Structure

a. Matrix

(1) The matrix is composed of an amorphous ground substance containing **proteoglycan aggregates** and **chondronectin,** in which **type II collagen** is embedded (see Tables 4.2 and 7.1).

Table 7.1. Cartilage Types, Characteristics, and Locations

Type of Cartilage	Identifying Characteristics	Perichondrium	Location
Hyaline	Type II collagen, basophilic matrix, chondrocytes usually arranged in groups (isogenous groups)	Perichondrium usually present except on articular surfaces	Articular ends of long bones, nose, larynx, trachea, bronchi, ventral ends of ribs, template for endochondral bone formation
Elastic	Type II collagen, elastic fibers	Perichondrium present	Pinna of ear, auditory canal and tube, epiglottis, some laryngeal cartilages
Fibrocartilage	Type I collagen, acidophilic matrix, chondrocytes arranged in parallel rows between bundles of collagen, always associated with dense collagenous connective tissue and/or hyaline cartilage	Perichondrium absent	Intervertebral discs, articular discs, pubic symphysis, insertion of tendons, meniscus of knee

Adapted with permission from Gartner LP, Hiatt JL. *Color Textbook of Histology.* Philadelphia, Saunders, 1997, p 111.

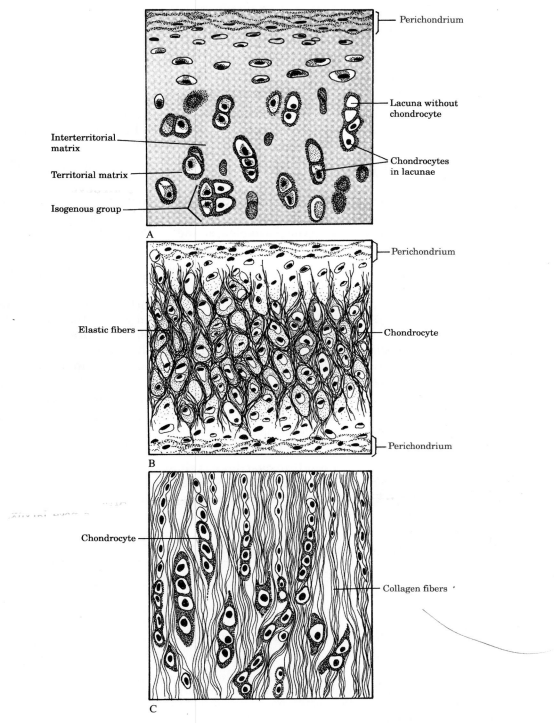

Figure 7.1. The three types of cartilage. *(A)* Hyaline cartilage. *(B)* Elastic cartilage. *(C)* Fibrocartilage. (Reprinted with permission from Borysenko M, Berringer T: *Functional Histology*, 2nd ed. Boston, Little, Brown, 1984.)

    (2) The matrix that is adjacent to chondrocytes is called the **territorial matrix.** This part of the matrix is poor in collagen but rich in proteoglycans and stains more deeply than does the **interterritorial matrix.**

  b. Perichondrium is a layer of dense irregular connective tissue that surrounds hyaline cartilage except at articular surfaces.

    (1) It consists of an **outer fibrous layer,** containing **type I collagen,** fibroblasts, and blood vessels, and an **inner cellular layer,** containing chondrogenic cells and chondroblasts.

    (2) It **provides the nearest blood supply** to the avascular cartilaginous tissue.

  c. Chondroblasts manufacture the cartilage matrix through which nutrients and waste materials pass to and from the cells. These cells contain an extensive Golgi complex, abundant rough endoplasmic reticulum (RER), lipid droplets, and glycogen.

  d. Chondrocytes are mature cartilage cells that are embedded within **lacunae** in the matrix.

    (1) They arise by differentiation of mesenchymal **chondrogenic cells** and from chondrogenic cells located within the inner layer of the perichondrium into chondroblasts, which are the earliest cells to produce cartilage matrix. Once these cells become totally enveloped by matrix, they are referred to as **chondrocytes** (see Figure 6.1A).

    (2) Those located **superficially** are ovoid and positioned with their longitudinal axis parallel to the cartilage surface. Those located **deeper** are more spherical in shape and may occur in groups of four to eight cells **(isogenous groups).**

2. Histogenesis of hyaline cartilage is similar to that of elastic cartilage and fibrocartilage and is affected by certain hormones and vitamins (Table 7.2). It occurs by the following two processes:

Table 7.2. Effects of Hormones and Vitamins on Hyaline Cartilage

	Effects on Cartilage
Hormones	
Thyroxine, testosterone, somatotrophin	Stimulates cartilage histogenesis
Cortisone, hydrocortisone, estradiol	Inhibits cartilage histogenesis
Vitamins	
Hypovitaminosis A	Diminishes thickness of epiphyseal plates
Hypervitaminosis A	Accelerates ossification of epiphyseal plates
Hypovitaminosis C	Stops matrix production, distorts cartilage columns in epiphyseal plates; scurvy develops.
Absence of vitamin D, resulting in deficient absorption of calcium and phosphorus	Epiphyseal cartilage cells proliferate, but matrix fails to calcify and growing bones become deformed; rickets develops

a. **Interstitial growth** results from cell **division of preexisting chondrocytes.** This type of growth occurs only during the early stages of cartilage formation and in articular cartilage and the epiphyseal plates of long bones.

b. **Appositional growth** results from differentiation of chondrogenic cells in the perichondrium. This type of growth results in the formation of chondroblasts and/or new chondrocytes, which elaborate a new layer of cartilage matrix at the periphery.

3. **Degeneration of hyaline cartilage** occurs when chondrocytes undergo hypertrophy and die and the matrix becomes calcified, a process that becomes more frequent with age. Degeneration of hyaline cartilage is a normal part of endochondral bone formation.

B. **Elastic cartilage** (see Figure 7.1 and Table 7.1) possesses a perichondrium and is nearly identical to hyaline cartilage except for a network of elastic fibers, which impart **a yellowish** color. Although it contains type II collagen, it is less prone to degeneration than hyaline cartilage and is located in areas where **flexible support** is required.

C. **Fibrocartilage** (see Figure 7.1 and Table 7.1) **lacks** an identifiable perichondrium. It is characterized by alternating rows of fibroblast-derived chondrocytes surrounded by scant matrix and thick parallel bundles of **type I collagen** fibers. Fibrocartilage is located in areas where **support and tensile strength** are required.

II. Bone

A. **Overview.** Bone is the primary constituent of the adult skeleton. It is a specialized type of connective tissue with a **calcified** extracellular matrix in which characteristic cells are embedded. Bone **functions** to protect vital organs, support fleshy structures, and provide a calcium reserve (bone contains about 99% of the body's calcium). It is a dynamic tissue that constantly undergoes changes in shape. **Applied pressure results in bone resorption,** whereas **applied tension results in bone formation.**

B. **Structure**

1. **Bone matrix**

   a. The **inorganic (calcified) portion of the bone matrix** (about 65% of the dry weight) is composed of calcium, phosphate, bicarbonate, citrate, magnesium, potassium, and sodium. It consists primarily of **hydroxyapatite crystals,** which have the composition $Ca_{10}(PO_4)_6(OH)_2$.

   b. The **organic portion of the bone matrix** (about 35% of the dry weight) consists primarily of **type I collagen** (95%). It has a ground substance that contains **chondroitin sulfate** and **keratan sulfate.**

      (1) **Osteocalcin,** a glycoprotein, binds to hydroxyapatite, which in turn binds to integrins on osteoblasts and osteoclasts.

      (2) **Bone sialoprotein** is a matrix protein that also binds to integrins of the osteoblasts and osteocytes, and is thus related to adherence of bone cells to bone matrix.

2. The **periosteum** is a layer of **noncalcified** connective tissue covering

bone on its **external** surfaces, except at synovial articulations and muscle attachments.

   a. It is composed of an outer, **fibrous,** dense collagenous layer and an inner cellular **osteoprogenitor (osteogenic)** layer.

   b. **Sharpey fibers** (type I collagen) attach the periosteum to the bone surface.

   c. The periosteum functions to distribute blood vessels to bone.

3. The **endosteum** is a thin specialized connective tissue that lines the **marrow cavities** and supplies **osteoprogenitor cells** and **osteoblasts** for bone growth and repair.

C. Bone cells

1. Osteoprogenitor cells

   a. These spindle-shaped cells are derived from embryonic mesenchyme and are **located in the periosteum and endosteum.**

   b. They are capable of differentiating into osteoblasts. However, at **low oxygen tensions, they may change into chondrogenic cells.**

2. Osteoblasts

   a. These cells are derived from osteoprogenitor cells and possess receptors for parathyroid hormone (PTH).

   b. They synthesize and secrete **osteoid** (uncalcified bone matrix).

   c. On bony surfaces, they resemble a layer of cuboidal, basophilic cells as they secrete organic matrix (see Figure 6.1A).

   d. They possess cytoplasmic processes with which they contact the processes of other osteoblasts and osteocytes.

   e. **When synthetically active,** they have a well-developed RER and Golgi complex.

   f. These cells become entrapped in **lacunae** but maintain contact with other cells via their cytoplasmic processes. Entrapped osteoblasts are known as osteocytes.

3. Osteocytes (Figure 7.2)

   a. Osteocytes are **mature bone cells** housed in their own lacunae.

   b. They have narrow cytoplasmic processes that extend through **canaliculi** in the calcified matrix (see Figures 6.1A and 7.2).

   c. They maintain communication with each other via **gap junctions** between their processes.

   d. They are nourished and maintained by nutrients and metabolites contained within the canaliculi.

   e. They contain abundant heterochromatin, a paucity of RER, and a small Golgi complex.

4. Osteoclasts

   a. **Overview. Osteoclasts** are large, motile, multinucleated cells (up to 50 nuclei) that are derived from cells in the bone marrow that are also precursors of **monocytes.** Their cytoplasm is usually **acidophilic.**

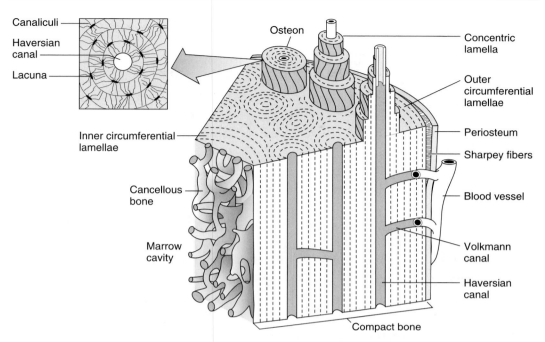

Canaliculi
Haversian canal
Lacuna
Osteon
Concentric lamella
Outer circumferential lamellae
Periosteum
Sharpey fibers
Inner circumferential lamellae
Blood vessel
Cancellous bone
Marrow cavity
Volkmann canal
Haversian canal
Compact bone

Figure 7.2. Histological aspects of a long bone. Inset: cross-section of an osteon. (Adapted with permission from Denhardt D, Guo X: A protein with diverse functions. *FASEB* 7:1475-1482, 1993.)

Osteoclasts function in the **resorption of bone** (osteolysis). They form and are located in depressions known as **Howship lacunae,** which represent areas of bone resorption.

b. Morphology. Osteoclasts display four regions in electron micrographs.

(1) The **ruffled border** is the site of active bone resorption. It is composed of irregular finger-like cytoplasmic projections extending into Howship lacunae.

(2) The **clear zone** surrounds the ruffled border. It contains **microfilaments,** which help osteoclasts maintain contact with the bony surface, and serves to isolate the region of osteolytic activity.

(3) The **vesicular zone** contains exocytotic vesicles that transfer lysosomal enzymes to Howship lacunae and endocytotic vesicles that transfer degraded bone products from Howship lacunae to the interior of the cell.

(4) The **basal zone** is located on the side of the cell opposite the ruffled border. It contains most of the cell organelles.

c. Bone resorption (Figure 7.3) involves the following events:

(1) Osteoclasts secrete **acid,** thus creating an acidic environment that decalcifies the surface layer of bone.

(2) Acid hydrolases, collagenases, and other proteolytic enzymes secreted by osteoclasts then **degrade the organic portion** of the bone.

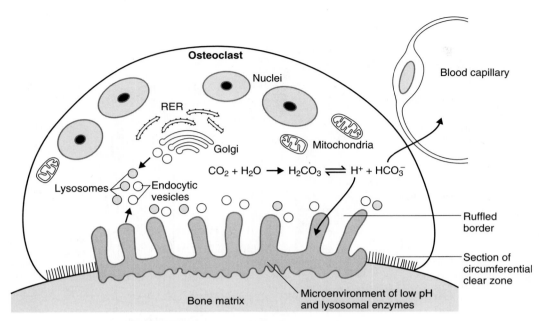

Figure 7.3. Osteoclastic function. RER = rough endoplasmic reticulum.

      (3) Osteoclasts resorb the organic and inorganic residues of the bone matrix and release them into connective tissue capillaries.

D. Classification of bone is based on both gross and microscopic properties.

  1. Gross observation (see Figure 7.2) of cross-sections of bone reveals two types:

    a. Spongy (cancellous) bone, which is composed of interconnected trabeculae. Trabeculae surround cavities filled with bone marrow. The trabeculae contain osteocytes and are lined by a single layer of osteoblasts.

    b. Compact bone, which has no trabeculae or bone marrow cavities

  2. Microscopic observation of bone reveals two types:

    a. Primary bone, also known as **immature** or **woven** bone

      (1) Primary bone contains many osteocytes and large, irregularly arranged type I collagen bundles.

      (2) It has a low mineral content.

      (3) It is the first compact bone produced during fetal development and bone repair.

      (4) It **is remodeled and replaced by secondary bone** except in a few places (e.g., tooth sockets, near suture lines in skull bones, and at insertion sites of tendons).

    b. Secondary bone, also known as **mature** or **lamellar** bone

      (1) Secondary bone is the compact bone of adults.

      (2) It has a **calcified matrix arranged in regular layers, or lamellae**. Each lamella is 3-7 micrometers (μm) thick.

(3) It contains osteocytes in lacunae, which are located between, and occasionally within, lamellae.

E. Organization of lamellae in compact bone (e.g., diaphysis of long bones) is characteristic and consists of the following elements:

  1. Haversian systems (osteons) are long cylinders that run approximately parallel to the long axis of the diaphysis.

   a. Haversian systems are composed of 4-20 lamellae surrounding a central haversian canal, which contains blood vessels, nerves, and loose connective tissue.

   b. They are often surrounded by an amorphous **cementing substance.**

   c. They are interconnected by **Volkmann canals,** which also connect to the periosteum and endosteum and **carry the neurovascular supply.**

  2. Interstitial lamellae are irregularly shaped lamellae located between haversian systems. They are remnants of remodeled haversian systems.

  3. Outer and inner circumferential lamellae are located at the external and internal surfaces of the diaphysis, respectively.

F. Histogenesis of bone occurs by two processes, **intramembranous** and **endochondral bone formation.** Both processes produce bone that appears histologically identical. Histogenesis is accompanied by bone resorption. The combination of bone formation and resorption, termed **remodeling,** occurs throughout life, although it is slower in secondary than in primary bone.

  1. Intramembranous bone formation is the process by which most of the flat bones (e.g., parietal bones of the skull) are formed. It involves the following events:

   a. Mesenchymal cells, condensed into **primary ossification centers,** differentiate into osteoblasts, which begin secreting **osteoid.**

   b. As calcification occurs, osteoblasts are trapped in their own matrix and become osteocytes. These centers of developing bone are called **trabeculae** (fused spicules).

   c. Fusion of the bony trabeculae produces **spongy bone** as blood vessels invade the area and other undifferentiated mesenchymal cells give rise to the bone marrow.

   d. The periosteum and endosteum develop from portions of the mesenchymal layer that do not undergo ossification.

   e. Mitotic activity of the mesenchymal cells gives rise to osteoprogenitor cells, which undergo cell division and form more osteoprogenitor cells or differentiate into osteoblasts within the inner layer of the developing periosteum.

  2. Endochondral bone formation is the process by which **long bones** are formed. It begins in a segment of **hyaline cartilage** that serves as a small **model** for the bone. The two stages of endochondral bone formation involve the development of primary and secondary centers of ossification.

    a. The **primary center of ossification** develops at the **midriff of the diaphysis** of the hyaline cartilage model by the following sequence of events:

        (1) Vascularization of the perichondrium at this site causes the transformation of chondrogenic cells to osteoprogenitor cells, which differentiate into osteoblasts. This region of the perichondrium is now called the **periosteum.**

        (2) Osteoblasts elaborate matrix deep to the periosteum, and via **intramembranous bone formation**, form the **subperiosteal bone collar.**

        (3) Chondrocytes within the core of the cartilaginous model undergo **hypertrophy** and degenerate, and their lacunae become confluent, forming large cavities (eventual marrow spaces).

        (4) Osteoclasts create perforations in the bone collar that permit the **periosteal bud** (blood vessels, osteoprogenitor cells, and mesenchymal cells) to enter the newly formed spaces in the cartilaginous model. The cartilage that constitutes the walls of these spaces then becomes calcified.

        (5) Newly developed osteoblasts elaborate bone matrix that becomes calcified on the surface of the calcified cartilage, forming a **calcified cartilage-calcified bone complex.** In histologic sections the calcified cartilage stains **basophilic,** whereas the calcified bone stains **acidophilic.**

        (6) The subperiosteal bone collar becomes thicker and elongates toward the epiphysis.

        (7) Osteoclasts begin to resorb the calcified cartilage-calcified bone complex, thus enlarging the primitive marrow cavity.

        (8) Repetition of this sequence of events results in bone formation spreading toward the **epiphyses.**

    b. Secondary centers of ossification develop at the **epiphyses** in a sequence of events similar to that described for the primary center, except a bone collar is not formed.

        (1) Development of these centers begins when osteoprogenitor cells invade the epiphysis and differentiate into osteoblasts, which elaborate bone matrix to replace the disintegrating cartilage. When the epiphyses are filled with bone tissue, cartilage remains in two areas, the articular surfaces and the epiphyseal plates.

        (2) Articular cartilage persists and does not contribute to bone formation.

        (3) Epiphyseal plates continue to grow by adding new cartilage at the epiphyseal end while it is being replaced at the diaphyseal end (lengthening the bone).

        (4) Ossification of the epiphyseal plates and cessation of growth occurs at about age 20.

  3. Zones of the epiphyseal plates are histologically distinctive and arranged in the following order:

    a. The **zone of reserve** cartilage is located at the epiphyseal side of the plate and possesses small, randomly arranged inactive chondrocytes.

    b. The **zone of proliferation** (of chondrocytes) is a region of rapid mitotic divisions, giving rise to rows of isogenous cell groups.

    c. The **zone of cell hypertrophy and maturation** is the region where the chondrocytes are greatly enlarged.

    d. The **zone of calcification** is the region where hypertrophied chondrocytes die and the cartilage becomes calcified.

    e. The **zone of ossification** is the area where newly formed osteoblasts elaborate bone matrix on the calcified cartilage, forming a calcified cartilage/calcified bone complex, which becomes resorbed.

  4. Calcification begins with the deposition of calcium phosphate on collagen fibrils and is stimulated by certain proteoglycans and **osteonectin,** a Ca^{2+}-binding glycoprotein. The mechanism by which calcification occurs is not well understood.

G. Bone remodeling. Bone is constantly being remodeled as necessary for growth and to alter its structural makeup to adapt to changing stresses in the environment throughout life. Several factors, including calcitonin and PTH, are responsible for this phenomenon.

  1. Early on, bone development outpaces bone resorption as new haversian systems are added and fewer are resorbed.

  2. Later when the epiphyseal plates are closed, ending bone growth, there is a balance between bone development and resorption.

H. Repair of a bone fracture. A bone fracture damages the matrix, bone cells, and blood vessels in the region and is accompanied by localized hemorrhaging and blood clot formation.

  1. Proliferation of osteoprogenitor cells occurs in the periosteum and endosteum in the vicinity of the fracture. As a result of this proliferation, a cellular tissue surrounds the fracture and penetrates between the ends of the damaged bone.

  2. Formation of a bony callus occurs both internally and externally at a fracture site.

    a. Fibrous connective tissue and hyaline cartilage are formed in the fracture zone.

    b. Endochondral bone formation replaces the cartilage with primary bone.

    c. Intramembranous bone formation also produces primary bone in the area.

    d. The irregularly arranged trabeculae of primary bone join the ends of the fractured bone, forming a **bony callus.**

    e. The primary bone is resorbed and replaced with secondary bone as the fracture heals.

I. Role of vitamins in bone formation

  1. Vitamin D is necessary for **absorption of calcium** from the small intestine. Vitamin D deficiency results in poorly calcified (soft) bone, a condition known as **rickets** in children and **osteomalacia** in adults. Vitamin D is also necessary for **bone formation** (ossification). An excess causes bone resorption.

2. **Vitamin A** deficiency inhibits proper bone formation and growth, whereas an excess accelerates ossification of the epiphyseal plates. Deficiency or excess results in small stature.

3. **Vitamin C** is necessary for **collagen formation**. Deficiency results in **scurvy,** characterized by poor bone growth and inadequate fracture repair.

J. **Role of hormones in bone formation**

1. **Parathyroid hormone (PTH)** activates osteoblasts to secrete **osteoclast-stimulating factor,** which then activates osteoclasts to **resorb bone,** thus **elevating blood calcium levels.** Excess PTH (hyperparathyroidism) renders bone **more susceptible to fracture** and subsequent deposition of calcium in arterial walls and certain organs such as the kidney.

2. **Calcitonin** is produced by parafollicular cells of the thyroid gland. It eliminates the ruffled border of osteoclasts and **inhibits bone-matrix resorption,** thus preventing the release of calcium.

3. **Pituitary growth hormone** is produced in the pars distalis of the pituitary gland and stimulates overall growth, especially that of epiphyseal plates. Excess during growing years causes **pituitary gigantism** and in adults causes acromegaly. Deficiency during growing years causes dwarfism.

III. Joints

A. **Synarthroses** are **immovable joints** composed of connective tissue, cartilage, or bone. These joints unite the first rib to the sternum and connect the skull bones to each other.

B. **Diarthroses (synovial joints)** permit **maximum movement** and generally unite long bones. These joints are surrounded by a two-layered **capsule,** enclosing and sealing the articular cavity. The articular cavity contains **synovial fluid,** a colorless, viscous fluid that is rich in hyaluronic acid and proteins.

1. The **external (fibrous) capsular layer** is a tough, fibrous layer of dense connective tissue.

2. The **internal (synovial) capsular layer** is also called the **synovial membrane.** It is lined by a layer of squamous to cuboidal epithelial cells on its internal surface. Two cell types are displayed in electron micrographs of this epithelium.

a. **Type A cells** are intensely phagocytic and have a well-developed Golgi complex, many lysosomes, and sparse RER.

b. **Type B cells** resemble fibroblasts and have a well-developed RER; these cells probably secrete synovial fluid.

IV. Clinical Considerations

A. Osteoporosis is a disease characterized by **low bone mass (bone mineral density)** and structural deterioration of bone tissue, making the bone more fragile and susceptible to fracture. Osteoporosis is associated with an abnormal ratio of mineral to matrix.

   1. It results from increased bone resorption, decreased bone formation, or both.

   2. This disease is most common in postmenopausal women because of diminished estrogen secretion, and in immobile patients because of lack of physical stress on the bone.

   3. Preventive measures include a balanced diet rich in calcium and vitamin D, and weight-bearing exercises.

B. Rickets occurs in children deficient in vitamin D, which results in calcium deficiency. It is characterized by **deficient calcification** in newly formed bone and is generally accompanied by deformation of the bone spicules in epiphyseal plates; as a result, bones grow more slowly than normal and are deformed.

C. Osteomalacia (rickets of adults) results from calcium deficiency in adults.

   1. It is characterized by **deficient calcification** in newly formed bone and decalcification of already calcified bone.

   2. This disease may be severe during pregnancy because the calcium requirements of the fetus may lead to calcium loss from the mother.

D. Acromegaly results from an **excess of pituitary growth hormone** in adults. It is characterized by **very thick bones** in the extremities and in portions of the facial skeleton.

Review Test

Directions: Each of the numbered items or incomplete statements in this section is followed by answers or by completions of the statement. Select the ONE lettered answer or completion that is BEST in each case.

1. Which of the following statements characterizes osteoclasts?

(A) They are enucleated cells.
(B) They produce collagen.
(C) They occupy Howship lacunae.
(D) They are derived from osteoprogenitor cells.

2. Which one of the following statements is correct concerning the periosteum?

(A) It is devoid of a blood supply.
(B) It produces osteoclasts.
(C) It is responsible for interstitial bone growth.
(D) Its inner layer contains osteoprogenitor cells.

3. Which one of the following statements is characteristic of osteocytes?

(A) They communicate via gap junctions between their processes.
(B) They contain large amounts of rough endoplasmic reticulum (RER).
(C) They are immature bone cells.
(D) They are housed as isogenous groups in lacunae.

4. Which one of the following statements is correct concerning hyaline cartilage?

(A) It is vascular.
(B) It contains type IV collagen.
(C) It undergoes appositional growth only.
(D) It is located at the articular ends of long bones.

5. A seven-year-old boy is seen by his pediatrician because the child broke his humerus as he tripped and fell while walking. The pediatrician asked about the child's diet and learned that he may have a possible dietary deficiency. Which of the following may be lacking in his diet?

(A) Potassium
(B) Calcium
(C) Iron
(D) Carbohydrates

6. A 22-year-old female patient is seen for the first time by her new physician, who notes that she has very thick bones in her extremities and face. The physician suspects acromegaly, which is a condition caused by which of the following?

(A) Hypervitaminosis A
(B) Excess growth hormone
(C) Hypovitaminosis A
(D) Hypervitaminosis D

7. Which of the following statements is characteristic of bone?

(A) Bone matrix contains primarily type II collagen.
(B) About 65% of the dry weight of bone is organic.
(C) Haversian canals are interconnected via Volkmann canals.
(D) Bone growth occurs via interstitial growth only.

The response options for the next four items are the same. Each item will state the number of options to select. Choose exactly this number.

Questions 8-11
For each physiologic effect, select the vitamin or hormone causing it.
(A) Thyroxine
(B) Hypervitaminosis A
(C) Absence of vitamin D
(D) Hydrocortisone
(E) Hypovitaminosis C

8. Inhibits histogenesis of cartilage (**SELECT 1 OPTION**)

9. Stimulates cartilage histogenesis (**SELECT 1 OPTION**)

10. Accelerates epiphyseal ossification (**SELECT 1 OPTION**)

11. Makes epiphyseal cartilage matrix fail to calcify (**SELECT 1 OPTION**)

Answers and Explanations

1-C. Osteoclasts are multinucleated cells that produce proteolytic enzymes and occupy Howship lacunae. They are not derived from osteoprogenitor cells but from monocyte precursors.

2-D. The inner layer of the periosteum possesses osteoprogenitor cells, whereas the outer layer of the periosteum is fibrous. The periosteum functions to distribute blood vessels to the bone; thus, appositional bone growth takes place here.

3-A. Osteocytes communicate with each other via gap junctions on narrow cytoplasmic processes that extend through canaliculi. They are mature bone cells that occupy individual lacunae as mature resting bone cells.

4-D. Hyaline cartilage is avascular, contains type II collagen, and grows both interstitially and appositionally. It is located at the articulating ends of long bones.

5-B. Because calcium must be maintained at a constant level in the blood and the tissues, a diet deficient in calcium leads to calcium loss from the bones. As a result, the bones become fragile.

6-B. Excessive growth hormone causes acromegaly. Excessive vitamin D causes bone resorption. Both an excess and a deficiency of vitamin A result in short stature.

7-C. Haversian canals run longitudinally, parallel to the long axis of bone. They are connected to one another by Volkmann canals that run perpendicular (or obliquely) to them.

8-D. Hydrocortisone inhibits cartilage growth and matrix formation.

9-A. Thyroxine, testosterone, and somatotropin stimulate cartilage growth and matrix formation.

10-B. Hypervitaminosis A accelerates ossification of epiphyseal plates, whereas hypovitaminosis A reduces the width of the epiphyseal plates.

11-C. In the absence of vitamin D, epiphyseal chondrocytes continue to proliferate but their matrix does not calcify, which leads to rickets.

8

Muscle

I. Overview—Muscle

A. Muscle is classified into three types: **skeletal, cardiac,** and **smooth** muscle.

B. Muscle cells possess **contractile filaments** containing **actin** and **myosin.**

C. Contraction may be **voluntary** (skeletal muscles) or **involuntary** (cardiac and smooth muscles).

II. Structure of Skeletal Muscle

A. Connective tissue investments include **aponeuroses** and **tendons,** which connect skeletal muscle to bone and other tissues and convey neural and vascular elements to muscle cells.

   1. Epimysium surrounds an entire muscle.

   2. Perimysium surrounds **fascicles** (small bundles) of muscle cells.

   3. Endomysium surrounds individual muscle cells and is composed of **reticular fibers** and an **external lamina.**[1]

B. Types of skeletal muscle cells

   1. Types of skeletal muscle cells (also known as muscle fibers) include **red** (slow), **white (fast),** and **intermediate.** All three types may be present in a given muscle.

   2. These three types vary in their content of **myoglobin** (binds O_2-similar to hemoglobin), **number of mitochondria, concentration of various enzymes,** and **rate of contraction** (Table 8.1).

   3. A change in **innervation** can change a fiber's type. If a red fiber is denervated and its innervation replaced with that of a white fiber, the red fiber will become a white fiber.

C. Skeletal muscle cells (Figure 8.1) are long, cylindrical, **multinucleated** cells that are enveloped by an external lamina and reticular fibers. Their cytoplasm is called **sarcoplasm,** and their plasmalemma is called the **sarcolemma** and forms deep tubular invaginations, or **T (transverse) tubules,** which extend into the cells. The muscle cells possess cylindrical collections of **myofibrils,** 1-2 micrometers (μm) in diameter, which extend the entire length of the cell.

   1. Myofibrils are longitudinally arranged, cylindrical bundles of **thick** and **thin myofilaments** observable by transmission electron microscopy (see Figure 8.1*D* and *E*).

---

[1]The term **external lamina** is synonymous with the term **basal lamina,** but is used in reference to nonepithelial cells.

Table 8.1. Characteristics of Red and White Muscle Fibers

Type	Myoglobin Content	Number of Mitochondria	Enzyme Content	Contraction	Primary Method of ATP Generation
Red (slow; type 1)	High	Many	High in oxidative enzymes; low in ATPase	Slow but repetitive; not easily fatigued	Oxidative phosphorylation
Intermediate (type 2A)	Intermediate	Intermediate	Intermediate in oxidative enzymes and ATPase	Fast but not easily fatigued	Oxidative phosphorylation and anaerobic glycolysis
White (fast; type 2B)	Low	Few	Low in oxidative enzymes; high in ATPase and phosphorylases	Fast and easily fatigued	Anaerobic glycolysis

ATP = adenosine triphosphate.

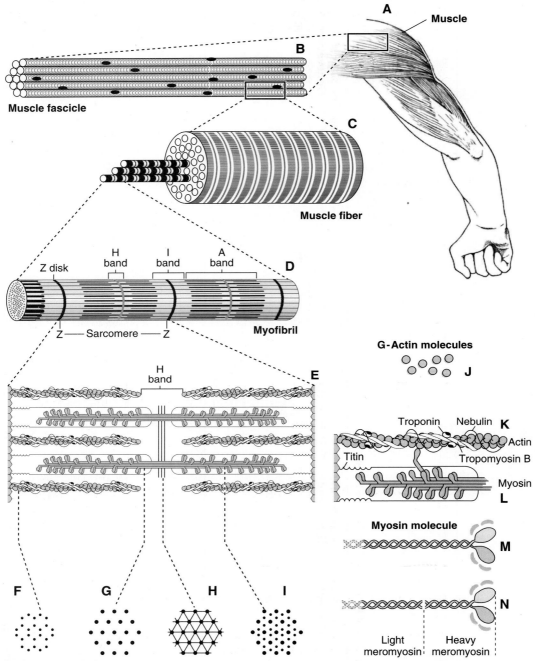

Figure 8.1. Diagram of skeletal muscle and its components as observed by light and electron microscopy. (Adapted with permission from Fawcett DW: *Bloom and Fawcett's Textbook of Histology*, 12th ed. New York, Chapman & Hall, 1994, p 279 and from Gartner LP, Hiatt JL: *Color Atlas of Histology*, 2nd ed. Baltimore, Williams & Wilkins, 1994, p 102.)

a. Precise alignment of myofibrils results in a characteristic banding pattern visible by light microscopy as alternating dark **A bands** and light **I bands;** the latter are bisected by **Z disks** (see Figure 8.1*D* and *E*).

b. Myofibrils are held in alignment by the intermediate filaments **desmin** and **vimentin,** which tether Z disks of adjacent myofibrils to one another.

2. The **sarcomere** is the regular repeating region between successive Z disks and constitutes the **functional unit of contraction** in skeletal muscle.

3. The **sarcoplasmic reticulum (SR)** is a modified smooth endoplasmic reticulum (SER) that surrounds myofilaments and forms a meshwork around each myofibril.

a. The SR forms a pair of dilated **terminal cisternae,** which encircle the myofibrils at the junction of each A and I band.

b. It **regulates muscle contraction** by sequestering calcium ions (leading to relaxation) or releasing calcium ions (leading to contraction).

4. Triads are specialized complexes consisting of a narrow central T tubule flanked on each side by terminal cisternae of the SR. They are located at the A-I junction in mammalian skeletal muscle cells and **help provide uniform contraction** throughout the muscle cell.

5. Myofilaments include **thick** filaments [15 nanometers (nm) in diameter and 1.5 μm long] and **thin** filaments (7 nm in diameter and 1.0 μm long). They lie parallel to the long axis of the myofibril in a precise arrangement that is responsible for the sarcomere banding pattern.

D. Satellite cells lie within the external lamina of skeletal muscle cells. These **regenerative cells** differentiate, fuse with one another, and form skeletal muscle cells when the need arises.

E. Skeletal muscle cross-striations (see Figure 8.1*D* and *E*)

1. A bands are anisotropic with polarized light; they usually stain dark. They contain **both thin and thick filaments,** which overlap and interdigitate. Six thin filaments surround each thick filament (see Figure 8.1*I*).

2. I bands are isotropic with polarized light and appear lightly stained in routine histologic preparations. They contain only **thin filaments.**

3. H bands are light regions transecting A bands; they consist of **thick filaments** only.

4. M lines are narrow, dark regions at the center of H bands formed by several cross-connections **(M-bridges)** at the centers of adjacent thick filaments.

5. Z disks (lines) are dense regions bisecting each I band.

a. Z disks contain α-actinin, which binds to thin filaments and anchors them to Z disks with the assistance of **nebulin.**

b. Desmin anchors Z disks to each other. Peripherally located Z disks are anchored to the sarcolemma by **vinculin,** which contains elements called **costameres.**

F. Molecular organization of myofilaments

1. **Thin filaments** are composed of F-actin, tropomyosin, troponin, and associated proteins.

    a. **F-actin** (see Figure 8.1*J* and *K*) is a polymer of G-actin monomers arranged in a double helix.

       (1) Each monomer possesses an **active site** that can interact with myosin.

       (2) F-actin is present as filaments (with a diameter of 5 nm) that exhibit **polarity,** having a (+) and (-) end.

    b. **Tropomyosin** molecules are 41 nm in length. They bind head-to-tail, forming filaments that are located in the grooves of the F-actin helix.

    c. **Troponin** is associated with each tropomyosin molecule and is composed of:

       (1) Troponin T (TnT), which forms the tail of the molecule and functions in binding the troponin complex to tropomyosin.

       (2) Troponin C (TnC), which possesses four binding sites for calcium. It may be related to calmodulin.

       (3) Troponin I (TnI), which binds to actin, inhibiting interaction of myosin and actin.

    d. **Nebulin** is a long inelastic protein. Two nebulin molecules wrap around each thin filament and assist in anchoring it to the Z disk (see Figure 8.1*K*).

       (1) Each nebulin molecule is embedded in the Z disk by its carboxy terminal, but does not span the entire Z disk.

       (2) The amino terminal of each nebulin molecule ends in the A band, at or near the free end of its thin filament.

       (3) Nebulin in skeletal muscle is thought to determine the length of its associated thin filament, although in cardiac muscle it extends only one quarter of the length of the thin filament.

2. **Thick filaments** each contain about 250 myosin molecules arranged in an antiparallel fashion and three associated proteins—myomesin, titin, and C protein.

    a. **Myosin** (see Figure 8.1*L* and *M*) is composed of two identical heavy chains and two pairs of light chains. The myosin molecule resembles a double-headed golf club.

       (1) Myosin **heavy chains** consist of a long rod-like "tail" and a globular "head." The tails of the heavy chains wind around each other in an α-helical configuration.

          (a) Tails function in the self-assembly of myosin molecules into bipolar thick filaments.

          (b) Actin-binding sites of the heads function in contraction.

       (2) Myosin **light chains** are of two types; one molecule of each type is associated with the globular head of each heavy chain.

(3) **Digestion of myosin**

    (a) Trypsin cleaves myosin into **light meromyosin** (part of the tail portion) and **heavy meromyosin** (the two heads and the remainder of the tail) [see Figure 8.1N].

    (b) Papain cleaves the heavy meromyosin, releasing the short tail (S2 fragment) and the two globular heads (S1 fragments). These S1 fragments have adenosine triphosphatase (ATPase) activity but require interaction with actin to release the non-covalently bound adenosine diphosphate (ADP) and P_i.

b. Myomesin is a protein at the M line that cross-links adjacent thick filaments to one another to maintain their spatial relations.

c. Titin is a large linear protein that displays axial periodicity. It forms an elastic lattice that parallels the thick and thin filaments, and two titin filaments anchor each thick filament to the Z disk, thus maintaining their architectural relationships to each other (see Figure 8.1L).

    (1) The amino terminal of the titin molecule spans the entire thickness of the Z disk and binds to α-actinin and Z proteins.

    (2) Within the Z disk, titin overlaps with other titin molecules from the neighboring sarcomere and probably forms bonds with them or with unidentified linker proteins.

    (3) The carboxyl terminal of the titin molecule spans the entire M line and overlaps with titin molecules from the other half of the same sarcomere, and binds to the protein **myomesin**.

    (4) Within the I band, in the vicinity of the Z disk, titin interacts with thin filaments.

    (5) Within the A band, titin interacts with **C protein**.

d. C protein binds to thick filaments in the vicinity of M lines along much of their lengths (between the M line and the end of the thin filament in the vicinity of the A-I junction). This region of the A band is referred to as the **C zone**.

III. Contraction of Skeletal Muscle. The contraction cycle (Figure 8.2) involves the binding, hydrolysis, and release of ATP.

A. Sliding-filament model (Table 8.2)

    1. During contraction, thick and thin filaments do not shorten but increase their overlap.

    2. Thin filaments slide past thick filaments and penetrate more deeply into the A band, which remains constant in length.

    3. I bands and H bands shorten as Z disks are drawn closer together.

B. Initiation and regulation of contraction

    1. Depolarization and Ca^{2+} release trigger binding of actin and myosin, leading to muscle contraction.

        a. The **sarcolemma** is depolarized at the myoneural junction.

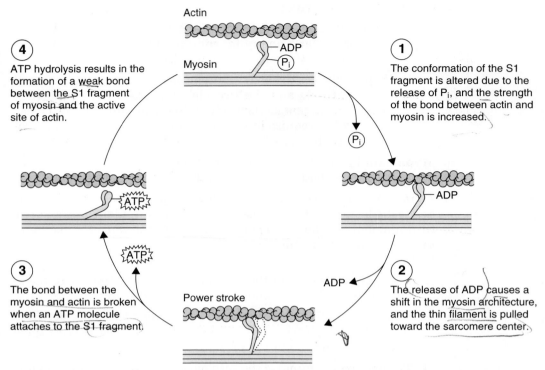

Actin

(4) ATP hydrolysis results in the formation of a weak bond between the S1 fragment of myosin and the active site of actin.

Myosin — ADP, Pᵢ

(1) The conformation of the S1 fragment is altered due to the release of Pᵢ, and the strength of the bond between actin and myosin is increased.

Pᵢ

ATP

— ADP

(3) The bond between the myosin and actin is broken when an ATP molecule attaches to the S1 fragment.

ATP

Power stroke

ADP

(2) The release of ADP causes a shift in the myosin architecture, and the thin filament is pulled toward the sarcomere center.

Figure 8.2. Contraction cycle in skeletal muscle cells. This sequence of steps is repeated many times, leading to extensive overlay of thick and thin filaments, which shortens the sarcomere and consequently the entire skeletal muscle fiber. ADP = adenosine diphosphate; ATP = adenosine triphosphate. (Adapted with permission from Alberts B, Bray D, Lewis J, et al: *Molecular Biology of the Cell*, 3rd ed. New York, Garland Publishing, 1994, p 852.)

   b. T tubules convey the wave of depolarization to the myofibrils. Voltage-sensitive dihydropyridine (DHP) receptors alter their conformation as a function of membrane depolarization.

   c. Ca²⁺ is released into the cytosol at the A-I junctions via **Ca²⁺-release channels** (junctional feet) of the SR terminal cisternae that are opened by activated DHP receptors. As long as the Ca²⁺ level is sufficiently high, the contraction cycle will continue.

  2. Activation of actin by Ca²⁺

   a. In the **resting state,** the myosin-binding sites on thin (actin) filaments are partially covered by tropomyosin. Also, Troponin I is bound to actin and hinders myosin-actin interaction.

Table 8.2. Effects of Contraction on Skeletal Muscle Cross-Bands

Bands	Myofilament Component	Change in Bands During Contraction
I	Thin only	Shorten
H	Thick only	Shorten
A	Thick and thin	No change in length
Z disks	Thin only (attached by α-actinin)	Move closer to each other

b. Ca^{2+} binding by troponin C results in a **conformational change** that breaks the TnI-actin bond; tropomyosin shifts its position slightly and uncovers the myosin-binding sites **(active state).**

C. Relaxation occurs when Ca^{2+} concentration in the cytosol is reduced enough that TnC loses its bound Ca^{2+}.

1. As a result, tropomyosin returns to its resting position, covering actin's binding sites and restoring the resting state.

2. Relaxation depends on a Ca^{2+} **pump** in the SR, which pumps Ca^{2+} from the cytosol to the inner surface of the SR membrane to be bound by **calsequestrin.**

D. A **motor unit** consists of a neuron and every muscle cell it innervates. A **muscle** may contract with varying degrees of strength, but a **muscle cell** obeys the **"all or none law"** (i.e., it either contracts or does not contract). All muscle cells of a single motor unit contract in unison.

IV. Innervation of Skeletal Muscle. Innervation consists of **motor** nerve endings (myoneural junctions) and two types of **sensory** nerve endings (muscle spindles and Golgi tendon organs). Both types of sensory nerve endings function in **proprioception.**

A. The **myoneural junction** (neuromuscular junction) is a **synapse** between a motor nerve and a skeletal muscle cell.

1. Structural components

a. The **axon terminal** lacks myelin but has a **Schwann cell** on its nonsynaptic surface.

(1) The membrane on the synaptic surface is called the **presynaptic membrane.**

(2) The axon terminal contains mitochondria, synaptic vesicles (containing the neurotransmitter **acetylcholine**), and SER elements.

b. The **synaptic cleft** is a narrow space between the presynaptic membrane of the axon terminal and the **postsynaptic membrane (motor end plate)** of the muscle cell. The synaptic cleft contains an amorphous external lamina derived from the muscle cell.

c. Muscle cell near the myoneural junction

(1) Sarcolemmal invaginations (of postsynaptic membrane), called **junctional folds,** are lined by an external lamina and extend inward from the synaptic cleft.

(2) Acetylcholine receptors are located in the postsynaptic membrane.

(3) The sarcoplasm is rich in mitochondria, ribosomes, and rough endoplasmic reticulum (RER).

2. Conduction of a nerve impulse across a myoneural junction

a. The presynaptic membrane is depolarized and **voltage-gated Ca^{2+} channels** open, permitting the entry of extracellular Ca^{2+} into the axon terminal.

b. The rise in cytosolic Ca^{2+} triggers the synaptic vesicles to release acetylcholine in multimolecular quantities **(quanta)** into the synaptic cleft.

    c. The released acetylcholine binds to receptors of the postsynaptic membrane, resulting in **depolarization** of the sarcolemma and generation of an **action potential.**

    d. Acetylcholinesterase located in the external lamina lining the junctional folds of the motor end plate degrades acetylcholine, thus ending the depolarizing signal to the muscle cell.

    e. Acetylcholine is recycled as **choline** and is returned to the axon terminal to be recombined with acetyl CoA (from mitochondria) and stored in synaptic vesicles.

    f. Membranes of the emptied synaptic vesicles are recycled via **clathrin-coated endocytic vesicles** (see Figure 3.5).

B. The **muscle (neuromuscular) spindle** is an elongated, fusiform sensory organ within skeletal muscle that functions primarily as a **stretch receptor.**

    1. Structure

        a. It is bounded by a connective tissue capsule enclosing the fluid-filled **periaxial space** and 8-10 modified skeletal muscle fibers **(intrafusal fibers).**

        b. Normal skeletal muscle fibers **(extrafusal fibers)** surround it.

        c. It is anchored via the capsule to the perimysium and endomysium of the extrafusal fibers.

    2. Function

        a. Stretching of a muscle also stretches the muscle spindle and thus stimulates the afferent nerve endings to send impulses to the central nervous system. The response is to both the **rate (phasic response)** and **duration (tonic response)** of stretching.

        b. Depolarization of γ-efferent neurons also stimulates the intrafusal nerve endings; the rate and duration of the stimulation are monitored in the same way as stretching.

        c. Muscle overstimulation results from stretching at too great a frequency or for too long a time. Overstimulation causes stimulation of **α-efferent neurons** to the muscle, initiating contraction and thus counteracting the stretching.

C. The **Golgi tendon organ**, located in tendons, counteracts the effects of muscle spindles.

    1. Structure. It is composed of encapsulated **collagen fibers** that are surrounded by terminal branches of **type Ib sensory nerves.**

    2. It is stimulated when the muscle contracts too strenuously, increasing tension on the tendon. Impulses from type Ib neurons then **inhibit** α-efferent (motor) neurons to the muscle, preventing further contraction.

V. Cardiac Muscle

A. General features—cardiac muscle cells (Table 8.3). Cardiac muscle cells have the following features:

    1. Contract spontaneously and display a rhythmic beat, which is modified by hormonal and neural (sympathetic and parasympathetic) stimuli.

Table 8.3. Comparison of Skeletal, Cardiac, and Smooth Muscle

Property	Skeletal Muscle	Cardiac Muscle	Smooth Muscle
Shape and size of cells	Long cylindrical	Blunt-ended, branched	Short, spindle-shaped
Number and location of nucleus	Many peripheral	One or two, central	One, central
Striations present	Yes	Yes	No
T tubules and sarcoplasmic reticulum	Has triads at A-I junctions	Has dyads at Z disks	Has caveolae (but no T tubules) and some SER
Gap junctions present	No	Yes (in intercalated disks)	Yes (in sarcolemma); known as the nexus
Sarcomere present	Yes	Yes	No
Regeneration	Restricted	None	Extensive
Voluntary contraction	Yes	No	No
Distinctive characteristics	Peripheral nuclei	Intercalated disks	Lack of striations

SER = smooth endoplasmic reticulum.

2. May branch at their ends to form connections with adjacent cells.

3. Contain **one** (occasionally **two**) centrally located nuclei.

4. Contain **glycogen granules,** especially at either pole of the nucleus.

5. Possess thick and thin filaments arranged in **poorly defined myofibrils.**

6. Exhibit a cross-banding pattern identical to that in skeletal muscle.

7. **Do not regenerate;** injuries to cardiac muscle are repaired by the formation of fibrous connective (scar) tissue by fibroblasts.

B. **Structural components of cardiac muscle cells** differ from those of skeletal muscle as follows:

1. **T tubules** are larger than those in skeletal muscle and are lined by external lamina. They invaginate at Z disks, not at A-I junctions as in skeletal muscle.

2. **SR** is poorly defined and contributes to the formation of **dyads,** each of which consists of one T tubule and one profile of SR. SR is also present in the vicinity of Z disks as small, basket-like saccules known as **corbular sarcoplasmic reticulum.**

3. **Calcium ions**

   a. During relaxation, Ca^{2+} **leaks** into the sarcoplasm at a slow rate, resulting in automatic rhythm. Ca^{2+} also enters cardiac muscle cells from the extracellular environment via voltage-gated Ca^{2+} channels of T tubules and sarcolemma.

   b. In response to calcium entering through the voltage-gated Ca^{2+} channels, Ca^{2+} is released from the SR (via **ryanodine receptors**) to cause contraction of cardiac muscle.

   c. The force of cardiac muscle contraction is directly dependent on the availability of Ca^{2+} in the sarcoplasm. During basal cardiac contraction, only 50% of the available calcium binding sites of TnC are occupied.

4. **Mitochondria** are more abundant than in skeletal muscle; they lie parallel to the I bands and often are adjacent to lipids.

5. **Atrial granules** are present in the atrial cardiac muscle cells and contain the precursor of **atrial natriuretic peptide,** which acts to **decrease resorption of sodium and water in the kidneys,** reducing body fluid volume and blood pressure.

6. **Intercalated disks** (Figure 8.3) are complex step-like junctions forming end-to-end attachments between adjacent cardiac muscle cells.

   a. The **transverse portion of intercalated disks** runs across muscle fibers at right angles and possesses three specializations: **fasciae adherentes** (analogous to zonula adherentes) to which actin filaments attach, **desmosomes** (macula adherentes), and **gap junctions** (see Chapter 5 II).

   b. The **lateral portion of intercalated disks** has desmosomes and numerous large gap junctions, which facilitate **ionic coupling** between cells and aid in coordinating contraction; thus, cardiac muscle behaves as a **functional syncytium.**

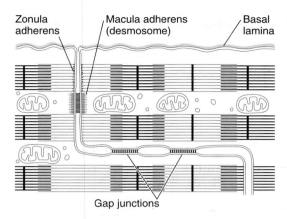

Figure 8.3. Schematic diagram of an intercalated disk, displaying its two regions: the transverse and lateral portions. (Adapted with permission from Junqueira LC, Carneiro J, and Kelley RO: *Basic Histology*, 8th ed. Norwalk, CT, Appleton & Lange, 1995, p 197.)

7. Their thin filaments are secured to the Z disk by α-actinin as well as by **nebulette**, a nebulin-like molecule that extends only as far as the proximal 25% of the length of the thin filament.

8. **Connective tissue elements** support a rich capillary bed that supplies sufficient nutrients and oxygen to maintain the high metabolic rate of cardiac muscle.

9. **Purkinje fibers** are **modified** cardiac muscle cells located in the **bundle of His.** They are specialized for **conduction** and contain a few peripheral myofibrils.

   a. These large, pale cells are rich in glycogen and mitochondria.

   b. They form gap junctions, fasciae adherentes, and desmosomes with cardiac muscle cells (but not through typical intercalated disks).

VI. Smooth Muscle

A. **Structure—smooth muscle cells** (see Table 8.3; Figure 8.4). Smooth muscle cells are **nonstriated, fusiform** cells that range in length from 20 μm in small blood vessels to 500 μm in the pregnant uterus. They contain a single nucleus and actively divide and **regenerate.** They are surrounded by an external lamina and a reticular fiber network and may be arranged in layers, small bundles, or helices (in arteries).

1. **Nucleus**

   a. The centrally located nucleus may not be visible in each cell in cross-sections of smooth muscle because some nuclei lie outside the plane of section.

   b. The nucleus in longitudinal sections of contracted smooth muscle has a **corkscrew shape** and is **deeply indented.**

2. **Cytoplasmic organelles**

   a. **Mitochondria, RER, and the Golgi complex** are concentrated near the nucleus and are involved in synthesis of type III collagen, elastin, glycosaminoglycans, external lamina, and growth factors.

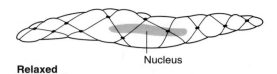

Relaxed

Nucleus

Contracted

Figure 8.4. A diagram of relaxed and contracted smooth muscle cells, displaying cytoplasmic and peripheral densities. Note that the nucleus of the smooth muscle cell assumes a corkscrew shape. (Adapted with permission from Gartner LP, Hiatt JL: *Color Textbook of Histology.* Philadelphia, PA, WB Saunders, 1997, p 151.)

    b. Sarcolemmal vesicles (caveolae), present along the periphery of smooth muscle cells, may function in the uptake and release of Ca^{2+}.

    c. SER is sparse and may be associated with caveolae.

  3. Filaments in smooth muscle

    a. Contractile filaments (actin and myosin) are **not** organized into myofibrils. They are attached to peripheral and cytoplasmic densities and aligned obliquely to the longitudinal axis of smooth muscle cells.

      (1) Thick filaments (composed of myosin) are each surrounded by as many as 15 thin filaments. In contrast to striated muscle, the heads of the myosin molecules all point in the same direction.

      (2) Thin filaments are composed of **actin, caldesmon, tropomyosin,** and **calponin.** Caldesmon functions similarly to TnT and TnI.

    b. Intermediate filaments are attached to cytoplasmic densities and include **vimentin** and **desmin** in **vascular** smooth muscle cells and **desmin** only in **nonvascular** smooth muscle cells.

  4. Cytoplasmic densities are believed to be **analogous to Z disks,** contain α-actinin, and function as **filament attachment sites.**

  5. Gap junctions between smooth muscle cells facilitate the spread of excitation. These gap junctions are collectively called a **nexus.**

B. Contraction of smooth muscle occurs more slowly and lasts longer than contraction of skeletal muscle because the rate of ATP hydrolysis is slower. Contraction of smooth muscle is regulated by a different mechanism from that of skeletal muscle contraction.

  1. The **contraction cycle** is stimulated by a transient increase in cytosolic Ca^{2+}.

    a. Ca^{2+} binds to **calmodulin,** altering its conformation.

b. The Ca²-calmodulin complex activates the enzyme myosin light-chain kinase, which catalyzes phosphorylation of one of the light chains of myosin.

c. In the presence of Ca^{2+}, the inhibitory effect of the caldesmon-tropomyosin complex on the actin-myosin interaction is eliminated. Another inhibitor of contraction is calponin, which, when phosphorylated, loses its inhibitory capability.

d. The globular head of phosphorylated myosin interacts with actin and stimulates myosin ATPase, resulting in contraction. As long as myosin is in its phosphorylated form, the contraction cycle continues.

e. Dephosphorylation of myosin disturbs the myosin-actin interaction and leads to relaxation.

2. Initiation of contraction

  a. In **vascular smooth muscle,** contraction is usually triggered by a **nerve impulse,** with little spread of the impulse from cell to cell.

  b. In **visceral smooth muscle,** it is triggered by stretching of the muscle itself **(myogenic);** the signal spreads from cell to cell.

  c. In the **uterus** during labor, it is triggered by **oxytocin.**

  d. In smooth muscle elsewhere in the body, it is triggered by **epinephrine.**

C. Innervation of smooth muscle is by **sympathetic** (noradrenergic) nerves and **parasympathetic** (cholinergic) nerves of the autonomic nervous system, which act in an antagonistic fashion to stimulate or depress activity of the muscle.

VII. Contractile Nonmuscle Cells

A. Myoepithelial cells

1. In certain glands, these cells share basal laminae of secretory and duct cells.

2. They arise from **ectoderm** and can **contract** to express secretory material from glandular epithelium into ducts and out of the gland.

3. Although generally similar in morphology to smooth muscle cells, they have a **basket-like shape** and several radiating **processes.**

4. They are attached to the underlying basal lamina via hemidesmosomes.

5. They contain **actin, myosin,** and intermediate filaments, as well as cytoplasmic and peripheral densities to which these filaments attach.

6. Contraction is similar to that of smooth muscle and occurs via a **calmodulin-mediated process.** In lactating **mammary glands,** they contract in response to **oxytocin.** In **lacrimal glands,** they contract in response to **acetylcholine.**

B. Myofibroblasts

1. Although they resemble fibroblasts, they possess higher amounts of **actin** and **myosin** and are capable of **contraction.**

2. They may contract during wound healing to decrease the size of the defect **(wound contraction).**

VIII. Clinical Considerations

A. Duchenne muscular dystrophy (DMD) is caused by a **sex-linked,** recessive genetic defect that results in the inability to synthesize **dystrophin,** an actin-binding protein normally present in small amounts in the sarcolemma. Dystrophin also stabilizes the sarcolemma and acts as a link between the cytoskeleton and the extracellular matrix.

   1. This common, serious degenerative disorder occurs in young men and results in death usually before age 20.

   2. DMD is characterized by the replacement of degenerating skeletal muscle cells by fatty and fibrous connective tissue, but it may also affect cardiac muscle.

B. Amyotrophic lateral sclerosis (ALS; Lou Gehrig disease) is marked by degeneration of motor neurons of the spinal cord, resulting in muscle atrophy. Death is usually due to respiratory muscle failure.

C. Myasthenia gravis is an **autoimmune disease** in which **antibodies block acetylcholine receptors** of myoneural junctions, reducing the number of sites available for initiation of sarcolemma depolarization.

   1. Myasthenia gravis is characterized by gradual weakening of skeletal muscles, especially the most active ones (e.g., muscles of the eyes, tongue, face, and extremities). Death may result from respiratory compromise and pulmonary infections.

   2. Clinical signs include thymic hyperplasia (thymoma) and the presence of circulating antibodies to acetylcholine receptors.

D. Myocardial infarct is an irreversible necrosis of cardiac muscle cells due to prolonged ischemia. It may result in death if the cardiac muscle damage is extensive.

E. Rigor mortis is a postmortem rigidity appearing as hardening of skeletal muscles caused by the inability of muscle cells to synthesize ATP. As a result, myosin remains bound to actin, and the muscles remain contracted.

F. Botulism is a form of **food poisoning** caused by ingestion of *Clostridium botulinum* toxin, which inhibits acetylcholine release at myoneural junctions. Botulism is marked by muscle paralysis, vomiting, nausea, and visual disorders and is fatal if untreated.

Review Test

Directions: Each of the numbered items or incomplete statements in this section is followed by answers or by completions of the statement. Select the ONE lettered answer or completion that is BEST in each case.

Questions 1-3

The response options for the next three items are the same. Each item will state the number of options to select. Choose exactly this number.

(A) T tubules are located at the Z disk
(B) T tubules are located at the A-I interface
(C) T tubules are lined by external lamina
(D) T tubules are absent
(E) Oxytocin triggers contraction
(F) Possess(es) dyads
(G) Possess(es) triads
(H) Possess caveolae
(I) Troponin is absent
(J) Myosin light-chain kinase participates in contraction
(K) Possess(es) intercalated disks

For each of the following muscle types, select the appropriate description(s).

1. Skeletal muscle (**SELECT 2 OPTIONS**)

2. Cardiac muscle (**SELECT 4 OPTIONS**)

3. Smooth muscle (**SELECT 5 OPTIONS**)

4. Contraction in all types of muscle requires calcium ions. Which of the following muscle components can bind or sequester calcium ions?

(A) Rough endoplasmic reticulum
(B) Tropomyosin
(C) Troponin
(D) Active sites on actin

5. Each smooth muscle cell

(A) has triads associated with its contraction
(B) has dyads associated with its contraction
(C) possesses a single centrally located nucleus
(D) is characterized by the absence of sarcolemmal vesicles

6. Thick filaments are anchored to Z disks by

(A) C protein
(B) nebulin
(C) titin
(D) myomesin
(E) α-actinin

7. The endomysium is a connective tissue investment that surrounds

(A) individual muscle fibers
(B) muscle fascicles
(C) individual myofibrils
(D) an entire muscle
(E) small bundles of muscle cells

8. Which of the following statements concerning triads in mammalian skeletal muscle is true?

(A) They are located in the Z disk.
(B) They consist of two terminal cisternae of the sarcoplasmic reticulum separated by a T tubule.
(C) They can be observed with the light microscope.
(D) They are characterized by a T tubule that sequesters calcium ions.
(E) They consist of two T tubules separated by a central terminal cisterna.

9. Which one of the following statements concerning cardiac muscle cells is true?

(A) They are spindle shaped.
(B) They require an external stimulus to undergo contraction.
(C) They are multinuclear cells.
(D) They are joined together end-to-end by intercalated disks.
(E) They possess numerous caveolae.

10. A 19-year-old male patient and his mother arrive in the emergency room, both suffering with nausea, vomiting, and visual disorders. The physician taking their history notes that they both had canned green beans that "tasted funny." Which of the following possibilities should the physician consider?

(A) Duchenne muscular dystrophy
(B) Amyotrophic lateral sclerosis
(C) Botulism
(D) Myasthenia gravis
(E) Myocardial infarct

Answers and Explanations

1-B and G.   The T tubules of skeletal muscle cells are positioned so that they form triads with the terminal cisternae of the sarcoplasmic reticulum at the interface of the A and I bands.

2-A, C, F, and K.   The T tubules of cardiac muscle cells are wider than those of skeletal muscle cells and are lined by external lamina (a basal lamina-like material). In contrast to skeletal muscle, the T tubules are located at the Z disk, where they often form dyads. Individual cardiac muscle cells are attached to one another by intercalated disks.

3-D, E, H, I, and J.   Smooth muscle cells do not have T tubules. Contraction may be initiated by stretching, neural impulses, the intercellular passage of small molecules via gap junctions, or the action of hormones such as oxytocin. The process of contraction is not dependent on troponin, which is absent from the thin filament of smooth muscle. Instead, calcium ions controlled by sarcolemmal vesicles known as caveolae are released into the cytosol, where they bind with calmodulin. The calcium-calmodulin complex activates myosin light chain kinase, which participates in the contraction process.

4-C.   Binding of calcium ions to the TnC subunit of troponin leads to the uncovering of myosin-binding sites on actin (thin filaments).

5-C.   Smooth muscle cells contain one centrally located nucleus.

6-C. Titin forms an elastic lattice that anchors thick filaments to Z disks.

7-A. The endomysium is a thin connective tissue layer, composed of reticular fibers and an external lamina, that invests individual muscle fibers (cells). The epimysium surrounds the entire muscle, and the perimysium surrounds bundles (fascicles) of muscle fibers.

8-B. A triad in skeletal muscle is composed of three components, a T tubule and two terminal cisternae of the sarcoplasmic reticulum (SR) that flank it. The SR, not the T tubules, sequesters calcium ions.

9-D. Cardiac muscle cells are joined together end-to-end by a unique junctional specialization called the intercalated disk.

10-C. Botulism is the only possible consideration, especially after learning that they both had canned food. Duchenne muscular dystrophy is most common in young men but very rare in older women. It would be highly unlikely that both mother and son would show symptoms of amyotrophic lateral sclerosis or myasthenia gravis or have myocardial infarct at the same time.

9

Nervous Tissue

I. Overview—Nervous Tissue. Nervous tissue can be organized by anatomical or functional divisions.

    A. Nervous tissue is divided **anatomically** into the **central nervous system** (CNS), which includes the brain and spinal cord, and the **peripheral nervous system** (PNS), which includes the nerves outside the CNS and their associated ganglia.

    B. Nervous tissue is divided **functionally** into a **sensory** component, which transmits electrical impulses (signals) to the CNS, and a **motor** component, which transmits impulses from the CNS to various structures of the body. The motor component is further divided into the **somatic** and **autonomic** systems.

    C. Nervous tissue contains two types of cells: **nerve cells (neurons),** which conduct electrical impulses, and **glial (neuroglial) cells,** which support, nurture, and protect the neurons.

II. Histogenesis of the Nervous System

    A. The **neuroepithelium** thickens and differentiates to form the neural plate.

    B. The **neural plate** invaginates and thickens to form the **neural groove.**

    C. The **neural tube,** a cylindrical structure that results from fusion of the edges of the neural groove, enlarges at its cranial end to form the **brain.** The remaining portion gives rise to the **spinal cord.**

    D. Neural crest cells stream off the edges of the neural groove before formation of the neural tube. These cells migrate throughout the body and give rise to the following structures.

        1. Sensory neurons of cranial and spinal sensory ganglia

        2. Most sensory neurons and Schwann cells of the PNS

        3. Enteric and autonomic ganglia and their postganglionic neurons and associated glia

        4. Most of the mesenchymal (ectomesenchymal) cells of the head and anterior portion of the neck

        5. Melanocytes of the skin and oral mucosa

        6. Odontoblasts (cells responsible for the production of dentin)

        7. Cells of the arachnoid and pia (rostral to the mesencephalon)

        8. Chromaffin cells of the adrenal medulla

III. Cells of Nervous Tissue

A. **Neurons** consist of a **cell body** and its processes, which usually include multiple **dendrites** and a single **axon.** Neurons comprise the smallest and largest cells of the body, ranging from 5 micrometers (μm) to 150 μm in diameter.

1. **Morphologic classification of neurons**

   a. **Unipolar neurons** possess a single process but are rare in vertebrates (see pseudounipolar neurons, III A 1 d).

   b. **Bipolar neurons** possess a single axon and a single dendrite. These neurons are present in some sense organs (e.g., the vestibular/cochlear mechanism).

   c. **Multipolar neurons** possess a single axon and more than one dendrite. These neurons are the **most common type** of neuron in vertebrates.

   d. **Pseudounipolar neurons** possess a single process that extends from the cell body and subsequently branches into an axon and dendrite. They are present in spinal and cranial ganglia.

      (1) These neurons originate embryologically as bipolar cells whose axon and dendrite later fuse into a single process functionally categorized as an axon.

      (2) They are frequently referred to as unipolar neurons.

2. **Functional classification of neurons**

   a. **Sensory neurons** receive stimuli from the internal and external environment. They conduct impulses **to the CNS** for processing and analysis.

   b. **Interneurons** connect other neurons in a chain or sequence. They most commonly connect sensory and motor neurons; they also regulate signals transmitted to neurons.

   c. **Motor neurons** conduct impulses **from the CNS** to other neurons, muscles, and glands.

3. **Neuron structure**

   a. **Neuronal cell body (soma, perikaryon)** is the region of a neuron containing the nucleus, various cytoplasmic organelles and inclusions, and cytoskeletal components.

      (1) The **nucleus** is large, spherical, and pale-staining and is **centrally located** in the soma of most neurons. It contains abundant euchromatin and a large nucleolus ("owl-eye" nucleus).

      (2) **Cytoplasmic organelles and inclusions**

         (a) **Nissl bodies** are composed of polysomes and rough endoplasmic reticulum (RER). They appear as clumps under light microscopy and are most abundant in large motor neurons.

         (b) The **Golgi complex** is located near the nucleus, and **mitochondria** are scattered throughout the cytoplasm.

         (c) **Melanin-containing granules** are present in some neurons in the CNS and in dorsal root and sympathetic ganglia.

(d) **Lipofuscin-containing granules** are present in some neurons and increase in number with age.

(e) **Lipid droplets** occasionally are present.

(3) **Cytoskeletal components**

(a) **Neurofilaments** [10 nanometers (nm) in diameter] are abundant and run throughout the soma cytoplasm. They are intermediate filaments composed of three intertwining polypeptide chains.

(b) **Microtubules** (24 nm in diameter) are also present in the soma cytoplasm.

(c) **Microfilaments [actin filaments** (6 nm in diameter)] are associated with the plasma membrane.

b. **Dendrites receive stimuli** (signals) from sensory cells, axons, or other neurons and convert these signals into small electrical impulses (action potentials) that are **transmitted toward** the soma.

(1) Dendrites have **arborized terminals** (except in bipolar neurons), which permit a neuron to receive stimuli simultaneously from many other neurons.

(2) The dendrite cytoplasm is similar to that of the soma except that it lacks a Golgi complex.

(3) Organelles become reduced or absent near the terminals except for mitochondria, which are abundant.

(4) **Spines** located on the surface of dendrites increase the area available for synapse formation with other neurons. These diminish with age and poor nutrition and exhibit altered configurations in individuals with trisomy 21 or trisomy 13.

c. **Axons** conduct impulses **away from** the soma to the axon terminals without any diminution in their strength.

(1) The diameter and length of axons in different types of neurons vary. Some axons are as long as 100 centimeters (cm).

(2) Axons originate from the **axon hillock,** a specialized region of the soma that lacks RER and ribosomes but contains many microtubules and neurofilaments; abundance of the latter may regulate neuron diameter.

(3) Axons may have **collaterals,** branching at right angles from the main trunk.

(4) Axon cytoplasm **(axoplasm)** lacks a Golgi complex but contains smooth endoplasmic reticulum (SER), RER, and elongated mitochondria.

(5) A plasma membrane surrounds the axon and is called the **axolemma.**

(6) Axons terminate in many small branches **(axon terminals)** from which impulses are passed to another neuron or other type of cell.

B. **Neuroglial cells** are located only in the CNS (Schwann cells are a PNS equivalent).

1. **General characteristics.** Neuroglial cells comprise several cell types and outnumber neurons by approximately 10 to 1. These cells are embedded in a web of tissue composed of modified ectodermal elements; the entire supporting structure is termed the **neuroglia.** They function to **support and protect neurons,** but they do not conduct impulses or form synapses with other cells. Neuroglia are revealed in histologic sections of the CNS only with special gold and silver stains.

2. **Types of neuroglial cells**

   a. **Astrocytes**

      (1) Astrocytes are the largest of the neuroglial cells. They have many processes, some of which possess expanded pedicles **(vascular feet)** that surround blood vessels, whereas others exhibit processes that contact the pia mater.

      (2) **Function**

         (a) Astrocytes scavenge ions and debris from neuron metabolism and supply energy for metabolism.

         (b) Along with other components of the neuroglia, astroglia form a protective **sealed barrier** between the pia mater and the nervous tissue of the brain and spinal cord.

         (c) They provide **structural support** for nervous tissue.

         (d) They proliferate to form **scar tissue** after injury to the CNS.

      (3) **Types of astrocytes**

         (a) **Protoplasmic astrocytes** reside mostly in gray matter and have branched processes that envelop blood vessels, neurons, and synaptic areas. They contain some intermediate filaments composed of **glial fibrillar acidic protein** (GFAP). These astrocytes help to establish the **blood-brain barrier** and may contribute to its maintenance.

         (b) **Fibrous astrocytes** reside mostly in **white matter** and have long, slender processes with few branches. They contain many intermediate filaments composed of GFAP.

   b. **Oligodendrocytes**

      (1) Oligodendrocytes are neuroglial cells that live **symbiotically** with neurons (i.e., each cell type is affected by the metabolic activities of the other). They are necessary for the survival of neurons.

      (2) They are located in both **gray matter and white matter.**

      (3) They possess a small, round, condensed nucleus and only a few short processes.

      (4) Their electron-dense cytoplasm contains ribosomes, numerous microtubules, many mitochondria, RER, and a large Golgi complex.

      (5) Oligodendrocytes produce **myelin,** a lipoprotein material organized into a sheath that insulates and protects axons in the CNS. Each oligodendrocyte produces myelin for several axons.

c. Schwann cells

   (1) Schwann cells are flattened cells with only a few mitochondria and a small Golgi region.

   (2) Although Schwann cells are derived from neural crest cells, they are still considered neuroglial cells.

   (3) These cells perform the same function in the PNS as oligodendrocytes in the CNS: they protect and insulate neurons. Schwann cells form either unmyelinated or myelinated coverings over neurons. A myelin sheath consists of several Schwann cell plasmalemmae wrapped around a **single axon** (see V).

d. Microglia are small, **phagocytic** neuroglial cells that are derived from the mononuclear phagocytic cell population in the bone marrow. They have a condensed, elongated nucleus and many short, branching processes. Activated microglial cells become antigen-presenting cells and secrete cytokines.

e. Ependymal cells, derived from the neuroepithelium, are the **epithelial cells** that line the neural tube and ventricles of the brain. In certain regions of the brain they possess **cilia,** which aid in moving the cerebrospinal fluid (CSF). Modified ependymal cells contribute to the formation of the **choroid plexus.**

IV. Synapses. Synapses are sites of **functional apposition** where signals are transmitted from one neuron to another, or from a neuron to another type of cell (e.g., muscle cell).

   A. Classification. Synapses are classified according to the site of synaptic contact and the method of signal transmission.

   1. Site of synaptic contact

   a. Axodendritic synapses are located between an axon and a dendrite.

   b. Axosomatic synapses are located between an axon and a soma. The CNS primarily contains axodendritic and axosomatic synapses.

   c. Axoaxonic synapses are located between axons.

   d. Dendrodendritic synapses are located between dendrites.

   2. Method of signal transmission

   a. Chemical synapses

   (1) These synapses involve the release of a chemical substance (**neurotransmitter** or **neuromodulator**) by the presynaptic cell, which acts on the postsynaptic cell to generate an action potential.

   (2) Chemical synapses are the most common neuron-neuron synapse and the only neuron-muscle synapse.

   (3) Signal transmission across these synapses is **delayed** by about 0.5 millisecond, which is the time required for secretion and diffusion of neurotransmitter from the presynaptic membrane of the first cell into the synaptic cleft and then to the postsynaptic membrane of the second cell.

      (4) Neurotransmitters do not effect the change, they only activate a response in the second cell.

  b. Electrical synapses

      (1) These synapses involve movement of ions from one neuron to another via **gap junctions,** which transmit the action potential of the presynaptic cell directly to the postsynaptic cell.

      (2) Electrical synapses are much less numerous than chemical synapses.

      (3) Signal transmission across these synapses is **nearly instantaneous.**

B. Synaptic morphology

  1. Axon terminals may vary morphologically depending on the site of synaptic contact.

    a. Boutons terminaux are bulbous expansions that occur singly at the end of axon terminals.

    b. Boutons en passage are swellings along the axon terminal; synapses may occur at each swelling.

  2. The **presynaptic membrane** is the thickened axolemma of the neuron that is transmitting the impulse. It contains **voltage-gated Ca²⁺** channels, which regulate the entry of calcium ions into the axon terminal.

  3. The **postsynaptic membrane** is the thickened plasma membrane of the neuron or other target cell that is receiving the impulse.

  4. The **synaptic cleft** is the narrow space (20-30 nm wide) between the presynaptic and postsynaptic membranes. Neurotransmitters diffuse across the synaptic cleft.

  5. Synaptic vesicles are small, membrane-bound structures (40-60 nm in diameter) located in the axoplasm of the transmitting neuron. They **discharge neurotransmitters** into the synaptic cleft by exocytosis.

C. Neurotransmitters (Table 9.1) are produced, stored, and released by presynaptic neurons. They diffuse across the synaptic cleft and bind to receptors in the postsynaptic membrane, leading to generation of an action potential.

V. Nerve Fibers. Nerve fibers are individual axons enveloped by a myelin sheath or by Schwann cells in the PNS (or oligodendrocytes in the CNS).

A. Myelin sheath

  1. The myelin sheath is produced by **oligodendrocytes** in the CNS and by **Schwann cells** in the PNS.

  2. It consists of several spiral layers of the plasma membrane of an oligodendrocyte or Schwann cell wrapping around the axon.

  3. It is not continuous along the length of the axon but is interrupted by gaps called **nodes of Ranvier.**

  4. Its thickness is constant along the length of an axon; however, thickness usually increases as axonal diameter increases.

Table 9.1. Common Neurotransmitters

Neuro-transmitter	Location	Function
Acetylcholine	Myoneural junctions; all parasympathetic synapses; preganglionic sympathetic synapses	Activates skeletal muscle, autonomic nerves; brain functions
Norepinephrine	Postganglionic sympathetic synapses	Increases cardiac output
Glutamic acid	CNS; presynaptic sensory and cortex	Most common excitatory neurotransmitter of CNS
γ-aminobutyric acid (GABA)	CNS	Most common inhibitory neurotransmitter of CNS
Dopamine	CNS	Inhibitory and excitatory, depending on receptor
Glycine	Brain stem and Spinal cord	Inhibitory
Serotonin	CNS	Pain inhibitor; mood control; sleep
Aspartate	CNS	Excitatory
Enkephalins	CNS	Analgesic; inhibits pain transmission?
Endorphins	CNS	Analgesic; inhibits pain transmission?

CNS = central nervous system.

5. The myelin sheath can be extracted by standard histologic techniques. Methods using **osmium tetroxide** preserve the myelin sheath and stain it black.

6. Under electron microscopy the myelin sheath displays the following features:

   a. **Major dense lines,** which represent **fusions** between the cytoplasmic surfaces of the plasma membranes of Schwann cells (or oligodendrocytes)

   b. **Intraperiod lines,** which represent **close contact,** but not fusion, of the extracellular surfaces of adjacent Schwann-cell (or oligodendrocyte) plasma membranes

   c. **Clefts (incisures) of Schmidt-Lanterman** (observed in both electron and light microscopy), which are cone-shaped, oblique, **discontinuities** of the myelin sheath due to the presence of Schwann cell (or oligodendrocyte) cytoplasm within the myelin

B. **Nodes of Ranvier** are regions along the axon that **lack myelin** and represent discontinuities between adjacent Schwann cells (or oligodendrocytes).

   1. In the PNS, the axon at the nodes is covered by Schwann cells. In the CNS, however, the axon may not be covered by oligodendrocytes.

   2. The axolemma at the nodes contains **many Na+** pumps and displays, in electron micrographs, a characteristic electron density.

C. **Internodes** are the segments of a nerve fiber **between adjacent nodes of Ranvier.** They vary in length from 0.08 to 1 millimeter (mm), depending on the size of the Schwann cells (or oligodendrocytes) associated with the fiber.

VI. **Nerves.** Nerves are cordlike bundles of nerve fibers surrounded by connective tissue sheaths (Figure 9.1). They are **visible to the unaided eye** and usually appear **whitish** due to the presence of myelin.

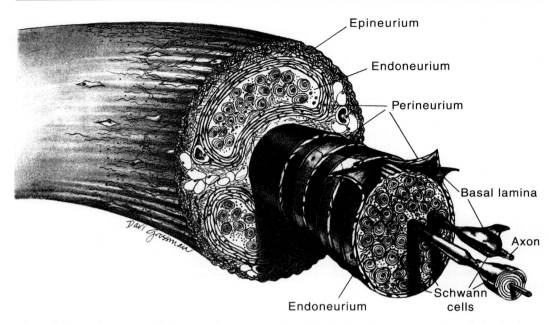

Figure 9.1. Drawing of a peripheral nerve in cross-section showing the various connective tissue sheaths. Each bundle of nerve fibers, or fascicle (one is extended in drawing), is covered by perineurium. (Reprinted with permission from Kelly DE, Wood RL, Enders AC: *Bailey's Textbook of Microscopic Anatomy*, 18th ed. Baltimore, Williams & Wilkins, 1984, p 353.)

A. Connective tissue investments

1. **Epineurium** is the layer of fibrous dense connective tissue **(fascia)** that forms the external coat of nerves.

2. **Perineurium** surrounds each bundle of nerve fibers **(fascicle).** Its inner surface consists of layers of flattened cells joined by **tight junctions** (zonulae occludentes) that prohibit passage of most macromolecules.

3. **Endoneurium** is a thin layer of reticular fibers, produced mainly by Schwann cells, that surrounds individual nerve fibers.

B. Functional classification of nerves

1. **Sensory nerves** contain **afferent** fibers and only carry sensory signals from the internal and external environment to the CNS.

2. **Motor nerves** contain **efferent** fibers and only carry signals from the CNS to effector organs.

3. **Mixed nerves,** which are the most common type of nerve, contain both afferent and efferent fibers and thus carry both sensory and motor signals.

VII. Ganglia. Ganglia are encapsulated aggregations of **neuronal cell bodies (soma)** located outside the CNS.

A. Autonomic ganglia are **motor** ganglia in which axons of preganglionic neurons synapse on postganglionic neurons.

B. Craniospinal ganglia are **sensory** ganglia associated with most cranial nerves and the dorsal roots of spinal nerves **(dorsal root ganglia).** Unlike autonomic ganglia, craniospinal ganglia **do not have synapses.** These ganglia contain the cell bodies of sensory neurons, which are **pseudounipolar (unipolar)** and transmit sensory signals from receptors to the CNS.

VIII. Histophysiology of Nervous Tissue

A. Resting membrane potential

1. The resting membrane potential exists across the plasma membrane of all cells.

2. It is established and maintained mostly by K+ leak channels and to a lesser extent by the **Na+-K+ pump,** which actively transports three Na+ ions out of the cell in exchange for two K+ ions (see Chapter 1 III B 1). The resting potential exists when there is no net movement of K+ ions (i.e., when outward diffusion of K+ ions is just balanced by the external positive charge acting against further diffusion).

B. An **action potential** is the electrical activity that occurs in a neuron as an impulse is propagated along the axon and is observed as a **movement of negative charges along the outside of an axon.** It is an **all-or-nothing event with a constant amplitude and duration.**

1. Generation of the action potential

a. An excitatory stimulus on a postsynaptic neuron partially **depolarizes** a portion of the plasma membrane (the potential difference is **less negative**).

b. Once the membrane potential reaches a critical **threshold, voltage-gated Na+ channels** in the membrane open, permitting Na+ ions to enter the cell (Figure 9.2).

c. The influx of Na+ ions leads to **reversal of the resting potential** in the immediate area (i.e., the external side becomes negative).

d. The Na+ channels close spontaneously and are inactivated for 1 to 2 milliseconds **(refractory period).**

e. Opening of **voltage-gated K+ channels** is also triggered by depolarization. Because these channels remain open longer than the Na+

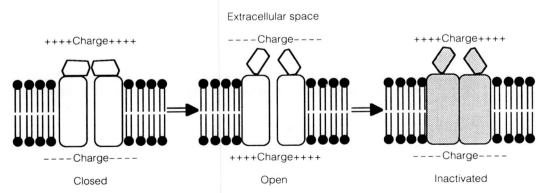

Figure 9.2. Model of the voltage-gated Na+ channel showing the transition between its closed, open, and inactivated states. In the resting state, the channel-blocking segment and gating keep the channel closed to entry of extracellular Na+ ions. Depolarization of the membrane causes a conformational change that opens the channel to influx of Na+ ions. The channel closes spontaneously and becomes inactive within a millisecond after opening.

channels, exit of K+ ions during the refractory period **repolarizes** the membrane back to its resting potential.

    f. The ion channels then return to their normal states. The cell is now ready to respond to another stimulus.

  2. Propagation of the action potential

    a. Propagation results from longitudinal diffusion of Na+ ions (which enter the cell at the initial site of excitation) toward the axon terminals **(orthodromic spread).** The longitudinal diffusion of Na+ ions depolarizes the adjacent region of membrane, giving rise to a new action potential at this site.

    b. Propagation does **not** result from diffusion of Na+ ions toward the soma **(antidromic spread)** because the Na+ channels are inactivated in this region.

    c. Action potentials are propagated most rapidly in myelinated fibers, which exhibit **saltatory conduction.** In this type of conduction, the action potential "jumps" from one node of Ranvier to the next.

C. Axonal transport of proteins, organelles, and vesicles occurs at high, intermediate, and low velocities depending on the nature of the transported materials.

  1. Anterograde transport carries material away from the soma.

  2. Retrograde transport carries material toward the soma for reutilization, recycling, or degradation.

D. Trophic function of nervous tissue

  1. Denervation of a muscle or gland leads to its atrophy.

  2. Reinnervation restores its structure and function.

IX. Somatic Nervous System and Autonomic Nervous System (ANS). "Somatic" and "autonomic" are functional concepts relating to all the neural elements involved in transmission of impulses from the CNS to the somatic and visceral components of the body, respectively.

A. The **somatic nervous system** contains sensory fibers that bring information to the CNS and the motor fibers that innervate voluntary **skeletal muscle.**

B. The **ANS** contains motor fibers that control and regulate **smooth muscle, cardiac muscle,** and some **glands.** It establishes and maintains **homeostasis** of the body's visceral functions. Anatomically and functionally it is divided into the **sympathetic** and **parasympathetic systems,** which generally **function antagonistically** in a given organ (i.e., when the sympathetic system stimulates an organ, the parasympathetic inhibits it, and vice versa).

  1. Autonomic nerve chains

    a. Cell bodies of preganglionic neurons are located in the CNS and extend their **preganglionic fibers** (axons) to an **autonomic ganglion** located outside of the CNS.

    b. In the ganglion, the preganglionic fibers synapse with postganglionic neurons, which typically are multipolar and surrounded by satellite cells.

      c. Postganglionic fibers leave the ganglion and terminate in the **effector organ** (smooth muscle, cardiac muscle, and glands).

  2. Sympathetic system (thoracolumbar outflow)

      a. The preganglionic cell bodies of this system are located in the thoracic and first two lumbar segments of the spinal cord.

      b. Function. The sympathetic system effects **vasoconstriction.** In general, it functions to prepare the body for "flight or fight" by increasing heart rate, respiration, blood pressure, and blood flow to skeletal muscles; dilating pupils; and decreasing visceral function.

  3. Parasympathetic system (craniosacral outflow)

      a. The preganglionic cell bodies of this system are located in certain cranial nerve nuclei within the brain and in some sacral segments of the spinal cord.

      b. Function. The parasympathetic system stimulates secretion (**secretomotor** function). In general, it functions to prepare the body for "rest or digest" by decreasing heart rate, respiration, and blood pressure; constricting pupils; and increasing visceral function.

X. CNS

A. White matter and gray matter are both present in the CNS.

  1. White matter contains mostly myelinated nerve fibers but also some unmyelinated fibers and neuroglial cells.

  2. Gray matter contains neuronal cell bodies, many unmyelinated fibers, some myelinated fibers, and neuroglial cells.

  3. Spinal cord gray matter appears in the shape of an **H** in cross-sections of the spinal cord.

      a. A small **central canal,** lined by ependymal cells, is at the center of the crossbar in the H. This canal is a remnant of the embryonic neural tube.

      b. The dorsal vertical bars of the H **(dorsal horns)** consist of **sensory** fibers extending from the dorsal root ganglia and cell bodies of interneurons.

      c. The ventral vertical bars of the H **(ventral horns)** consist of the fibers of large multipolar motor neurons.

  4. Brain gray matter is located at the periphery (cortex) of the cerebrum and cerebellum. White matter lies beneath the gray matter in these structures.

      a. The **Purkinje cell layer** (cerebellar cortex only) consists of flask-shaped Purkinje cells. These cells have a centrally located nucleus, highly branched (arborized) dendrites, and a single myelinated axon. These cells may receive several hundred thousand excitatory and inhibitory impulses to sort and integrate.

      b. Brain gray matter also forms the **basal ganglia,** which are located deep within the cerebrum and are surrounded by white matter.

B. Meninges are **membranous coverings of the brain and spinal cord.** They are formed from connective tissue. There are three layers of meninges:

the outermost **dura mater;** the intermediate **arachnoid mater;** and the innermost, highly vascular **pia mater.**

C. Cerebrospinal fluid (CSF)

1. CSF is a clear fluid produced primarily by cells of the **choroid plexus,** located in the ventricles of the brain. The choroid plexus is composed of folds of pia mater and capillaries that are surrounded by cuboidal ependymal cells.

2. CSF circulates through the ventricles, subarachnoid space, and central canal, bathing and nourishing the brain and spinal cord; it also acts as a shock-absorbing cushion to protect these structures.

3. CSF is about 90% water and ions; it contains little protein, occasional white blood cells, and infrequent desquamated cells.

4. CSF is continuously produced and is reabsorbed by **arachnoid granulations** that transport it into the superior sagittal sinus. If reabsorption is blocked, **hydrocephalus** may occur.

XI. Degeneration and Regeneration of Nerve Tissue

A. Death of neurons occurs as the result of injury to or disease affecting the somata.

1. Neuronal death results in degeneration and permanent loss of nerve tissue because **neurons of the CNS cannot divide.**

2. In the CNS, neuronal death may be followed by proliferation of the neuroglia, which fills in areas left by dead neurons.

B. Transection of axons induces changes in the soma including **chromatolysis** (disruption of Nissl bodies with a concomitant loss of cytoplasmic basophilia), increase in soma volume, and movement of the nucleus to a peripheral position.

1. **Degeneration of distal axonal segment (anterograde changes)**

   a. The axon and its myelin sheath, which are separated from the soma, degenerate completely **(wallerian degeneration),** and the remnants are removed by macrophages.

   b. Schwann cells proliferate, forming a **solid cellular column** that is distal to the injury and that remains attached to the effector cell.

2. **Regeneration of proximal axonal segment** (retrograde changes) (Figure 9.3)

   a. The distal end, closest to the wound, initially degenerates, and the remnants are removed by macrophages.

   b. Growth at the distal end then begins (0.5-3 mm/day) and progresses toward the columns of Schwann cells.

   c. Regeneration is successful if the sprouting axon penetrates a Schwann-cell column and re-establishes contact with the effector cell.

XII. Clinical Considerations. Hirschsprung disease (congenital megacolon) is the result of abnormal organogenesis, in which neural crest cells fail to migrate into the wall of the gut. The disease is characterized by the **absence of Auerbach plexus,** a part of the parasympathetic system innervating the

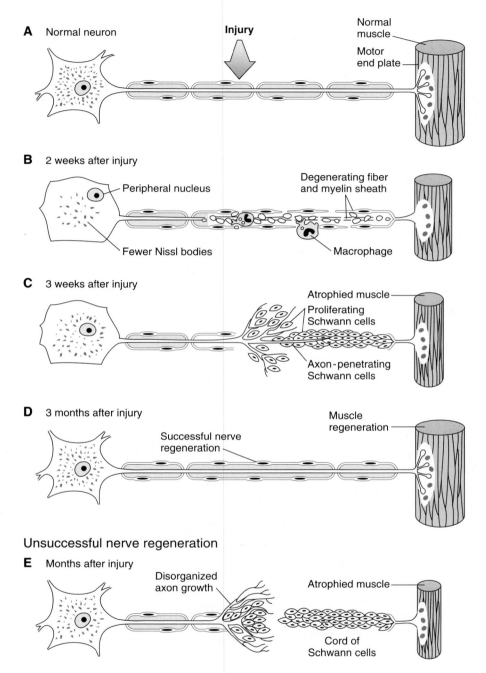

A Normal neuron

Injury

Normal muscle

Motor end plate

B 2 weeks after injury

Peripheral nucleus

Degenerating fiber and myelin sheath

Fewer Nissl bodies

Macrophage

C 3 weeks after injury

Atrophied muscle

Proliferating Schwann cells

Axon-penetrating Schwann cells

D 3 months after injury

Successful nerve regeneration

Muscle regeneration

Unsuccessful nerve regeneration

E Months after injury

Disorganized axon growth

Atrophied muscle

Cord of Schwann cells

Figure 9.3. Schematic diagram of peripheral nerve regeneration. (Adapted with permission from Gartner LP, Hiatt JL: *Color Textbook of Histology*. WB Saunders, Philadelphia, 1997, p 185.)

distal segment of the colon. This loss of motor function leads to dilation of the colon. **Neuroglial tumors** constitute 50% of all intracranial tumors. They are derived from astrocytes, oligodendrocytes, and ependymocytes. The tumors range in severity from slow-growing **benign oligodendrogliomas** to rapid-growing fatal **malignant astrocytomas.**

Review Test

Directions: Each of the numbered items or incomplete statements in this section is followed by answers or by completions of the statement. Select the ONE lettered answer or completion that is BEST in each case.

1. Neural crest cells give rise to which of the following?

(A) Dorsal horns of the spinal cord
(B) Adrenal cortex
(C) Sympathetic ganglia
(D) Preganglionic autonomic nerves

2. Which one of the following neurotransmitters functions to increase cardiac output?

(A) Dopamine
(B) Serotonin
(C) Norepinephrine
(D) Glutamic acid
(E) 3-Aminobutyric acid

3. Which of the following statements regarding nerve cell membrane potentials is true?

(A) Membrane potentials are maintained at rest by Na^+ ions entering the cell.
(B) Entrance of K^+ ions causes the membrane to return to its resting potential.
(C) Depolarization triggers the opening of voltage-gated K^+ channels.
(D) Voltage gated Na^+ channels become activated during the refractory period.

4. Which of the following statements is characteristic of the perineurium?

(A) It is a fascia surrounding many bundles of nerve fibers.
(B) It consists in part of epithelioid cells that surround a bundle (fascicle) of nerve fibers.
(C) It is a thin layer of reticular fibers covering individual nerve fibers.
(D) It is a fascia that excludes macromolecules and forms the external coat of nerves.

5. Acetylcholine is the neurotransmitter in which of the following regions of the nervous system?

(A) Central nervous system
(B) Presynaptic sensory cortex

(C) Myoneural junctions
(D) Postganglionic sympathetic synapses
(E) Motor cortex

6. Nissl bodies are composed of

(A) synaptic vesicles and acetylcholine
(B) polyribosomes and rough endoplasmic reticulum
(C) lipoprotein and melanin
(D) neurofilaments and microtubules
(E) smooth endoplasmic reticulum and mitochondria

7. The axon hillock contains

(A) rough endoplasmic reticulum
(B) ribosomes
(C) microtubules
(D) Golgi complex
(E) synaptic vesicles

8. Synaptic vesicles possess which of the following characteristics?

(A) Manufacture neurotransmitter
(B) Enter the synaptic cleft
(C) Become incorporated into the presynaptic membrane
(D) Become incorporated into the postsynaptic membrane

9. A patient with Hirschsprung disease will present with which of the following symptoms?

(A) Absent cranial vault
(B) Exposed spinal cord
(C) Headache
(D) Large, dilated colon

10. Myelination of peripheral nerves is accomplished by

(A) astrocytes
(B) oligodendrocytes
(C) Schwann cells
(D) neural crest cells
(E) basket cells

Answers and Explanations

1-C. Neural crest cells migrate throughout the body and give rise to ganglia and other structures, including portions of the adrenal medulla, but they do not contribute to the development of preganglionic autonomic nerves, adrenal cortex, or the dorsal horns of the spinal cord.

2-C. Norepinephrine increases cardiac output, whereas dopamine and 3-aminobutyric acid are central nervous system (CNS) inhibitors. Glutamic acid is the most common excitatory neurotransmitter of the CNS. Serotonin functions as a pain inhibitor, in mood control, and in sleep.

3-C. Once the critical threshold is reached, voltage-gated Na^+ channels open and Na^+ ions enter the cell, which depolarizes the cell. Depolarization triggers the opening of voltage-gated K^+ channels, and K^+ ions then exit the cell.

4-B. Each bundle of nerve fibers is surrounded by the perineurium, which consists primarily of several layers of epithelioid cells. Tight junctions between these cells exclude most macromolecules. The external coat of nerves, the epineurium, surrounds many fascicles but does not exclude macromolecules. The layer of reticular fibers that covers individual nerve fibers is the endoneurium; it also does not exclude macromolecules.

5-C. Acetylcholine is the neurotransmitter for myoneural junctions as well as for preganglionic sympathetic and preganglionic and postganglionic parasympathetic synapses.

6-B. Nissl bodies are large granular basophilic bodies composed of polysomes and rough endoplasmic reticulum. They are found only in neurons (in the soma cytoplasm).

7-C. The axon hillock is devoid of organelles, but it does contain microtubules, which are arranged in bundles.

8-C. Synaptic vesicles release neurotransmitter into the synaptic cleft by exocytosis. In this process, the vesicle membrane is incorporated into the presynaptic membrane. Although these vesicles contain neurotransmitter, they do not manufacture it.

9-D. Hirschsprung disease is characterized by a dilated colon caused by the absence of the parasympathetic myenteric ganglia known as Auerbach plexus.

10-C. Schwann cells produce myelin in the peripheral nervous system, whereas oligodendrocytes produce myelin in the central nervous system. Astrocytes, neural crest cells, and basket cells do not produce myelin.

10

Blood and Hemopoiesis

I. Overview—Blood

A. Blood is a specialized connective tissue that consists of formed elements **(erythrocytes, leukocytes, and platelets)** and a fluid component called **plasma.**

B. The volume of blood in an average human adult is approximately **5 liters.**

C. Blood circulates in a closed system of vessels and **transports** nutrients, waste products, hormones, proteins, ions, etc.

D. It also **regulates body temperature** and assists in regulation of **osmotic** and **acid-base balance.**

E. Blood cells have short life spans and are replaced continuously by a process called **hemopoiesis.**

II. Blood Constituents

A. Plasma consists of 90% **water,** 9% **organic compounds** (such as proteins, amino acids, and hormones), and 1% **inorganic salts.**

1. Main plasma proteins

a. Albumin, a small protein [60,000 molecular weight (MW)], preserves osmotic pressure in the vascular system and helps transport some metabolites.

b. γ-Globulins are antibodies (immunoglobulins) (see Chapter 12).

c. α-Globulins and β-globulins transport metal ions (e.g., iron and copper) and lipids (in the form of lipoproteins).

d. Fibrinogen is converted into fibrin during blood clotting.

e. Complement proteins (C1 through C9) function in nonspecific host defense and initiate the inflammatory process.

2. Serum is the **yellowish fluid** that remains after blood has clotted. It is similar to plasma but lacks fibrinogen and clotting factors.

B. Formed elements of blood (Table 10.1)

1. Erythrocytes [red blood cells (RBCs)]

a. General features

(1) RBCs are round, **anucleate,** biconcave cells that **stain light salmon pink** with either Wright or Giemsa stain.

(2) The average life span of an RBC is 120 days. Aged RBCs are fragile and express membrane surface factors recognized by splenic macrophages, which destroy them.

Table 10.1. Size and Number of Formed Elements in Human Blood

Cell Type	Diameter (μm) Smear	Diameter (μm) Section	Cells per mm³	Leukocytes (%)
Erythrocyte	7–8	6–7	5×10^6 (men) 4.5×10^6 (women)	–
Agranulocytes				
Lymphocyte	8–10	7–8	1500–2500	20–25
Monocyte	12–15	10–12	200–800	3–8
Granulocytes				
Neutrophil	9–12	8–9	3500–7000	60–70
Eosinophil	10–14	9–11	150–400	2–4
Basophil	8–10	7–8	50–100	0.5–1
Platelet	2–4	1–3	250,000–400,000	–

Reprinted from Gartner LP, Hiatt JL: *Color Atlas of Histology*, 2nd ed. Baltimore, Williams & Wilkins, 1994, p 84.

(3) Carbohydrate determinants for the **A, B, and O blood groups** are located on the external surface of their plasmalemmae.

(4) Several **cytoskeletal proteins** (ankyrin, band 4.1 and band 3 proteins, spectrin, and actin) function in maintaining the shape of RBCs (see Chapter 1 V A).

(5) Mature erythrocytes possess no organelles but are filled with **hemoglobin (Hb)**.

(6) Erythrocytes contain **soluble enzymes** that are responsible for glycolysis and the hexose monophosphate pathway and the production of adenosine triphosphate (ATP).

b. The **hematocrit** is an estimation of the **volume of packed erythrocytes per unit volume of blood.**

(1) The hematocrit is expressed as a percentage.

(2) Normal values are 40%-50% in adult men, 35%-45% in adult women, 35% in children up to 10 years of age, and 45%-60% in newborns.

c. **Hb** is a protein composed of four polypeptide chains, each covalently linked to a heme group. The four chains that normally occur in humans are α, β, γ, and δ. Each chain differs in its amino acid sequence.

(1) Hb occurs in several normal forms that differ in their **chain composition.**

 (a) The predominant form of adult Hb is **HbA₁** ($\alpha_2\beta_2$).

 (b) A minor form is **HbA₂** ($\alpha_2\delta_2$).

 (c) Fetal Hb is designated **HbF** ($\alpha_2\gamma_2$).

(2) Abnormal forms include **HbS,** which occurs as a result of a point mutation in the β chain (substitution of the amino acid **valine** for **glutamate**). Erythrocytes containing HbS are sickle-shaped and fragile and cause **sickle cell anemia.**

    d. Transport of O$_2$ and CO$_2$ to and from the tissues of the body is carried out by erythrocytes.

      (1) In the lungs, where the partial pressure of O$_2$ is high, Hb preferentially binds O$_2$, forming **oxyhemoglobin.**

      (2) In the tissues (low O$_2$ and high CO$_2$ levels), oxyhemoglobin releases O$_2$ and binds CO$_2$, forming **carbaminohemoglobin** (also known as **carbamylhemoglobin).**

      (3) Because CO binds avidly to Hb, it can block binding of O$_2$ and cause **CO asphyxiation** if it is inhaled in sufficient amounts.

      (4) Nitric oxide (**NO**), a neurotransmitter substance, also binds to Hb, and, in oxygen-poor areas, facilitates dilation of blood vessels and a more efficacious exchange of O$_2$ for CO$_2$.

2. Leukocytes [white blood cells (WBCs)] [see Table 10.1] possess varying numbers of **azurophilic granules.** These are lysosomes containing various hydrolytic enzymes.

    a. Granulocytes (Table 10.2) include **neutrophils, eosinophils,** and **basophils.**

      (1) Granulocytes possess **specific granules** with type-specific contents.

Table 10.2. Selected Characteristics of Granulocytes

Characteristic	Neutrophils	Eosinophils	Basophils
Nuclear shape	Lobulated (3 or 4 lobes)	Bilobed	S-shaped
Number of azurophilic granules	Many	Few	Few
Specific granules			
Size	Small	Large	Large
Color*	Light pink	Dark pink	Dark blue to black
Contents	Alkaline phosphatase	Acid phosphatase	Eosinophil chemotactic factor (ECF)
	Collagenase	Arylsulfatase	
	Lactoferrin	β-Glucuronidase	Heparin
	Lysozyme	Cathepsin	Histamine
	Phagocytin	Major basic protein	Peroxidase
		Peroxidase	
		Phospholipase	
		RNase	
Life span	1 week	Few hours in blood, 2 weeks in connective tissue	Very long (1-2 years in mice)
Main functions	Phagocytose, kill, and digest bacteria	Moderate inflammatory reactions by inactivating histamine and leukotriene C	Mediate inflammatory responses in a manner similar to mast cells
Special properties	Form H$_2$O$_2$ during phagocytosis	Are decreased in number by corticosteroids	Have receptors for immunoglobulin E on their plasma membrane

*Cells stained with Giemsa or Wright stain.

(2) These cells generate ATP via the glycolytic pathway, Krebs cycle (basophils), and anaerobic pathways (neutrophils).

(3) Destruction of phagocytosed microorganisms by neutrophils occurs in two ways.

(a) Azurophilic granules release **hydrolytic enzymes** into phagosomes to destroy microorganisms.

(b) Reactive oxygen compounds **superoxide (O_2^-)**, hydrogen peroxide (H_2O_2), and hypochlorous acid (HOCl) formed within phagosomes (catalyzed by **myeloperoxidase**) destroy microorganisms.

b. **Agranulocytes** (Table 10.3) lack specific granules.

(1) They include **lymphocytes** and **monocytes.**

(2) They also include **null cells** (about 5% of the circulating lymphocytes), pluripotential hemopoietic stem cells (PHSCs), and natural killer (NK) cells. Null cells resemble lymphocytes but lack their characteristic surface determinants.

3. **Platelets (thrombocytes)** [see Table 10.1 and Table 10.4] are anucleated, disk-shaped cell fragments that arise from megakaryocytes in bone marrow.

a. A clear, peripheral region, the **hyalomere,** and a region containing purple granules, the **granulomere,** are visible in stained blood smears.

b. Platelets are surrounded by a **glycocalyx,** which coats the plasmalemma. **Calcium ions** and **adenosine diphosphate** (ADP) increase the "stickiness" of the glycocalyx and enhance platelet adherence.

c. Platelets function in **blood coagulation** by aggregating at lesions in vessel walls and producing various factors that aid in clot formation.

d. They are also responsible for **clot retraction** and contribute to **clot removal.**

Table 10.3. Selected Characteristics of Agranulocytes

Characteristic	Monocytes	T Lymphocytes	B Lymphocytes
Plasma membrane	Form filopodia and pinocytic vesicles	Have T-cell receptors	Have Fc receptors and antibodies
Number of azurophilic granules	Many	Few	Few
Life span	Less than 3 days in blood	Several years	Few months
Main functions	Become macrophages in connective tissue	Generate cell-mediated immune response, secrete numerous growth factors	Generate humoral immune response

Table 10.4. Platelet Components

Hyalomere		Granulomere			
Structure	Function	Granules	Size (nm)	Contents	Function
Actin and myosin	Platelet contraction	α	300–500	Fibrinogen, platelet thromboplastin, factors V and VIII, PDGF	Repair of vessel, platelet aggregation, coagulation
Microtubule bundles	Maintains platelet shape	β (dense bodies)	250–300	Pyrophosphate, ADP, ATP, histamine, serotonin, Ca^{2+}	Vasoconstriction, platelet aggregation and adhesion
Surface opening tubule system	Facilitates exocytosis and endocytosis in activated platelets	λ	200–250	Lysosomal enzymes	Clot removal
Dense tubular system	Prevents platelet stickiness by sequestering Ca^{2+}				

ADP = adenosine diphosphate; ATP = adenosine triphosphate; PDGF = platelet-derived growth factor.

III. Blood Coagulation

A. Blood coagulation contributes to hemostasis and is normally tightly controlled so that it occurs **only in regions where the endothelium is damaged.**

B. The activation of at least 13 plasma proteins, **coagulation factors,** is necessary for blood coagulation. The coagulation factors participate in a cascade of reactions.

C. **Platelet membranes** and **Ca^{2+}** (factor IV) are also required for blood coagulation.

D. Blood coagulation occurs via two interrelated pathways, the **intrinsic** and **extrinsic pathways.** The final steps in both pathways involve the transformation of prothrombin to **thrombin,** an enzyme that catalyzes the conversion of fibrinogen (factor I) to **fibrin monomers,** which coalesce to form a **reticulum of clot.**

   1. The **extrinsic pathway** occurs in response to damaged blood vessels. It is initiated within seconds **(rapid onset)** after trauma that releases **tissue thromboplastin** (factor III).

   2. The **intrinsic pathway** is initiated within several minutes **(slow onset)** after trauma to blood vessels or when platelets or factor XII are exposed to collagen in the vessel wall. This pathway depends on **von Willebrand factor** and **factor VIII.** These factors form a complex that binds to subendothelial collagen and to receptors on platelet membranes, thus promoting **platelet aggregation** and **adherence to collagen** in the vessel wall.

IV. Bone Marrow

A. Yellow marrow is located in the long bones of adults and is highly infiltrated with fat. It is **not** hemopoietic, but has the potential to become so if necessary.

B. In adults, **red marrow** is located in the epiphyses of long bones and in flat, irregular, and short bones. It is highly vascular and composed of a **stroma,** irregular **sinusoids,** and islands of **hemopoietic cells.** Red marrow is the site of blood cell **differentiation** and **maturation.**

   1. Sinusoids are large vessels with highly attenuated (very thin) endothelia. They are associated on their extravascular surfaces with reticular fibers and **adventitial reticular cells,** which manufacture these fibers.

   2. Stromal cells include macrophages, adventitial reticular cells, fibroblasts, and endothelial cells. These cells produce and release various **hemopoietic growth factors.**

     a. Macrophages are located in extravascular areas near sinusoids and extend processes between endothelial cells into sinusoidal lumina.

     b. Adventitial reticular cells are believed to subdivide the bone-marrow cavity into smaller compartments, which are occupied by **is-**

lands of hemopoietic cells. They may accumulate fat (instead of fat cells), thus transforming red marrow into yellow marrow.

V. Prenatal Hemopoiesis. This process occurs successively in the yolk sac, liver, spleen, and bone marrow.

A. The bone marrow first participates in hemopoiesis at about 6 months' gestation and assumes an increasingly larger role thereafter.

B. The liver and spleen cease hemopoiesis at about the time of birth.

VI. Postnatal Hemopoiesis. This process involves three classes of cells: stem, progenitor, and precursor.

A. Comparison of stem, progenitor, and precursor cells

1. **Stem cells** are capable of **self-renewal** and can undergo enormous proliferation.

   a. These cells can differentiate into **multiple** cell lineages.

   b. They are present in circulation (as null cells) and bone marrow.

2. **Progenitor cells** have reduced potentiality and are committed to a single cell lineage.

   a. They **proliferate** and **differentiate** into precursor cells in the presence of appropriate growth factors.

   b. They are morphologically indistinguishable from stem cells and both appear similar to small lymphocytes.

3. **Precursor cells** are all the cells in each lineage that display **distinct morphologic characteristics.**

B. Initial steps in blood formation (stem cells)

1. **Pluripotential hemopoietic stem cells (PHSC)** give rise to multipotential hemopoietic stem cells in the bone marrow.

2. **Multipotential hemopoietic stem cells** are of two types: colony-forming unit—spleen **(CFU-S)** and colony-forming unit—lymphocyte **(CFU-Ly)**. These cells divide and differentiate in bone marrow to form progenitor cells.

   a. **CFU-S,** the **myeloid stem cell,** is the multipotential stem cell that gives rise to erythrocytes, granulocytes, monocytes, and platelets. Probably **Hox 2 genes** are active in the early stages of differentiation of the erythroid lines, and **Hox 1 genes** may be active in the early stages of differentiation of granulocytes, monocytes, and platelets.

   b. **CFU-Ly,** the **lymphoid stem cell,** is the multipotential stem cell that gives rise to T and B lymphocytes and NK cells.

C. Erythrocyte formation (erythropoiesis) begins with formation of two types of progenitor cells: **burst-forming unit—erythroid (BFU-E),** derived from CFU-S, and **colony-forming unit—erythroid (CFU-E),** which arises from BFU-E. Erythropoiesis yields about 1 trillion RBCs daily in a normal adult.

Table 10.5. Selected Characteristics of Erythrocyte Precursor Cells

Characteristic	Proerythroblast	Basophilic Erythroblast	Polychromatophilic Erythroblast (Normoblast)	Orthochromatophilic Erythroblast	Reticulocyte
Nucleus Shape	Round	Round	Round and small	Round	None
Color*	Burgundy red	Burgundy red	Dense blue	Dark, may be extruding	None
Chromatin network	Very fine	Fine	Coarse	Pyknotic	None
Number of nucleoli	3-5 (very pale gray)	1-2	None	None	None
Mitosis	Yes	Yes	Yes	No	No
Cytoplasmic color*	Pale gray with blue clumps	Grayish pink with intensely blue clumps	Yellowish pink with bluish background	Pink with trace of blue	Pink†
Hemoglobin	None (ferritin is present)	Some	Abundant	Abundant	Abundant

*Cells stained with Giemsa or Wright stain.
†Cells stained supravitally with brilliant cresyl blue display a reticulum.

1. **Erythroid progenitor cells**

   a. **BFU-E** has a high rate of mitotic activity and responds to high concentrations of **erythropoietin,** a hormone that stimulates erythropoiesis.

   b. **CFU-E** responds to low concentrations of erythropoietin and gives rise to the first histologically recognizable erythrocyte precursor, the **proerythroblast.**

2. **Erythrocyte precursor cells** include a series of cell types (the **erythroid series**) that differentiate sequentially to form mature erythrocytes (Table 10.5).

D. **Granulocyte formation** begins with production of three unipotential or bipotential cells, all of which are descendants of CFU-S. Granulocyte formation yields about 1 million granulocytes daily in a normal adult.

   1. **Granulocyte progenitor cells** give rise to histologically identical **myeloblasts** and **promyelocytes** in all three cell lineages.

      a. **CFU-Eo** is the progenitor of the **eosinophil** lineage.

      b. **CFU-Ba** is the progenitor of the **basophil** lineage.

      c. **CFU-NM,** the common progenitor of **neutrophils** and **monocytes,** gives rise to **CFU-N** (neutrophil) and **CFU-M** (monocyte).

   2. **Granulocyte precursor cells** are histologically similar in the early stages of all three lineages (myeloblasts and promyelocytes). They develop characteristic granules unique to each cell type during the myelocyte stage and a distinctive nuclear shape during the stab (band) stage (Table 10.6).

E. **Monocyte formation (CFU-M)** begins with the common progenitor CFU-NM and involves only two precursor cells: **monoblasts** and **promonocytes.** Monocyte formation yields about 10 trillion monocytes daily in a normal adult.

   1. **Promonocytes** are reported to be large cells [16-18 micrometers (μm) in diameter] and contain a kidney-shaped, acentric nucleus; numerous azurophilic granules; an extensive rough endoplasmic reticulum (RER) and Golgi complex; and many mitochondria. They undergo cell division and subsequently develop into **monocytes.**

   2. **Monocytes** leave the bone marrow to enter the circulation. From the bloodstream, they enter connective tissue where they **differentiate into macrophages.**

F. **Platelet formation** begins with the progenitor **CFU-Meg,** which arises from CFU-S, and involves a single precursor cell, the **megakaryoblast,** and the mature **megakaryocyte,** which remains in the bone marrow and sheds platelets.

   1. **Megakaryoblasts**

      a. These large cells (25-40 μm in diameter) have a single large nucleus that may be indented or lobed and displays a fine chromatin network.

      b. Their basophilic, nongranular cytoplasm contains large mitochondria, many polysomes, some RER, and a large Golgi complex.

Table 10.6. Selected Characteristics of Neutrophil Precursor Cells

Characteristic	Myeloblast	Promyelocyte	Neutrophilic Myelocyte	Neutrophilic Metamyelocyte	Neutrophilic Stab Cell
Cell diameter (μm)	12-14	16-24	10-12	10-12	11-12
Nucleus shape	Large, round	Large, round	Flat (acentric)	Kidney-shaped (acentric)	Horseshoe-shaped
Color*	Reddish blue	Reddish blue	Blue to dark blue	Dark blue	Dark blue
Chromatin network	Very fine	Fine	Coarse	Very coarse	Very coarse
Number of nucleoli	2 or 3 (pale gray)	1 or 2 (pale gray)	1 (?)	None	None
Mitosis	Yes	Yes	Yes	No	No
Cytoplasmic appearance*	Blue clumps in pale blue background, cytoplasmic blebs at cell periphery	Bluish hue, no cytoplasmic blebs	Pale blue	Blue	Similar to mature neutrophils
Granules	None	Azurophilic	Azurophilic and specific	Azurophilic and specific	Azurophilic and specific

*Cells stained with Giemsa or Wright stain.

   c. They divide endomitotically (i.e., no daughter cells are formed) and enlarge; the ploidy of the nucleus increases to as much as 64N, giving rise to megakaryocytes.

  2. Megakaryocytes

    a. These extremely large cells (40-100 μm in diameter) have a single large **polypoid** nucleus that is highly indented.

    b. They possess a well-developed Golgi complex associated with the formation of α-granules, lysosomes, and dense bodies (δ-granules); they also contain many mitochondria and an extensive RER.

    c. Megakaryocytes lie just outside the sinusoids in the bone marrow and form **platelet demarcation channels,** which fragment into pro-platelets (clusters of adhering platelets) or single platelets that are released into the sinusoidal lumen.

 G. Lymphocyte formation (lymphopoiesis) begins with differentiation of CFU-Ly, the lymphoid stem cell, into the **immunoincompetent** progenitor cells, **CFU-LyB** and **CFU-LyT.** These **prelymphocytes** are processed and become mature, immunocompetent cells.

  1. B-lymphocyte (B cell) maturation

    a. Pre-B lymphocytes acquire cell-surface markers, including membrane-bound **antibodies,** which confer **immunocompetence.**

    b. In mammals, B-cell maturation occurs in the bone marrow, whereas in birds it occurs in the bursa of Fabricius (hence, **B** lymphocytes).

  2. T-lymphocyte (T cell) maturation involves migration of progenitor T lymphocytes to the **thymus,** where they acquire cell-surface markers, including **T-cell receptors,** which confer **immunocompetence.** Most of the newly formed T lymphocytes are destroyed in the thymus and do not enter into circulation.

  3. Mature B and T lymphocytes leave the bone marrow and thymus, respectively, and circulate to peripheral organs (e.g., the lymph nodes and spleen) to establish **clones** of immunocompetent lymphocytes (see Chapter 12).

VII. Hemopoietic Growth Factors [Colony Stimulating Factors (CSFs)]

 A. Hemopoiesis is modulated by several **growth factors** and **cytokines,** including CSFs, stem cell factor (steel factor), interleukins, and macrophage inhibiting protein-α (Table 10.7).

 B. These factors may circulate in the bloodstream, thus acting as hormones, or they may act as local factors produced in the bone marrow that facilitate and stimulate formation of blood cells in their vicinity.

 C. They act at low concentrations and bind to specific membrane receptors on single target cells.

 D. Their various effects on target cells include control of mitotic rate, enhancement of cell survival, control of the number of times the cells divide before they differentiate, and promotion of cell differentiation.

Table 10.7. Hemopoietic Growth Factors

Factors	Principal Action of the Factor	Site of Origin of the Factor
Stem cell factor	Facilitates hemopoiesis	Stromal cells of bone marrow
GM-CSF	Facilitates CFU-GM mitosis and differentiation as well as granulocyte activity	T cells, endothelial cells
G-CSF	Induces mitosis and differentiation of CFU-G, facilitates neutrophil activity	Macrophages, endothelial cells
M-CSF	Facilitates mitosis and differentiation of CFU-M	Macrophages, endothelial cells
IL-1 (together with IL-3 and IL-6)	Facilitates proliferation of PHSC, CFU-S, and CFU-Ly; also suppresses erythroid precursors	Monocytes, macrophages, endothelial cells
IL-2	Promotes proliferation of activated T-cells and B-cells, also facilitates NK cell differentiation	Activated T cells
IL-3	See IL-1; also facilitates proliferation of all unipotential precursors (except for LyB and LyT)	Activated T and B cells
IL-4	Promotes activation of T cells and B cells, also facilitates development of mast cells and basophils	Activated T cells
IL-5	Facilitates proliferation of CFU-Eo, also activates eosinophils	T cells
IL-6	See IL-1, also promotes differentiation of CTLs and B cells	Monocytes and fibroblasts
IL-7	Stimulates CFU-LyB and NK cell differentiation	Adventitial reticular cells?
IL-8	Promotes migration and degranulation of neutrophils	Leukocytes, endothelial cells, and smooth muscle cells
IL-9	Promotes activation and proliferation of mast cells, modulates IgE synthesis, stimulates proliferation of T helper cells	T helper cells
IL-10	Inhibits the synthesis of cytokines by NK cells, macrophages, and T cells; promotes CTL differentiation and B cell and mast cell proliferation	Macrophages and T cells
IL-12	Stimulates NK cells, promotes CTL and NK cell function	Macrophages
γ-Interferons	Activates monocytes and B cells, promotes CTL differentiation, enhances the expression of class II HLA	T cells and NK cells
Erythropoietin	Promotes CFU-E differentiation and proliferation of BFU-E	Endothelial cells of the peritubular capillary network of kidney, hepatocytes
Thrombopoietin	Enhances mitosis and differentiation of CFU-Meg and megakaryoblasts	Not known

Modified with permission from Gartner LP, Hiatt JL: *Color Textbook of Histology,* 2nd ed. Philadelphia, PA, Saunders, 2001.
CTL = cytotoxic lymphocyte; CFU = colony-forming unit (Eo = eosinophil; G = granulocyte; GM = granulocyte-monocyte; Ly = lymphocyte; S = spleen); CSF = colony-stimulating factor (G = granulocyte; GM = granulocyte-monocyte; M = monocyte); HLA = human leukocyte antigen; IL = interleukin; NK = natural killer; PHSC = pluripotential hemopoietic stem cells.

E. Hemopoietic stem cells that do not contact growth factors usually enter **apoptosis** and are eliminated by **macrophages.**

VIII. Clinical Considerations

A. Erythrocyte disorders

1. **Sickle cell anemia** is caused by a point mutation in the deoxyribonucleic acid (DNA) encoding the Hb molecule, leading to production of an **abnormal Hb** (HbS).

   a. This disease occurs almost exclusively among people of African descent (1 in 600 are affected in the United States).

   b. Crystallization of Hb under low O_2 tension gives RBCs the characteristic sickle shape. **Sickled RBCs** are fragile and have a **higher rate of destruction** than normal cells.

   c. Signs include hypoxia, increased bilirubin levels, low RBC count, and capillary stasis.

2. **Pernicious anemia** is caused by a severe **deficiency of vitamin B_{12},** resulting from impaired production of **gastric intrinsic factor** by the parietal cells of the stomach. This factor is required for the proper absorption of vitamin B_{12}.

B. Leukocyte disorders

1. **Infectious mononucleosis** is caused by **Epstein-Barr virus (EBV),** which is related to the herpes virus.

   a. This disease mostly affects young individuals of high school and college age.

   b. Signs and symptoms include fatigue, swollen and tender lymph nodes, fever, sore throat, and an increase in circulating lymphocytes.

   c. EBV may be transmitted by saliva (as in kissing).

   d. Infectious mononucleosis may be life-threatening in immunosuppressed or immunodeficient individuals, whose B cells can undergo intense proliferation leading to death.

2. **Leukemias** are characterized by the replacement of normal hemopoietic cells of the bone marrow by neoplastic cells and are classified according to the **type** and **maturity** of the cells involved.

   a. **Acute leukemias** occur mostly in children.

      (1) These leukemias involve **immature cells.**

      (2) **Rapid onset** of the following signs and symptoms occur: anemia; high WBC count and/or many circulating immature WBCs; low platelet count; tenderness in bones; enlarged lymph nodes, spleen, and liver; vomiting; and headache.

   b. **Chronic leukemias** occur mainly in adults.

      (1) These leukemias initially involve relatively **mature cells.**

      (2) Early signs include **slow onset** of a mild leukocytosis and enlarged lymph nodes; later signs and symptoms include anemia, weakness, enlarged spleen and liver, and reduced platelet count.

C. Coagulation disorders result from inherited or acquired defects of coagulation factors.

1. **Factor VIII deficiency (hemophilia A)** is an X chromosome-linked disorder that affects mostly men.

   a. Severity varies and depends on the extent of reduction in the level of factor VIII (produced by hepatocytes).

   b. Hemophilia A results in excessive bleeding (into joints, in severe cases).

   c. Affected individuals have a normal platelet count, normal bleeding time, and an absence of petechiae, but thromboplastin time is increased.

2. **Von Willebrand disease** is an autosomal-dominant genetic defect resulting in a **decrease in the amount of von Willebrand factor,** which is required in the intrinsic pathway of coagulation.

   a. Most cases are mild and do not involve bleeding into the joints.

   b. Severe cases are characterized by excessive and/or spontaneous bleeding from mucous membranes and wounds.

Review Test

Directions: Each of the numbered items or incomplete statements in this section is followed by answers or by completions of the statement. Select the ONE lettered answer or completion that is BEST in each case.

1. Which of the following proteins associated with the erythrocyte plasma membrane is responsible for maintaining the cell's biconcave disk shape?

(A) Hemoglobin (Hb) A$_1$
(B) HbA$_2$
(C) Porphyrin
(D) Spectrin
(E) α-Actinin

Questions 2-10

The response options for the next nine items are the same. Each item will state the number of options to select. Choose exactly this number.

(A) Erythrocyte
(B) Lymphocyte
(C) Platelet
(D) Neutrophil

For each of the following descriptions, select the formed element that is most closely associated with it.

2. Is immunocompetent (**SELECT 1 OPTION**)

3. Is derived from colony-forming unit megakaryocyte (CFU-Meg) (**SELECT 1 OPTION**)

4. Is derived from CFU-E (**SELECT 1 OPTION**)

5. Is derived from CFU-NM (**SELECT 1 OPTION**)

6. Is associated with demarcation channels (**SELECT 1 OPTION**)

7. Is derived from myeloblasts (**SELECT 1 OPTION**)

8. Is associated with antibody production (**SELECT 1 OPTION**)

9. Possesses specific and azurophilic granules (**SELECT 1 OPTION**)

10. Is derived from reticulocytes (**SELECT 1 OPTION**)

11. A 4-year-old male child is brought by his parents to the pediatrician because of vomiting, headaches, and tenderness in the bones of his arms and legs. On palpation the physician notes that many lymph nodes are enlarged, as is the liver. The pediatrician should order a complete blood count in order to determine whether or not the child may have

(A) chronic leukemia
(B) infectious mononucleosis
(C) von Willebrand disease
(D) acute leukemia
(E) pernicious anemia

Answers and Explanations

1-D. Spectrin is associated with the erythrocyte cell membrane and assists in maintaining its biconcave disk shape.

2-B. Lymphocyte

3-C. Platelet ⎯ Sarm ~platelet demarcation channel~

4-A. Erythrocyte

5-D. Neutrophil

6-C. Platelet

7-D. Neutrophil

8-B. Lymphocyte

9-D. Neutrophil

10-A. Erythrocyte

11-D. Acute leukemia is a disease of children with symptoms that include headaches; vomiting; swollen lymph nodes, liver, and spleen; and the sensation of tenderness in bones. Chronic leukemia is a disease that usually affects adults. Von Willebrand disease is a coagulation disorder and does not have the same symptoms as acute leukemia. Infectious mononucleosis affects mostly young adults of high school and college age. Pernicious anemia is caused by vitamin B deficiency, and its symptoms do not resemble those of acute leukemia.

11

Circulatory System

I. Overview—Blood Vascular System. The blood vascular system consists of the heart, arteries, veins, and capillaries. This system transports **oxygen and nutrients to tissues,** carries **carbon dioxide and waste products from tissues,** and circulates **hormones** from the site of synthesis to their target cells.

A. The **heart** is a four-chambered pump composed of two **atria** and two **ventricles** and is surrounded by a fibroserous sac called the **pericardium.** It receives **sympathetic** and **parasympathetic** nerve fibers, which **modulate the rate of the heartbeat** but do not initiate it. The heart produces **atrial natriuretic peptide,** a hormone that increases secretion of sodium and water by the kidneys, inhibits renin release, and decreases blood pressure.

1. **Cardiac layers**

   a. **Endocardium** lines the lumen of the heart and is composed of simple squamous epithelium **(endothelium)** and a thin layer of loose connective tissue. It is underlain by **subendocardium,** a connective tissue layer that contains veins, nerves, and Purkinje fibers.

   b. **Myocardium** consists of layers of **cardiac muscle cells** arranged in a spiral fashion about the heart's chambers and inserted into the fibrous skeleton. The myocardium contracts to propel blood into arteries for distribution to the body. Specialized cardiac muscle cells in the atria produce several peptides including **atrial naturietic polypeptide, atriopeptin**, **cardiodilatin, and cardionatrin**, hormones that maintain fluid and electrolyte balance and decreases blood pressure.

   c. **Epicardium** is the outermost layer of the heart and constitutes the **visceral layer of the pericardium.** It is composed of simple squamous epithelium **(mesothelium)** on the external surface, fibroelastic connective tissue containing nerves and the coronary vessels, and adipose tissue.

2. The **fibrous skeleton of the heart** consists of thick bundles of **collagen fibers** oriented in various directions. It also contains occasional foci of fibrocartilage.

3. **Heart valves** [atrioventricular (AV)] are composed of a skeleton of fibrous connective tissue, arranged like an aponeurosis, and lined on both sides by endothelium. They are attached to the **annuli fibrosi** of the fibrous skeleton. These valves function in preventing regurgitation of ventricular blood into the atria.

4. The **impulse-generating and impulse-conducting system** of the heart comprises several specialized structures that have coordinated functions that act to **initiate and regulate the heartbeat.**

   a. The **sinoatrial (SA) node,** the **pacemaker** of the heart, is located within the wall of the right atrium. It **generates impulses** that initiate contraction of atrial muscle cells; the impulses are then conducted to the AV node.

   b. The **AV node** is located in the wall of the right atrium, adjacent to the tricuspid valve.

   c. The **AV bundle of His** is a band of conducting tissue radiating from the AV node into the interventricular septum, where it divides into two branches and continues as Purkinje fibers.

   d. **Purkinje fibers** are large, modified cardiac muscle cells (see Chapter 8 V B 9) that make contact with cardiac muscle cells at the apex of the heart via gap junctions, desmosomes, and fasciae adherentes.

   e. The autonomic nervous system modulates the heart rate and stroke volume. Sympathetic innervation accelerates the heart rate, whereas parasympathetic stimulation slows the heart rate.

B. **Arteries** conduct blood **away from the heart** to the organs and tissues. Arterial walls are composed of three layers (tunicae): the **tunica intima** (inner), **tunica media** (middle), and **tunica adventitia** (outer). Components of these layers and variations among types of arteries are summarized in Table 11.1.

   1. **Types of arteries**

      a. **Elastic arteries (conducting arteries)** are large and include the **aorta** and its **major branches.**

         (1) Elastic arteries help to **reduce changes in blood pressure** associated with the heartbeat.

         (2) Small vessels **(vasa vasorum)** and nerves are located in their tunicae adventitia and media.

         (3) Thick, concentric sheaths of **elastic membranes,** known as **fenestrated membranes,** of elastin are located in the tunica media.

      b. **Muscular arteries (distributing arteries)** distribute blood to various organs.

         (1) They include most of the **named** arteries of the human body.

         (2) These medium-sized arteries are smaller than elastic arteries but larger than arterioles. They possess vasa vasorum and nerves in their tunica adventitia.

         (3) The tunica adventitia contains vasa vasorum and nerves, whereas the tunica media is thick, composed of layers of smooth muscle cells.

         (4) The tunica intima is characterized by a prominent internal elastic lamina.

      c. **Arterioles regulate** blood pressure and are the terminal arterial vessels. They are the **smallest** arteries with diameters less than 0.1

Table 11.1. Comparison of Tunicae in Different Types of Arteries

Tunica Components	Elastic Arteries	Muscular Arteries	Arterioles	Metarterioles
Intima				
Endothelium	+	+	+	+
Factor VIII in endothelium	+	+	+	−
Basal lamina	+	+	+	+
Subendothelial layer*	+	+	±	−
Internal elastic lamina	Incomplete	Thick, complete	Some elastic fibers	−
Media				
Fenestrated elastic membranes	40-70	−	−	−
Smooth muscle cells	Interspersed between elastic membranes	Up to 40 layers	1 or 2 layers	Discontinuous layer
External elastic lamina	Thin	Thick	−	−
Vasa vasorum	±	−	−	−
Adventitia				
Fibroelastic connective tissue	Thin layer	Thin layer	−	−
Loose connective tissue	−	−	+	±
Vasa vasorum	+	±	−	−
Lymphatic vessels	+	+	−	−
Nerve fibers	+	+	+	−

The symbol + indicates that the component is present and prominent; ± indicates the component is present but is not prominent; − indicates the component is absent.
*In elastic arteries, the subendothelial layer is composed of loose connective tissue containing fibroblasts, collagen, and elastic fibers. In arterioles, this layer is less prominent; the connective tissue is sparse and contains a few reticular fibers.

millimeter (mm) and a narrow lumen; their luminal diameters usually equal the wall thickness.

(1) The tunica adventitia is scant, whereas the tunica media consists of up to two layers of smooth muscle.

(2) The tunica intima consists of an endothelium, basal lamina, and scant connective tissue.

   d. **Metarterioles** are narrow vessels arising from arterioles that give rise to **capillaries.**

(1) They are surrounded by incomplete rings of smooth muscle cells and possess individual smooth muscle cells (**precapillary sphincters**) that surround capillaries at their origin.

(2) Constriction of precapillary sphincters prevents blood from entering the capillary bed.

  2. **Vasoconstriction** primarily involves **arterioles** and reduces blood flow to a specific region. Vasoconstriction is stimulated by **sympathetic nerve fibers** (see Chapter 9 IX B).

  3. **Vasodilation** is induced by **parasympathetic nerve fibers** as follows:

a. Acetylcholine released from these nerve terminals stimulates the endothelium to release **nitric oxide (NO),** previously known as endothelial-derived relaxing factor (EDRF).

b. **NO** diffuses to smooth muscle cells in the vessel wall and activates their cyclic guanosine monophosphate (cGMP) system, resulting in relaxation.

C. Capillaries are small vessels [about 8-10 micrometers (μm) in diameter and usually less than 1 mm long]. Capillaries exhibit **selective permeability,** permitting the exchange of oxygen, carbon dioxide, metabolites, and other substances between the blood and tissues. They form **capillary beds** interposed between arterioles and venules.

1. Capillary endothelial cells—General features. Capillaries consist of a **single layer of endothelial cells** arranged as a cylinder, which is surrounded by a basal lamina and occasional **pericytes** (see Chapter 6 III B). Endothelial cells:

a. Are nucleated, polygonal cells with an attenuated cytoplasm.

b. Possess a Golgi complex, ribosomes, mitochondria, and some rough endoplasmic reticulum (RER).

c. Contain intermediate filaments of either **desmin, vimentin**, or both in the perinuclear zone; these filaments probably have a supportive function.

d. Generally are joined by **fasciae occludentes** (tight junctions); some desmosomes and gap junctions also are present. Characteristically they contain pinocytotic vesicles.

2. Classification of capillaries. They are classified into three types, depending on the structure of their endothelial cells and the continuity of the basal lamina.

a. Continuous (somatic) capillaries (Figure 11.1) contain numerous **pinocytic vesicles** except in the central nervous system (CNS), where they contain only a limited number of pinocytic vesicles (a property that partly is responsible for the blood-brain barrier).

(1) Continuous capillaries lack fenestrae and have a **continuous** basal lamina.

(2) They are located in nervous tissue, muscle, connective tissue, exocrine glands, and the lungs.

b. Fenestrated (visceral) capillaries are formed from endothelial cells that are perforated with **fenestrae.** These openings are 60-80 nanometers (nm) in diameter and are **bridged by a diaphragm** thinner than a cell membrane; in the renal glomerulus, the fenestrae are larger and lack a diaphragm.

(1) Fenestrated capillaries have a **continuous** basal lamina and few pinocytic vesicles.

(2) They are located in endocrine glands, the intestine, the pancreas, and the glomeruli of kidneys.

c. Sinusoidal capillaries possess many large **fenestrae that lack diaphragms.**

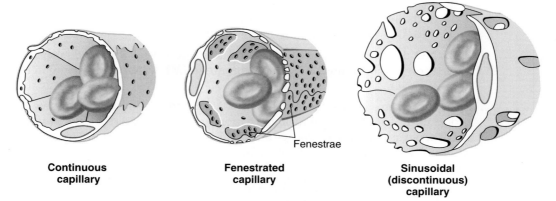

Fenestrae

| Continuous
capillary | Fenestrated
capillary | Sinusoidal
(discontinuous)
capillary |

Figure 11.1. Drawing of the three types of capillaries. (Reprinted with permission from Gartner LP, Hiatt JL: *Color Atlas of Histology.* Baltimore, Williams & Wilkins, 1994, p 147.)

    (1) Sinusoidal capillaries are 30-40 μm in diameter, thus they are much larger than continuous and fenestrated capillaries.

    (2) Sinusoidal capillaries have a **discontinuous** basal lamina and lack pinocytic vesicles.

    (3) Gaps may be present at the cell junctions, permitting leakage between endothelial cells.

    (4) They are located in the liver, spleen, bone marrow, lymph nodes, and adrenal cortex.

  3. Permeability of capillaries is dependent on the morphology of their endothelial cells as well as the size, charge, and shape of the traversing molecules. Permeability is altered during the inflammatory response by **histamine** and **bradykinin.**

    a. Some substances **diffuse** whereas others are **actively transported** across the plasma membrane of capillary endothelial cells.

    b. Other substances move across capillary walls via **small pores** (intercellular junctions) or **large pores** (fenestrae and pinocytic vesicles).

    c. Leukocytes leave the bloodstream to enter the tissue spaces by penetrating intercellular junctions. This process is called **diapedesis.**

  4. Metabolic functions of capillaries are carried out by the endothelial cells and include the following:

    a. Conversion of inactive angiotensin I to active angiotensin II (especially in the lung); this powerful vasoconstrictor stimulates secretion of aldosterone, a hormone that promotes water retention

    b. Deactivation of various pharmacologically active substances (e.g., bradykinin, serotonin, thrombin, norepinephrine, prostaglandins)

    c. Breakdown of lipoproteins to yield triglycerides and cholesterol

    d. Release of prostacyclin, a potent vasodilator and inhibitor of platelet aggregation

  5. Blood flow into capillary beds occurs either from **metarterioles** (with precapillary sphincters) or from **terminal arterioles.**

       a. Central channels are vessels that traverse a capillary bed and connect arterioles to small venules. Their proximal portion is the **metarteriole** (possessing precapillary sphincters) and their distal portion is the **thoroughfare channel** (with no precapillary sphincter).

       b. Metarterioles supply blood to the capillary bed, whereas thoroughfare channels receive blood from capillary beds.

    6. Bypassing a capillary bed

       a. Contraction of precapillary sphincters forces the blood flow from the **metarteriole** directly into the **thoroughfare channel,** thus bypassing the capillary bed and draining into a small vein.

       b. Arteriovenous anastomoses are small vessels that directly connect arterioles to venules, bypassing the capillary bed. They function in **thermoregulation** and also control blood pressure and flow.

D. Veins conduct blood away from the organs and tissues **to the heart** and contain about 70% of the body's total blood volume at any given time. Their walls are composed of three layers: the **tunica intima** (inner), **tunica media** (middle), and **tunica adventitia** (outer), which is the thickest and most prominent. The components of these layers and the variations among different types of veins are summarized in Table 11.2.

    1. Comparison with arteries. Veins have thinner walls and larger, more irregular lumina than the companion arteries. They may have valves in their lumina that prevent retrograde flow of the blood.

    2. Types of veins

       a. Large veins include the vena cava and pulmonary veins. These veins possess **cardiac muscle** in the tunica adventitia for a short distance as they enter the heart. This layer also contains vasa vasorum and nerves.

Table 11.2. Comparison of Tunicae in Different Types of Veins

Tunica Components	Large Veins	Medium and Small Veins	Venules
Intima			
Endothelium	+	+	+
Basal lamina	+	+	+
Valves	In some	In some	−
Subendothelial layer	+	+	−
Media			
Connective tissue	+	Reticular and elastic fibers	±
Smooth muscle cells	+	+	±
Adventitia			
Smooth muscle cells	Longitudinally oriented bundles	−	−
Collagen layers with fibroblasts	+	+	+

The symbol + indicates that the component is present and prominent; ± indicates the component is present but is not prominent; - indicates the component is absent.

    b. Small and medium-sized veins include the external jugular vein. These veins have a diameter of 1 to 9 mm.

    c. Venules have a diameter of 0.2 to 1 mm and are involved in **exchange of metabolites** with tissues and in **diapedesis.**

II. Overview—Lymphatic Vascular System. This system consists of peripheral lymphatic capillaries, lymphatic vessels of gradually increasing size, and lymphatic ducts. The lymphatic vascular system **collects excess tissue fluid (lymph) and returns it to the venous system.** It drains most tissues with the exception of the nervous system and bone marrow.

    A. Lymphatic capillaries are thin-walled vessels that begin as **blind-ended channels** (e.g., **lacteals**).

        1. They are composed of a single layer of **attenuated endothelial cells** that lack fenestrae and fasciae occludentes. They possess a sparse basal lamina.

        2. These capillaries are leaky; lymph enters via spaces between overlapping endothelial cells.

        3. Small **microfibrils** hold open these capillaries and also attach them to the surrounding connective tissue.

    B. Large lymphatic vessels possess valves and are similar in structure to small veins, except that they have larger lumina and thinner walls.

        1. Lymph nodes that filter the lymph are interposed along their routes.

        2. These vessels converge to form the **thoracic duct** and **right lymphatic duct.**

III. Clinical Considerations

    A. Aneurysm is a **ballooning out of an artery.**

        1. An aneurysm occurs because of a weakness in the arterial wall, which may result from an age-related displacement of elastic fibers by collagen.

        2. Aneurysms may be associated with **atherosclerosis, syphilis, or connective tissue disorders** such as Marfan syndrome or Ehlers-Danlos syndrome.

        3. This condition can be life threatening because the weakness of the wall may cause the artery to rupture.

    B. Atherosclerosis is characterized by deposits of yellowish plaques **(atheromas)** in the intima of large and medium-sized arteries. The plaques may block blood flow to the region supplied by the affected artery.

    C. Rheumatic heart-valve disease is a **sequel to childhood rheumatic fever** (subsequent to streptococcal infection), which causes scarring of the heart valves.

        1. The disease is characterized by reduced elasticity of the heart valves, making them unable to close **(incompetence)** or open **(stenosis)** properly.

        2. It most commonly affects the **mitral valve,** followed by the aortic valve.

    D. Ischemic (coronary) heart disease is usually caused by **coronary atherosclerosis,** which results in decreased blood flow to the myocardium. It may result (depending on its severity) in angina pectoris, myocardial infarct, chronic ischemic cardiopathy, or sudden cardiac death.

E. **Tetralogy of Fallot** is a **congenital malformation** consisting of a defective interventricular septum, hypertrophy of the right ventricle (due to a narrow pulmonary artery or valve), and transposed (dextroposed) aorta. It should be **repaired surgically** early in life before the pulmonary constriction becomes exacerbated.

F. **Varicose veins** are abnormally tortuous, **dilated veins,** usually of the **leg.**

    1. They are caused by a decline in muscle tone, degenerative alteration of the vessel wall, and valvular incompetence. They generally occur in older people.

    2. When they occur in the region of the anorectal junction, they are known as **hemorrhoids.**

Review Test

Directions: Each of the numbered items or incomplete statements in this section is followed by answers or by completions of the statement. Select the ONE lettered answer or completion that is BEST in each case.

1. The epicardium is one of the three layers of the heart. It is

(A) continuous with the endocardium
(B) also known as the visceral pericardium
(C) composed of modified cardiac muscle cells
(D) capable of increasing intraventricular pressure
(E) capable of decreasing the rate of contraction

2. The atrial muscle of the heart produces a hormone that

(A) decreases blood pressure
(B) increases blood pressure
(C) causes vasoconstriction
(D) facilitates the release of renin
(E) facilitates sodium resorption in the kidneys

3. The generation of impulses in the normal heart is the responsibility of which of the following structures?

(A) Atrioventricular (AV) node
(B) AV bundle of His
(C) Sympathetic nerves
(D) Sinoatrial (SA) node
(E) Purkinje fibers

4. Metarterioles, vessels interposed between arterioles and capillary beds

(A) function to control blood flow into arterioles
(B) possess a complete layer of smooth muscle cells in their tunica media
(C) possess precapillary sphincters
(D) receive blood from thoroughfare channels
(E) possess valves to regulate the direction of blood flow

5. Which of the following statements concerning innervation of blood vessels is true?

(A) Vasoconstriction is controlled by parasympathetic nerve fibers.
(B) Acetylcholine acts directly on smooth muscle cells.
(C) Acetylcholine acts directly on endothelial cells.
(D) Vasodilation is controlled by sympathetic nerve fibers.

6. Which of the following characteristics distinguishes somatic capillaries from visceral capillaries?

(A) Presence or absence of fenestrae
(B) Size of the lumen
(C) Thickness of the vessel wall
(D) Presence or absence of pericytes
(E) Thickness of the basal lamina

7. The blood-brain barrier is thought to occur because capillaries in the central nervous system (CNS) have which of the following characteristics?

(A) Discontinuous basal lamina
(B) Fenestrae with diaphragms
(C) Fenestrae without diaphragms
(D) A few pinocytic vesicles
(E) No basement membrane

8. Which of the following statements about healthy, intact capillaries is true?

(A) They control blood pressure.
(B) They are lined by a simple columnar epithelium.
(C) They have a smooth muscle coat.
(D) They inhibit clot formation.
(E) Satellite cells share their basal lamina.

9. A patient complains of shortness of breath even after only mild exercise. She states that she has had this condition for 2 years but recently has noticed that it has become more pronounced. Her medical history indicates that she had rheumatic fever when she was a child. Auscultation indicates an enlarged heart. What may the physician expect to find with other diagnostic tests?

(A) Mitral valve stenosis
(B) Tetralogy of Fallot
(C) Pulmonary artery aneurysm
(D) Coronary heart disease

10. Diagnostic tests for ischemic heart disease usually reveal

(A) malformed heart valves
(B) atherosclerosis of coronary arteries
(C) irregular heartbeat
(D) faulty sinoatrial valve

Answers and Explanations

1-B. The pericardium is a fibroserous sac that encloses the heart. The innermost layer of the pericardium, the epicardium, is also known as the visceral pericardium.

2-A. Atrial natriuretic peptide, which decreases blood pressure, is produced mainly by cardiac muscle cells of the right atrium. It inhibits the release of renin and causes the kidneys to decrease the resorption of sodium and water.

3-D. Impulses are generated in the sinoatrial (SA) node, which is the pacemaker of the heart. They are then conducted to the atrioventricular (AV) node. The bundle of His and Purkinje fibers conduct impulses from the AV node to the cardiac muscle cells of the ventricles. Sympathetic nerves can increase the rate of the heartbeat but do not originate it.

4-C. The proximal portion of a central channel is known as a metarteriole, whereas its distal portion is the thoroughfare channel. Blood from metarterioles may enter the capillary bed if their precapillary sphincters are relaxed. If the precapillary sphincters of metarterioles are constricted, blood bypasses the capillary bed and flows directly into thoroughfare channels and from there into a venule.

5-C. Acetylcholine stimulates the endothelial cells of a vessel to release nitric oxide (endothelial-derived relaxing factor), which causes relaxation of smooth muscle cells. Thus, acetylcholine does not act directly on smooth muscle cells.

6-A. Somatic (continuous) capillaries lack fenestrae, whereas visceral (fenestrated) capillaries are characterized by their presence. Both types of capillary possess a continuous basal lamina and are surrounded by occasional pericytes.

7-D. Capillaries in the central nervous system (CNS) are of the continuous type and thus lack fenestrae but have a continuous basal lamina. In contrast to continuous capillaries in other parts of the body, they contain only a few pinocytic vesicles; this characteristic is thought to be partly responsible for the blood-brain barrier.

8-D. The smooth endothelial lining of intact, healthy capillaries inhibits clot formation. Capillaries do not control blood pressure.

9-A. A person who has had rheumatic fever as a child may develop heart-valve disease later in life. Although the mitral valve is the one most commonly affected, the other valves may be also be involved. The mitral valve becomes inflamed, fibrotic, eventually becoming incompetent or stenotic. This condition leads to respiratory hypertension and eventual edema, thus restricting respiratory function. The key here is a history of rheumatic fever, because those individuals are predisposed to developing heart-valve diseases.

10-B. Diagnostic tests for ischemic heart disease reveal atherosclerosis of the coronary vessels. Over time, excessive plaque composed of cholesterol and fats is layered beneath the intima of these vessels, thus restricting blood flow to the myocardium of the heart, leading to angina pectoris, an infarction, or perhaps sudden death.

12

Lymphoid Tissue

I. Overview—The Lymphoid (Immune) System

A. The lymphoid system consists of **capsulated lymphoid tissues** (thymus, spleen, tonsils, and lymph nodes); **diffuse lymphoid tissue**; and lymphoid cells, primarily **T lymphocytes** (T cells), **B lymphocytes** (B cells), and **macrophages**.

B. The **immune system** has two components, the innate immune system (non-specific) and the adaptive immune system (specific).

1. The **innate immune system** consists of complement, neutrophils and macrophages, and natural killer (NK) cells.

2. The **adaptive immune system** possesses four characteristics: ability to recognize self/nonself, memory, specificity, and diversity. The cells of the adaptive immune system, namely T lymphocytes, B lymphocytes, and antigen presenting cells, communicate with one another by the use of signaling molecules (**cytokines**), thus relaying information to each other in response to antigenic invasion.

C. This system functions primarily to defend the organism by mounting **humoral immune responses** against foreign substances (**antigens**) and **cell-mediated immune responses** against microorganisms, tumor and transplanted cells, and virus-infected cells.

II. Cells of the Immune System

A. Overview—Cells of the Immune System

1. Cells of the immune system include **clones of T and B lymphocytes**. A clone is a small number of **identical** cells that can recognize and respond to a single or a small group of related antigenic determinants (**epitopes**). Exposure to antigen and one or more **cytokines** induces **activation** of resting T and B cells, leading to their proliferation and differentiation into **effector cells** (Figure 12.1).

2. Antigen-presenting cells (APCs) (e.g., macrophages, lymphoid dendritic cells, Langerhans cells, follicular dendritic cells, M cells, and B cells[1]), **mast cells**, and **granulocytes** are also cells of the immune system (see Chapter 6 III G and H).[1]

B. T lymphocytes

1. Overview—T lymphocytes

---

[1]Although B cells can present epitopes to T cells, and are thus antigen-presenting cells, their role in this function is probably limited to the secondary (anamnestic) immune response rather than the primary immune response.

    a. T lymphocytes include several functionally distinct subtypes and are responsible for **cell-mediated immune responses**. They assist B cells in developing humoral responses to **thymic-dependent antigens**.

    b. **T cell receptors** (TCR) are present on the surfaces of T lymphocytes. TCRs recognize only **protein** antigens.

    c. Because they only recognize **epitopes** that are bound to **major histocompatibility complex (MHC) molecules** on the surface of APCs, T cells are **MHC restricted.**

2. Maturation of T lymphocytes occurs in the thymus and involves the following events:

    a. Immunoincompetent progenitor T lymphocytes migrate from the bone marrow to the thymus. Once in the thymus, they are also referred to as **thymocytes.**

    b. Within the thymic **cortex,** thymocytes undergo **gene rearrangements** and begin to express antigen-specific T-cell receptors, which are integral membrane proteins. Thus the cells become **immunocompetent.**

    c. Cortical T cells express thymus-induced **CD** (cluster of differentiation) **markers** (CD2, CD3, CD4, CD8, and CD28) on their surfaces.

    d. In the thymic **medulla,** some thymocytes lose CD4 and develop into CD8+ cells; others lose CD8 and develop into CD4+ cells.

3. T-lymphocyte subtypes

    a. **T helper (T_H1 and T_H2) cells** are **CD4+** cells. After their activation, these cells synthesize and release numerous growth factors known as **cytokines (lymphokines).** T_H1 cells regulate responses against viral or bacterial invasions and instruct macrophages in killing bacteria, and T_H2 cells regulate humoral responses against parasitic or mucosal attacks.

        (1) The cytokines **interleukin 4 (IL-4), interleukin 5 (IL-5),** and **interleukin 6 (IL-6)** released by T_H2 cells induce B cells to proliferate and mature and thus respond to an antigenic stimulus.

        (2) Other cytokines produced by T_H1 cells [such as IL-2, interferon (IFN)-γ, and others] **modulate the immune response** in diverse ways (Table 12.1).

    b. **T cytotoxic (T_C) cells** are **CD8+** cells. After priming by an antigenic stimulus via an APC, T cells are induced by **interleukin 2** (IL-2) to proliferate, forming new **cytotoxic T lymphocytes** (CTLs), which mediate (via **perforins** and **granzymes**) **apoptosis** of foreign cells and virally altered self-cells (see Figure 12.1).

    c. **T suppressor (T_S) cells** are **CD8+** cells. These cells can modulate the extent of the immune response by suppressing the activity of other immunocompetent cells and may be important in **preventing autoimmune responses.** Some investigators do not recognize T_S cells as a separate subtype of T cells.

    d. **T memory cells** are **long-lived, committed, immunocompetent cells** that are formed during proliferation in response to an antigenic challenge. They do not react against the antigen but remain in the cir-

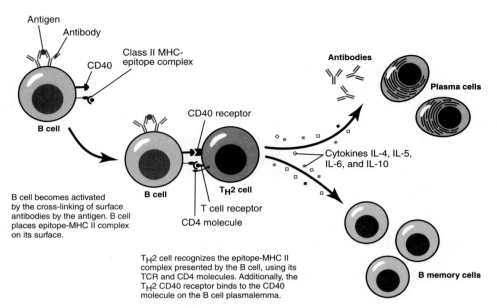

(A)

Antigen
Antibody
Class II MHC-epitope complex
CD40
B cell
CD40 receptor
B cell
T$_H$2 cell
T cell receptor
CD4 molecule
Antibodies
Plasma cells
Cytokines IL-4, IL-5, IL-6, and IL-10
B memory cells

B cell becomes activated by the cross-linking of surface antibodies by the antigen. B cell places epitope-MHC II complex on its surface.

T$_H$2 cell recognizes the epitope-MHC II complex presented by the B cell, using its TCR and CD4 molecules. Additionally, the T$_H$2 CD40 receptor binds to the CD40 molecule on the B cell plasmalemma.

Binding of CD40 to CD40 receptor causes proliferation of B cells. The T$_H$2 cell releases cytokines IL-4, IL-5, IL-6, and IL-10.

IL-4, IL-5, and IL-6 facilitate the activation and differentiation of B cells into B memory cells and antibody-forming plasma cells. IL-10 inhibits the proliferation of T$_H$1 cells.

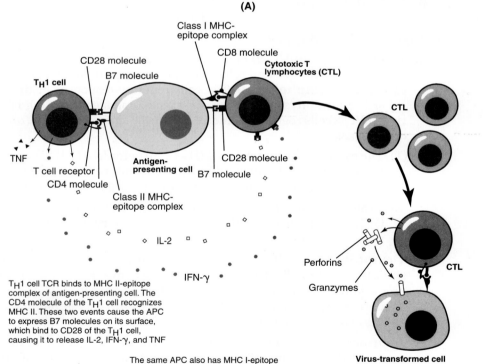

(B)

Class I MHC-epitope complex
CD28 molecule
B7 molecule
CD8 molecule
Cytotoxic T lymphocytes (CTL)
T$_H$1 cell
CTL
TNF
T cell receptor
CD4 molecule
Class II MHC-epitope complex
Antigen-presenting cell
CD28 molecule
B7 molecule
IL-2
IFN-γ
Perforins
Granzymes
CTL
Virus-transformed cell

T$_H$1 cell TCR binds to MHC II-epitope complex of antigen-presenting cell. The CD4 molecule of the T$_H$1 cell recognizes MHC II. These two events cause the APC to express B7 molecules on its surface, which bind to CD28 of the T$_H$1 cell, causing it to release IL-2, IFN-γ, and TNF

The same APC also has MHC I-epitope complex expressed on its surface that is bound by a CTL's CD8 molecule and T cell receptor. Additionally, the CTL has CD28 molecules bound to the APC's B7 molecule. The CTL also pocesses IL-2 receptors, which bind the IL-2 released by the T$_H$1 cell, causing the CTL to undergo proliferation, and IFN-γ causes its activation.

The newly formed CTLs attach to the MHC I-epitope complex via their TCR and CD8 molecules and secrete perforins and granzymes, killing the virus-transformed cells. Killing occurs when granzymes enter the cell through the pores established by perforins and act on the intracellular components to drive the cell into apoptosis.

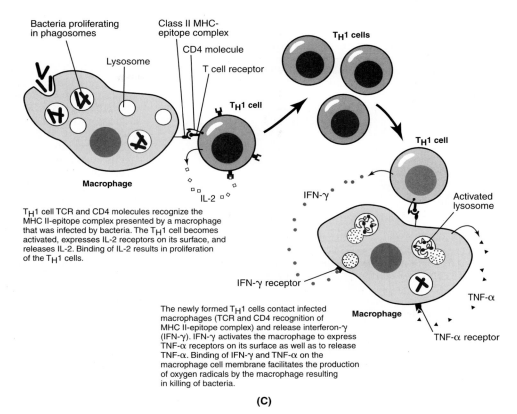

The following labels appear in the figure:

Bacteria proliferating in phagosomes

Class II MHC-epitope complex

CD4 molecule

Lysosome

T cell receptor

T$_H$1 cells

T$_H$1 cell

T$_H$1 cell

Activated lysosome

Macrophage

IL-2

IFN-γ

T$_H$1 cell TCR and CD4 molecules recognize the MHC II-epitope complex presented by a macrophage that was infected by bacteria. The T$_H$1 cell becomes activated, expresses IL-2 receptors on its surface, and releases IL-2. Binding of IL-2 results in proliferation of the T$_H$1 cells.

IFN-γ receptor

TNF-α

Macrophage

TNF-α receptor

The newly formed T$_H$1 cells contact infected macrophages (TCR and CD4 recognition of MHC II-epitope complex) and release interferon-γ (IFN-γ). IFN-γ activates the macrophage to express TNF-α receptors on its surface as well as to release TNF-α. Binding of IFN-γ and TNF-α on the macrophage cell membrane facilitates the production of oxygen radicals by the macrophage resulting in killing of bacteria.

(C)

Figure 12.1. Schematic overview of the interactions among the various cells of the immune system. *(A)* Thymic-dependent antigen-induced B-memory and plasma-cell formation. *(B)* Cytotoxic T-lymphocyte activation and cell killing. *(C)* Macrophage activation by T$_H$1 cells. IL = interleukin; MHC = major histocompatibility complex; TCR = T cell receptors. (Adapted with permission from Gartner LP, Hiatt JL: *Color Textbook of Histology,* 2nd ed. Philadelphia, WB Saunders, 2001, p 282-284.)

culation or in specific regions of the lymphoid system. Because they increase the size of the original clone, they **provide a faster and greater secondary response** (anamnestic response) against a future challenge by the same antigen.

C. B lymphocytes

1. Overview—B lymphocytes

a. B lymphocytes originate and mature into **immunocompetent** cells within the bone marrow. They are responsible for the **humoral immune response.**

b. Immunoglobulins (IgD and the monomeric form of IgM) are attached to the external aspect of their plasma membranes; all of the immunoglobulin molecules on a given B cell recognize and bind to the **same antigenic determinant** (epitope).

c. CD40 molecules are present on their plasmalemma. They interact with CD40 receptors on T$_H$2 cells, causing release of cytokines that facilitate proliferation and transformation of B cells into B memory and plasma cells, and inhibit T$_H$1 cell proliferation.

Table 12.1. Biological Activities of Selected Cytokines in the Immune Response*

Cytokine	Secreted by	Targeted Cell	Action
IL-1a and IL-1b	Macrophages and epithelial cells	T cells and macrophages	Activates both T lymphocytes and macrophages
IL-2	T_H1 cells	Activated T and B cells	Induces mitosis of activated T and B cells
IL-4	T_H2 cells	B cells	Induces mitosis of B cells as well as their transformation into plasma cells, also promotes isotype switching from IgM to IgG and IgE
IL-5	T_H2 cells	B cells	Induces mitosis and maturation of B cells, also promotes isotype switching from IgM to IgE
IL-6	Antigen presenting cells and T_H2 cells	T cells and activated B cells	Activates T cells, induces maturation of B cells to IgG forming plasma cells
IL-10	T_H2 cells	T_H1 cells	Inhibits the formation of T_H1 cells and retards their ability to manufacture cytokines
IL-12	B cells and macrophages	NK cells and T cells	Activates NK cells, facilitates the formation of T_H1-like cells
TNF-α	Macrophage	Macrophages	Macrophages self-activate to manufacture and release IL-12
	T_H1 cells	Activated macrophages	Promotes the production of oxygen radicals facilitating bacterial killing within endosomes of activated macrophages
α-IFN	Virally attacked cells	NK cells and macrophages	Activates macrophages and NK cells
β-IFN	Virally attacked cells	NK cells and macrophages	Activates macrophages and NK cells
γ-IFN	T_H1 cells	Macrophages and T cells	Activates cytotoxic T cells to kill altered and/or foreign cells, promotes phagocytosis by macrophages

CTL = cytotoxic T lymphocyte; IFN = interferon; Ig = immunoglobulin; IL = interleukin; NK = natural killer; T_H = T helper cell; TNF = tumor necrosis factor.
*See Chapter 10 for discussion about cytokines involved in hemopoiesis.

    d. The cytokines released by T helper cells depend on the **invading pathogen**.

        (1) IL-4 and IL-5 are released by T helper cells in response to parasitic worms, and B cells switch to IgE formation.

        (2) IL-6 and IFN-γ are released by T helper cells in response to bacteria and viruses in the connective tissue, and B cells switch to IgG formation.

> **(3)** TGF-β is released by T helper cells in response to the presence of bacteria and viruses on mucosal surfaces, and B cells switch to IgA formation.

> **e.** B lymphocytes can present epitopes [complexed with class II human leukocyte antigen (HLA)] to T_H1 cells.

> **f.** When activated, B lymphocytes release **interleukin 12 (IL-12)** to induce T_H1 cell formation and natural killer (NK) cell activation.

> **g.** During a humoral immune response, B lymphocytes proliferate and differentiate after an antigenic challenge to form plasma cells and B memory cells (see Figure 12.1*A*).

2. Plasma cells lack surface antibody and actively synthesize and secrete **antibody specific for the challenging antigen.**

3. B memory cells have the same properties as T memory cells.

D. Natural Killer (NK) cells belong to a category of **null cells,** a small group of peripheral-blood lymphocytes that **lack the surface determinants** characteristic of T and B lymphocytes.

1. As soon as they are formed, these cells are immunocompetent. They do not have to enter the thymic environment.

2. NK cells are not MHC restricted and exhibit an apparently **nonspecific cytotoxicity** against tumor cells and virus-infected cells. The mechanism by which NK cells recognize these target cells is not yet understood.

3. NK cells can also kill specific target cells that have antibodies bound to their surface antigens in a process known as **antibody-dependent cell-mediated cytotoxicity (ADCC);** macrophages, neutrophils, and eosinophils also exhibit ADCC.

4. NK cells use **perforins** and **granzymes** to drive the virally altered cells or tumor cells into **apoptosis.**

E. Macrophages function both as **APCs** and as **cytotoxic effector cells** in ADCC.

1. When acting as APCs, macrophages phagocytose antigens, fragment them into small components, and present them to T cells. The most antigenic of these small compounds are known as **epitopes.**

2. Macrophages produce **IL-1,** which helps activate T_H cells and self-activate macrophages. Moreover, they produce **TNF-α (tumor necrosis factor-alpha),** which also self-activates macrophages, and in conjunction with IFN-γ, facilitates their killing of endocytosed bacteria (see Table 12.1). Additionally, macrophages secrete **prostaglandin E_2 (PGE_2),** which decreases certain immune responses.

III. Antigen Presentation and the Role of MHC Molecules

A. Major histocompatibility complex (MHC)

1. The MHC is a large genetic complex with many loci that encode two main classes of membrane molecules: **class I molecules (MHC I),** which are expressed by nearly all **nucleated cells,** and **class II molecules (MHC II),** which are expressed by the various cells that function as **APCs.**

2. In humans, the MHC is referred to as the **HLA complex.** Therefore, MHC I = class I HLA, and MHC II = class II HLA.

B. **Immunogens** are molecules that are capable of **inducing an immune response.** Immunogens are **antigens,** that is, molecules that can react with an **antibody** or a **T-cell receptor.** Most antigens are immunogens.

1. **Exogenous immunogens** are endocytosed or phagocytosed by APCs and degraded intracellularly, yielding antigenic peptides containing an epitope that enter the trans Golgi network (TGN).

2. In the TGN the complex is sorted into specialized antigenic peptides (containing an epitope) that associate with **class II HLA molecules.**

   a. These epitopes are relatively long, composed of 13 to 25 amino acids.

   b. Class II HLA molecules are synthesized on the rough endoplasmic reticulum (RER), and are loaded within the RER cisternae with a protein known as **CLIP** (**cl**ass II associated **i**nvariant **p**rotein).

   c. The class II HLA-CLIP complex enters the Golgi apparatus, where it is delivered to the trans-Golgi vesicles [**MHC class II compartment (MIIC) vesicles** that already contain epitopes derived from exogenous immunogens], where the CLIP is exchanged for the epitope.

   d. The epitope-class II HLA complexes are transported to and displayed on the cell surface, where they are **presented** to T cells.

3. **Endogenous immunogens** are produced **within** host cells (these may be **viral proteins** synthesized in virus-infected cells or **tumor proteins** synthesized in cancerous cells).

   a. **Class I HLA molecules**, synthesized on the RER surface, enter the RER cisternae.

   b. Endogenous immunogens are degraded by organelles of the host cells, known as proteasomes, into short polypeptide fragments. These fragments are antigenic peptides (8 to 12 amino acids in length) known as **epitopes**.

   c. The epitopes are transported by **TAP1** and **TAP2** (**t**ransporter **p**roteins 1 and 2) into the RER cisternae.

   d. Within the RER cisternae, the epitopes derived from endogenous immunogens are loaded on the **class I HLA.**

   e. The peptide-class I HLA complexes are transported to the Golgi complex for sorting and eventual delivery within clathrin-coated vesicles to the cell surface, where they are presented to T cells.

C. **HLA restriction—T lymphocytes.** Each subtype of T lymphocytes (except T memory cells) recognizes only those epitopes that are associated with either class I or class II HLA molecules as follows:

1. T_H1, T_H2, and T_s cells recognize class II HLA molecules.

2. Cytotoxic T cells recognize class I HLA molecules.

3. T memory cells recognize both class I and class II HLA molecules.

IV. Immunoglobulins. Immunoglobulins are glycoproteins that are synthesized and secreted by **plasma cells.** They constitute the active agents of the humoral immune response and have **specific antibody activity** against one antigen or a few closely related antigens. Immunoglobulins bind antigens to form antigen-antibody complexes, which are cleared from the body by various means, some of which involve the **complement system,** whereas others involve **eosinophils.**

    A. Structure—immunoglobulins

        1. Immunoglobulins are composed of monomers containing two **heavy chains** and two **light chains.**

        2. Each immunoglobulin possesses a **constant region** that is identical in all immunoglobulin molecules.

        3. Each immunoglobulin also possesses a **variable region** that differs in the antibody molecules that recognize different antigens. Thus the **variable regions determine the specificity** of an antibody molecule (i.e., its ability to bind to a particular antigenic determinant). Large antigens may have multiple antigenic determinants, thus inducing production of antibodies with different specificities.

    B. Immunoglobulin classes

        1. Human serum contains five classes **(isotypes)** of immunoglobulins, which differ in the amino acid composition of their **heavy-chain constant regions.**

        2. The different isotypes exhibit functional differences.

            a. IgA are secretory immunoglobulins.

            b. IgD and **IgE** are reaginic antibodies; IgE binds to IgE receptors of mast cells and basophils. IgD binds to B-cell plasma membranes.

            c. IgG is the most abundant serum immunoglobulin and crosses the placental barrier.

            d. IgM forms pentamers and is the first isotype to be formed in an immune response. It activates the **complement system.** Its monomeric form binds to the B-cell plasma membrane.

V. Diffuse Lymphoid Tissue. Diffuse lymphoid tissue is especially prominent in the mucosa of the gastrointestinal and respiratory systems. It is organized as nonencapsulated clusters of lymphoid cells or as lymphoid (lymphatic) nodules. Diffuse lymphoid tissue is collectively called **mucosa-associated lymphoid tissue (MALT).**

    A. MALT consists of two major types, **bronchus-associated lymphoid tissue (BALT)** and **gut-associated lymphoid tissue (GALT).** Both types possess lymphoid nodules that are isolated from one another, except in the case of Peyer patches.

    B. Peyer patches are aggregates of lymphoid nodules found in the **ileum.** They are components of the GALT.

    C. Lymphoid (lymphatic) nodules are transitory, dense, spherical **accumulations of lymphocytes** (mostly B cells). The dark, peripheral region of nodules **(corona)** is composed mainly of small, newly formed lymphocytes.

Lymphoid nodules of the GALT are isolated from the lumina of their respective tracts by **microfold (M) cells,** which transfer antigens from the lumen and present them (**without** processing them into epitopes) to lymphocytes and macrophages lying in deep invaginations of their basal cell surfaces. From here an appropriate immune response is mounted by lymphoid tissue in the underlying lamina propria.

1. **Secondary nodules** have a central, lightly staining area, called the **germinal center,** which is composed of lymphocytes (**lymphoblasts** called centroblasts and centrocytes).

    a. **Centroblasts** do not display surface immunoglobulins (SIGs), whereas centrocytes have expressed SIGs.

    b. Centrocytes that express SIGs against self are forced into apoptosis.

    c. Surviving centrocytes become B memory cells or plasma cells. The darker, denser-staining area around the germinal center is the corona (mantle).

2. **Primary nodules** lack germinal centers.

VI. Lymphoid Organs

A. Lymph nodes

1. **Overview—lymph nodes**

    a. A lymph node is a small, encapsulated, ovoid to kidney-shaped organ with a capsule that sends **trabeculae** into the substance of the node.

    b. The **convex** surface of a lymph node receives afferent lymphatic vessels, whereas the **concave** surface (the **hilum**) is the site where arterioles enter and efferent lymphatic vessels and venules exit.

    c. Lymph nodes possess a **stroma** composed of a supportive framework rich in **reticular fibers.**

    d. **Function.** Lymph nodes filter lymph, maintain and produce T and B cells, and possess memory cells (especially T memory cells). Antigens delivered to lymph nodes by antigen presenting cells are recognized by T cells, and an immune response is initiated.

2. **Structure—lymph nodes.** Lymph nodes are divided into **cortical** and **medullary** regions.

    a. The **cortex of lymph nodes:**

        (1) Lies deep to the capsule from which it is separated by a subcapsular sinus

        (2) Is incompletely subdivided into compartments by connective tissue septa derived from the capsule

        (3) Contains lymphoid nodules, sinusoids, and the paracortex

            (a) **Lymphoid nodules** are composed mainly of B cells but also of T cells, follicular dendritic cells, macrophages, and reticular cells. They may possess a germinal center.

            (b) **Sinusoids** are endothelium-lined spaces that extend along the capsule and trabeculae and are known as **subcapsular** and **cortical sinusoids,** respectively.

(c) The **paracortex** is located between the cortex and the medulla. It is composed of a non-nodular arrangement of **mostly T lymphocytes** (the thymus-dependent area of the lymph node). The paracortex is the region where circulating lymphocytes gain access to lymph nodes via **postcapillary (high endothelial) venules.**

b. The **medulla of a lymph node** lies deep to the paracortex and cortex except at the region of the hilum. It is composed of medullary sinusoids and medullary cords.

   (1) **Medullary sinusoids** are endothelium-lined spaces supported by reticular fibers and reticular cells and frequently contain macrophages. Medullary sinusoids receive lymph from the cortical sinuses.

   (2) **Medullary cords** are composed of lymphocytes and plasma cells.

B. Thymus

1. Overview—thymus

a. The thymus is derived from both **endoderm (epithelial reticular cells)** and **mesoderm** (lymphocytes). It begins to involute near the time of puberty.

b. A connective tissue **capsule** surrounds the thymus. The septa of this capsule divide the parenchyma into incomplete lobules, each of which contains a **cortical** and **medullary region.** The thymus does not contain lymphoid nodules.

2. Structure—thymus

a. **The thymic cortex** is supplied by arterioles in the septa; these arterioles provide capillary loops that enter the substance of the cortex. The cortex is the region in which **T-cell maturation** occurs.

   (1) **Epithelial reticular cells**

   (a) These cells originate from endoderm and form a meshwork with interstices in which T cells are tightly packed.

   (b) They are pale cells (derived from the third and perhaps fourth pharyngeal pouch) and have a large, ovoid, lightly staining nucleus, which often displays a nucleolus.

   (c) They possess **long processes** that surround the thymic cortex, isolating it from both the connective tissue septa and the medulla. These processes, which are filled with bundles of **tonofilaments,** form desmosomal contacts with each other.

   (d) They manufacture **thymosin, serum thymic factor, and thymopoietin;** these function in the transformation of immature T lymphocytes into immunocompetent T cells.

   (e) Epithelial reticular cells are of six types, each with different surface markers and presumably different functions.

   (f) Some epithelial reticular cells present HLA I and HLA II molecules to the developing T cells.

(2) Thymocytes

    (a) Thymocytes are **immature T lymphocytes** present in large numbers within the thymic cortex in different stages of differentiation.

    (b) They are surrounded by processes of epithelial reticular cells, which help segregate thymocytes from antigens during their maturation.

    (c) They migrate toward the medulla as they mature; however, those T cells that are unable to recognize the presented HLA I or HLA II molecules or whose TCRs recognize self proteins are forced into apoptosis and never reach the medulla. Most T cells die in the cortex and are phagocytosed by macrophages.

    (d) Surviving T cells are naive. They leave the thymus and are distributed to secondary lymphoid organs by the vascular system.

(3) Blood-thymus barrier

    (a) This barrier exists in the **cortex only,** making it an immunologically protected region.

    (b) It ensures that antigens escaping from the bloodstream do not reach developing T cells in the thymic cortex.

    (c) It consists of the following layers: **endothelium** of the thymic capillaries and the associated basal lamina, **perivascular connective tissue** and **cells** (e.g., pericytes and macrophages), and **epithelial reticular cells** and their basal laminae.

b. Thymic medulla

    (1) The thymic medulla is continuous between adjacent lobules and contains large numbers of **epithelial reticular cells** and **mature T cells,** which are loosely packed, causing the medulla to stain lighter than the cortex.

    (2) It also contains whorl-like accretions of epithelial reticular cells, called **Hassall corpuscles** (thymic corpuscles). These structures display various stages of keratinization and increase in number with age. Their function is unknown.

    (3) Mature T cells exit the thymus via venules and efferent lymphatic vessels from the thymic medulla. The T cells then migrate to secondary lymphoid structures.

    (4) Hormones acting on the thymus

        (a) Thymosin, thymopoietin, thymulin, somatotropin, and **thymic humoral factor** promote the formation of T cells.

        (b) Thyroxin encourages thymulin production by epithelial reticular cells.

        (c) Adrenocorticosteroids depress T cell formation in the thymus.

C. Spleen

1. Overview—spleen

a. A simple squamous epithelium (peritoneum) covers the **capsule** of the spleen. The capsule consists of dense irregular collagenous connective tissue, which sends **trabeculae** into the substance of the spleen to form a supportive framework. The spleen possesses a hilum.

b. The spleen differs from the thymus and lymph nodes in that it **lacks a cortex and medulla.** It additionally differs from lymph nodes by the absence of **afferent lymphatics.**

c. It is divided into **red pulp** and **white pulp;** the latter contains lymphoid elements. These two regions are separated from each other by the **marginal zone.**

d. **Function.** The spleen filters blood, stores erythrocytes, phagocytoses damaged and aged erythrocytes, and is a site of proliferation of B and T lymphocytes as well as the production of antibodies by plasma cells.

2. Vascularization of the spleen is derived from the **splenic artery,** which enters the hilum and gives rise to trabecular arteries.

a. **Trabecular arteries** leave the trabeculae, become invested by a **periarterial lymphatic sheath (PALS),** and are known as central arteries.

b. **Central arteries** branch but maintain their lymphatic sheath until they leave the white pulp to form several straight penicillar arteries.

c. **Penicillar arteries** enter the red pulp. They have three regions: pulp arterioles, macrophage-sheathed arterioles, and **terminal arterial capillaries.** The latter either drain directly into the splenic sinusoids **(closed circulation)** or terminate as open-ended vessels within the splenic cords of the red pulp **(open circulation).**

d. **Splenic sinusoids** are drained by pulp veins, which are tributaries of the trabecular veins; these, in turn, drain into the splenic vein, which exits the spleen at the hilum.

3. Structure—spleen

a. **White pulp of the spleen** includes **all** of the organ's lymphoid tissue (diffuse and nodular), such as **lymphoid nodules** (mostly B cells) and **PALS** (mostly T cells) around the central arteries. It also contains macrophages and other APCs.

b. The **marginal zone of the spleen:**

(1) Is a **sinusoidal region** between the red and white pulps, located at the periphery of the PALS

(2) Receives blood from capillary loops derived from the central artery and thus is the **first site where blood contacts the splenic parenchyma**

(3) Is richly supplied by avidly phagocytic macrophages and other APCs

(4) Is the region where circulating T and B lymphocytes **enter the spleen** before becoming segregated to their specific locations within the organ

c. **Red pulp of the spleen** is composed of an interconnected network of sinusoids supported by a loose type of reticular tissue **(splenic cords).**

(1) **Sinusoids:**

(a) Are lined by long, fusiform endothelial cells separated by relatively large intercellular spaces

(b) Have a discontinuous basal lamina underlying the endothelium and circumferentially arranged "ribs" of reticular fibrils

(2) **Splenic cords (cords of Billroth)** contain plasma cells, reticular cells, blood cells, and macrophages enmeshed within the spaces of the reticular fiber network. Processes of the macrophages enter the lumina of the sinusoids through the spaces between the endothelial cells.

D. **Tonsils** are **aggregates of lymphoid tissue,** which sometimes lack a capsule. All tonsils are located in the upper section of the digestive tract, lying beneath, but in contact with, the epithelium. Tonsils assist in combating antigens entering via the nasal and oral epithelia.

1. **Palatine tonsils:**

a. Possess **crypts,** deep invaginations of the stratified squamous epithelium covering of the tonsils; the crypts frequently contain debris

b. Possess lymphoid nodules, some (secondary nodules) with germinal centers

c. Are separated from subjacent structures by a connective tissue **capsule**

2. The **pharyngeal tonsil** is a **single** tonsil located in the posterior wall of the nasopharynx.

a. It is covered by a pseudostratified ciliated columnar epithelium.

b. Instead of crypts, it has longitudinally disposed pleats (infoldings).

3. **Lingual tonsils**

a. Lingual tonsils are located on the dorsum of the posterior third of the tongue and are covered by a stratified squamous nonkeratinized epithelium.

b. Each lingual tonsil possesses a deep **crypt,** which frequently contains cellular debris. Ducts of mucous glands often open into the base of these crypts.

VII. Clinical Considerations

A. **DiGeorge syndrome,** also called **congenital thymic aplasia,** is characterized by the congenital absence of the thymus and parathyroid glands resulting from abnormal development of the third and fourth pharyngeal pouches.

1. This syndrome is associated with **abnormal cell-mediated immunity** but relatively normal humoral immunity.

2. It usually results in **death** from **tetany** or uncontrollable **infection.**

B. **Acquired immunodeficiency syndrome (AIDS)** is caused by infection with **human immunodeficiency virus (HIV),** which preferentially invades T_H cells, causing a severe depression in their number and thus **suppressing both cell-mediated and humoral immune responses.**

1. AIDS is characterized by **secondary infections** by opportunistic microorganisms that cause pneumonia, toxoplasmosis, candidiasis, and other diseases.

2. It is also characterized by development of certain malignancies such as **Kaposi sarcoma** and **non-Hodgkin lymphoma.**

C. **Hodgkin disease** is a malignancy involving **neoplastic transformation of lymphocytes,** occurring mostly in young adult men.

1. It is characterized by the presence of **Reed-Sternberg cells,** which are giant cells with two large, vacuolated nuclei, each with a dense nucleolus.

2. Signs and symptoms include painless, progressive **enlargement of the lymph nodes, spleen,** and **liver;** anemia; fever; weakness; anorexia; and weight loss.

Review Test

Directions: Each of the numbered items or incomplete statements in this section is followed by answers or by completions of the statement. Select the ONE lettered answer or completion that is BEST in each case.

1. Which of the following statements concerning T helper cells is true?

(A) They possess membrane-bound antibodies.
(B) They can recognize and interact with antigens in the blood.
(C) They produce numerous cytokines.
(D) They function only in cell-mediated immunity.
(E) Their activation depends on interferon γ.

2. Which of the following statements concerning T cytotoxic (T_C) cells is true?

(A) They assist macrophages in killing microorganisms.
(B) They possess antibodies on their surfacs.
(C) They possess CD8 surface markers.
(D) They possess CD28 surface markers.
(E) They secrete interferon γ.

3. Which of the following cell types is thought to function in preventing immune responses against self-antigens?

(A) T_S cells
(B) B cells
(C) T memory cells
(D) T_H cells
(E) Mast cells

4. Which of the following statements concerning interferon γ is true?

(A) It is produced by T memory cells.
(B) It is produced by T suppressor cells.
(C) It activates macrophages.
(D) It inhibits macrophages.
(E) It induces viral proliferation.

5. A patient who was given penicillin has an adverse reaction to the antibiotic. Although the reaction is due to the actions of mast cells, the response occurred because mast cells have immunoglobulin E (IgE) receptors in their cell membranes. Which of the following cells produced the IgE decorating the plasma cell's surface?

(A) T memory cells
(B) B memory cells
(C) T helper cells
(D) Plasma cells
(E) T cytotoxic cells

6. Which of the following statements concerning the thymus is true?

(A) Lymphoid nodules form much of the thymic cortex.
(B) Epithelial reticular cells form Hassall corpuscles.
(C) T cells migrate into the medulla, where they become immunologically competent.
(D) Most T cells that enter the thymus are killed in the medulla.
(E) Macrophages are essential components of the blood-thymus barrier.

7. Which of the following statements concerning Hassall corpuscles is true?

(A) They are located in the thymic cortex of young individuals.
(B) They are located in the thymic cortex of old individuals.
(C) They are derived from mesoderm.
(D) They are located in the thymic medulla.
(E) They are derived from T memory cells.

8. After their maturation in the thymus and release into the circulation, T lymphocytes migrate preferentially to which of the following sites?

(A) Paracortex of lymph nodes
(B) Cortical lymphoid nodules of lymph nodes
(C) Hilus of lymph nodes
(D) Lymphoid nodules of the tonsils
(E) Lymphoid nodules of the spleen

9. In which of the following sites do lymphocytes become immunocompetent?

(A) Germinal center of secondary lymphoid nodules
(B) White pulp of the spleen
(C) Thymic cortex
(D) Red pulp of the spleen
(E) Paracortex of lymph nodes

10. Which of the following statements about immunoglobulin G (IgG) is true?

(A) It is located in the serum and on the membrane of B cells.
(B) It can cross the placental barrier.
(C) It is involved in allergic reactions.
(D) It exists as a pentamer.
(E) It binds to antigens on the body surface and in the lumen of the gastrointestinal tract.

Answers and Explanations

1-C. T helper cells produce a number of cytokines that affect other cells involved in both the cell-mediated and the humoral immune responses. T helper cells possess antigen-specific T-cell receptors (not antibodies) on their membranes. These cells recognize and interact with antigenic determinants that are associated with class II human leukocyte antigen (HLA) molecules on the surface of antigen-presenting cells. IL-1 is necessary for activation of T helper cells.

2-C. T cytotoxic cells are CD8+. CD28 molecules are present on T_H1 cells. Interferon (IFN)-γ is released by T_H1 cells, and they also assist macrophages in killing microorganisms.

3-A. The immune response is decreased by T suppressor (T_S) cells. Their activity is thought to help prevent autoimmune responses against self-antigens.

4-C. Interferon γ activates macrophages, as well as natural killer (NK) cells and T cytotoxic cells, enhancing their phagocytic or cytotoxic activity or both.

5-D. Individuals allergic to penicillin produce immunoglobulin E (IgE) antibodies. The cells that manufacture IgE are plasma. After an antigenic challenge, proliferation and differentiation of B cells give rise to plasma cells and B memory cells.

6-B. Epithelial reticular cells of the medulla congregate to form Hassall (thymic) corpuscles.

7-D. Hassall corpuscles are concentrically arranged accretions of epithelial reticular cells (derived from endoderm) found only in the medulla of the thymus.

8-A. T lymphocytes are preferentially located in the paracortex of lymph nodes, whereas B lymphocytes are found in lymphoid nodules located in lymph nodes, tonsils, and the spleen.

9-C. T lymphocytes mature and become immunocompetent in the cortex of the thymus, whereas B lymphocytes do so in the bone marrow. After an antigenic challenge, lymphocytes proliferate and differentiate in various lymphoid tissues.

10-B. Immunoglobulin G (IgG) is the most abundant immunoglobulin isotype in the serum. It can cross the placental barrier but does not bind to the B-cell plasma membrane. It exists as a monomer, functions to activate complement, and acts as an opsonin.

13

Endocrine System

I. Overview—The Endocrine System

A. The endocrine system is composed of several **ductless glands, clusters of cells** located within certain organs, and isolated **endocrine cells** [so-called diffuse neuroendocrine system cells (DNES)] in the epithelial lining of the gastrointestinal and respiratory systems.

B. Glands of the endocrine system include the **pituitary, thyroid, parathyroid, adrenal,** and **pineal** glands.

C. Function. The endocrine system secretes hormones into nearby capillaries and interacts with the nervous system to modulate and control the body's metabolic activities.

II. Hormones. Hormones are **chemical messengers** that are carried via the bloodstream to distant **target cells.** Hormones include low-molecular-weight **water-soluble** proteins and polypeptides [e.g., insulin, glucagon, follicle-stimulating hormone (FSH)] and **lipid-soluble** substances, principally the steroid hormones (e.g., progesterone, estradiol, testosterone).

A. Water-soluble hormones interact with specific cell-surface receptors on target cells, which communicate a message that generates a biological response by the cell.

1. G-protein-linked receptors are used by some hormones [e.g., epinephrine, thyroid-stimulating hormone (TSH), serotonin]. Binding of the hormone to the G-protein-linked receptor leads to production of a second messenger that evokes a target cell response.

2. Catalytic receptors are used by insulin and growth hormone. Binding of the hormone to the catalytic receptor activates protein kinases that phosphorylate target proteins.

B. Lipid-soluble hormones diffuse across the plasma membrane of target cells and bind to specific receptors in the cytosol or nucleus, forming hormone-receptor complexes that regulate transcription of deoxyribonucleic acid (DNA).

III. Overview—Pituitary Gland (Hypophysis). The pituitary gland lies below the hypothalamus, to which it is structurally and functionally connected. It is divided into two major subdivisions, the **adenohypophysis** and the **neurohypophysis.** Each subdivision is derived from a distinct embryonic analog, which is reflected in its unique cellular constituents and functions.

A. The **adenohypophysis** is also called the **anterior pituitary gland** (Figure 13.1). It originates from an ectodermal diverticulum of the stomodeum **(Rathke pouch).** It is subdivided into the **pars distalis, pars intermedia,** and **pars tuberalis.**

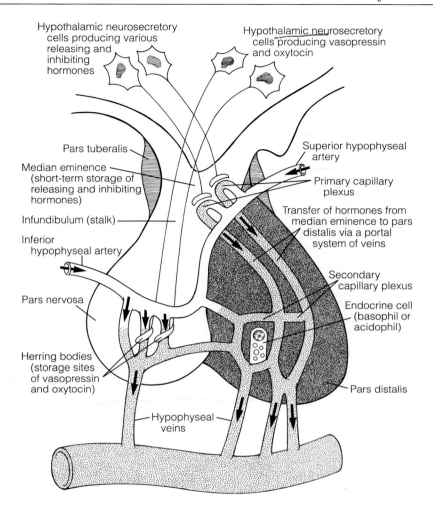

Figure 13.1. Diagram of the pituitary gland showing its connections to the hypothalamus, sites of hormone synthesis and storage, and vascularization. The adenohypophysis lies to the right and consists of the pars distalis, pars tuberalis, and pars intermedia (not shown). The neurohypophysis consists of the infundibulum (stalk) and pars nervosa. Various releasing and inhibiting hormones stored in the median eminence are transferred, via the hypophyseal portal system, to the pars distalis. (Adapted with permission from Junqueira LC, Carneiro J, Kelley RO: *Basic Histology*, 9th ed. Stamford, CT, Appleton & Lange, 1998, p 380.)

1. The **pars distalis** is supported by a connective tissue capsule and framework. It consists of irregular cords of parenchymal cells lying adjacent to **fenestrated** capillaries.

    a. **Chromophils**

    (1) **Overview.** Chromophils are parenchymal cells that stain intensely due to their hormone-containing secretory granules. They synthesize, store, and release several hormones. They are regulated by specific stimulatory and inhibitory hormones produced by neurosecretory cells in the **hypothalamus** and are conveyed to the pars distalis via a system of portal blood vessels originating in the median eminence.

    (2) **Types.** Chromophils are classified into two types, depending on the dyes they bind using special histologic stains. With hema-

toxylin-eosin stain, the distinction between the two cell types is much less obvious.

(a) **Acidophils** (Table 13.1) bind acid dyes and often stain **orange or red.** They are small cells of two subtypes: somatotrophs and mammotrophs.

(i) **Somatotrophs,** which produce **somatotropin (growth hormone),** are stimulated by **somatotropin-releasing hormone (SRH)** and are inhibited by **somatostatin.**

(ii) **Mammotrophs** produce **prolactin,** which is stored in small secretory granules. They are stimulated by **prolactin-releasing hormone (PRH)** and are inhibited by **prolactin-inhibiting hormone (PIH).**

Table 13.1. Physiologic Effects of Pituitary Hormones

Cell	Hormone	Major Function
Hormones released by the pars distalis		
Acidophils		
Somatotroph	Somatotropin (growth hormone)	Increases metabolism in most cells, indirectly stimulates epiphyseal plate and growth of long bones via production of liver somatomedins (insulin-like growth factors I and II)
Mammotroph	Prolactin	Development of mammary gland during pregnancy, milk synthesis during lactation
Basophils		
Corticotroph	ACTH (adrenocorticotrophic hormone)	Stimulates glucocorticoid secretion by zona fasciculata cells of the adrenal cortex
Gonadotroph	FSH (follicle-stimulating hormone)	Stimulates growth of secondary ovarian follicles and estrogen secretion in women; stimulates spermatogenesis via production of androgen-binding protein in Sertoli cells in men
	LH (luteinizing hormone) or ICSH (interstitial cell stimulating hormone)	Ovulation, formation of corpus luteum, and progesterone secretion in women; testosterone synthesis by Leydig cells of testis in men
Thyrotroph	TSH (thyroid-stimulating hormone)	Stimulates synthesis and release of thyroid hormones (T_3, T_4) by follicular cells
Hormones released by the pars nervosa		
Neurosecretory cells of hypothalamus (in supraoptic and paraventricular nuclei)	Oxytocin	Induces contraction of smooth muscle in wall of uterus at parturition and in myoepithelial cells of mammary gland during nursing
	Vasopressin [ADH (antidiuretic hormone)]	Renders kidney collecting tubules permeable to water, which is resorbed to produce a concentrated urine; constricts smooth muscle in wall of arterioles

(b) **Basophils** (see Table 13.1) bind basic dyes and typically stain **blue.** They include three subtypes: corticotrophs, thyrotrophs, and gonadotrophs.

(i) **Corticotrophs** produce **adrenocorticotropic hormone (ACTH)** and **lipotropic hormone (LPH),** a precursor of β-endorphin. They are stimulated by **corticotropin-releasing hormone (CRH).**

(ii) **Thyrotrophs** produce **TSH** and are stimulated by **thyrotropin-releasing hormone (TRH).**

(iii) **Gonadotrophs** produce **FSH** and **luteinizing hormone (LH)** in both sexes, although in men the latter is sometimes referred to as **interstitial cell-stimulating hormone (ICSH).** Gonadotrophs are stimulated by **gonadotropin-releasing hormone (GnRH).**

b. **Chromophobes:**

(1) Are parenchymal cells that **do not stain intensely**

(2) Appear as small cells under the light microscope; the cells lack (or have only a few) secretory granules and are arranged close to one another in clusters

(3) Sometimes resemble degranulated chromophils in the electron microscope, suggesting that they may represent different stages in the life cycle of various acidophil and basophil populations

(4) May also represent undifferentiated cells that are capable of differentiating into various types of chromophils

c. **Folliculostellate cells:**

(1) Are numerous in the pars distalis and lie between the chromophils and chromophobes

(2) Possess long processes that form gap junctions with processes of other folliculostellate cells

(3) Produce many peptides that are thought to regulate the production of pars distalis hormones via a paracrine effect

2. The **pars intermedia** lies between the pars distalis and pars nervosa.

a. It is characterized by the presence of many **colloid-containing cysts (Rathke cysts)** that are lined by cuboidal cells.

b. It possesses **basophilic cells,** which sometimes extend into the pars nervosa. These cells secrete the **prohormone** proopiomelanocortin **(POMC),** which is cleaved to form **melanocyte-stimulating hormone (MSH).** In humans, MSH acts in various ways to modulate inflammatory responses throughout the body, and it may play a role in controlling stores of body fat.

3. The **pars tuberalis** surrounds the cranial part of the infundibulum (hypophyseal stalk).

a. It is composed of cuboidal **basophilic cells,** arranged in cords along an abundant capillary network.

b. Its cells may secrete FSH and LH, but this has not been confirmed.

B. The **neurohypophysis** (see Figure 13.1 and Table 13.1) is also called the **posterior pituitary gland.** It originates from an evagination of the hypothalamus and is divided into the **infundibulum,** which is continuous with the hypothalamus, and the pars nervosa or main body of the neurohypophysis.

1. **Hypothalamohypophyseal tract:**

   a. Contains the unmyelinated axons of **neurosecretory cells** whose cell bodies are located in the **supraoptic** and **paraventricular nuclei** of the hypothalamus

   b. Transports **oxytocin, antidiuretic hormone (ADH; vasopressin), neurophysin** (a binding protein specific for each hormone), and adenosine triphosphate (ATP) to the pars nervosa

2. **Pars nervosa:**

   a. Contains the distal ends of the hypothalamohypophyseal axons and is the site where the neurosecretory granules in these axons are stored in accumulations known as **Herring bodies**

   b. Releases oxytocin and ADH into fenestrated capillaries in response to nerve stimulation

3. **Pituicytes:**

   a. Occupy approximately 25% of the volume of the pars nervosa

   b. Are glial-like cells that support axons in this region

   c. Possess numerous cytoplasmic processes and contain lipid droplets, intermediate filaments, and pigments

C. **Vascularization of the pituitary gland**

1. **Arterial supply** is from two pairs of blood vessels derived from the internal carotid artery.

   a. The right and left **superior** hypophyseal arteries serve the pars tuberalis, infundibulum, and median eminence.

   b. The right and left **inferior** hypophyseal arteries serve mostly the pars nervosa.

2. **Hypophyseal portal system** (see Figure 13.1)

   a. The **primary capillary plexus** consists of fenestrated capillaries coming off the superior hypophyseal arteries.

      (1) This plexus is located in the **median eminence** where stored hypothalamic neurosecretory hormones enter the blood.

      (2) It is drained by **hypophyseal portal veins,** which descend through the infundibulum into the adenohypophysis.

   b. The **secondary capillary plexus** consists of **fenestrated** capillaries coming off the hypophyseal portal veins. This plexus is located in the **pars distalis** where neurosecretory hormones leave the blood to stimulate or inhibit the parenchymal cells.

D. **Regulation of the pars distalis**

1. Neurosecretory cells in the hypothalamus synthesize specific hormones that enter the hypophyseal portal system and stimulate or inhibit the parenchymal cells of the pars distalis.

2. The hypothalamic neurosecretory cells in turn are regulated by the level of hormones in the blood (**negative feedback**) or by other physiologic (or psychologic) factors.

3. Some hormones (e.g., thyroid hormones, cortisol) exert negative feedback on the pars distalis directly.

IV. Overview—Thyroid Gland.
The thyroid gland is composed of two lobes connected by an **isthmus.** It is surrounded by a dense irregular collagenous connective tissue capsule, in which (posteriorly) the **parathyroid glands** are embedded. The thyroid gland is subdivided by capsular septae into lobules containing **follicles.** These septae also serve as conduits for blood vessels, lymphatic vessels, and nerves.

A. **Thyroid follicles** are spherical structures filled with **colloid,** a viscous gel consisting mostly of **iodinated thyroglobulin.**

1. Thyroid follicles are enveloped by a layer of epithelial cells, called **follicular cells,** which in turn are surrounded by **parafollicular cells.** These two parenchymal cell types rest on a basal lamina, which separates them from the abundant network of **fenestrated capillaries** in the connective tissue.

2. **Function.** Thyroid follicles synthesize and store thyroid hormones.

B. **Follicular cells**

1. **Structure**

a. Follicular cells are normally cuboidal in shape but become columnar when stimulated and squamous when inactive.

b. They possess a distended rough endoplasmic reticulum (RER) with many ribosome-free regions, a supranuclear Golgi complex, numerous lysosomes, and rod-shaped mitochondria.

c. Follicular cells contain many small **apical vesicles,** which are involved in the transport and release of thyroglobulin and enzymes into the colloid.

d. They possess short, blunt microvilli, which extend into the colloid.

2. **Synthesis and release of the thyroid hormones thyroxine (T_4) and triiodothyronine (T_3)** occur by the sequence of events illustrated in Figure 13.2. These processes are promoted by **TSH,** which binds to G-protein-linked receptors on the basal surface of follicular cells.

C. **Parafollicular cells** are also called **clear (C) cells** because they stain less intensely than thyroid follicular cells.

1. Parafollicular cells are present singly or in small clusters of cells located between the follicular cells and basal lamina.

2. These cells belong to the population of **DNES cells** (**d**iffuse **n**euro**e**ndocrine **s**ystem), also known as **APUD cells** (**a**mine **p**recursor **u**ptake and **d**ecarboxylation cells), or enteroendocrine cells.

3. They possess elongated mitochondria, substantial amounts of RER, a well-developed Golgi complex, and many membrane-bound secretory granules.

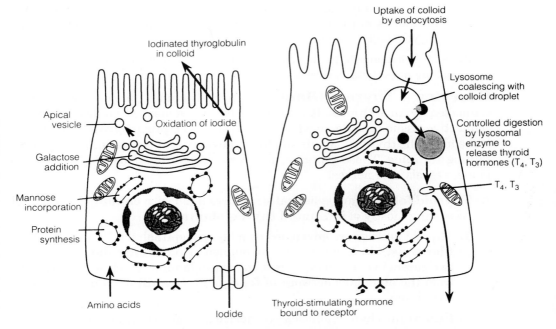

Figure 13.2. Synthesis and release of thyroxine (T_4) and triiodothyronine (T_3) by follicular cells of the thyroid gland. *(A)* Thyroglobulin is synthesized like other secretory proteins. Circulating iodide is actively transported into the cytosol, where a thyroid peroxidase oxidizes it and iodinates tyrosine residues on the thyroglobulin molecule; iodination occurs mostly at the apical plasma membrane. A rearrangement of the iodinated tyrosine residues of thyroglobulin in the colloid produces the iodothyronines T_4 and T_3. *(B)* Binding of thyroid-stimulating hormone to receptors on the basal surface stimulates follicular cells to become columnar and to form apical pseudopods, which engulf colloid by endocytosis. After the colloid droplets fuse with lysosomes, controlled hydrolysis of iodinated thyroglobulin liberates T_3 and T_4 into the cytosol. These hormones move basally and are released basally to enter the bloodstream and lymphatic vessels. (Adapted with permission from Junqueira LC, Carneiro J, Kelley RO: *Basic Histology*, 9th ed. Stamford, CT, Appleton & Lange, 1998, p 403, and from Fawcett DW: *Bloom and Fawcett: A Textbook of Histology,* 12th ed. New York, Chapman and Hall, 1994, p 496.)

    4. They synthesize and release **calcitonin**, a polypeptide hormone, in response to high blood calcium levels.

  D. Physiologic effects of thyroid hormones

    1. T_4 and T_3 act on a variety of target cells. These hormones **increase the basal metabolic rate** and thus promote heat production. They have broad effects on gene expression and induction of protein synthesis.

    2. Calcitonin functions primarily to lower blood calcium levels by inhibiting bone resorption by osteoclasts.

V. Parathyroid Glands

  A. Overview

    1. The parathyroid glands are four small glands that lie on the posterior surface of the thyroid gland, embedded in its connective tissue capsule.

    2. They have a parenchyma composed of two types of cells, **chief cells** and **oxyphil cells.**

    3. They are supported by septae from the capsule, which penetrate each gland and also convey blood vessels into its interior.

4. They become infiltrated with fat cells in older persons; the number of oxyphil cells also increases.

B. **Chief cells** are small, basophilic cells, arranged in clusters.

1. Chief cells form anastomosing cords, surrounded by a rich, fenestrated capillary network.

2. These cells possess a centrally located nucleus, a well-developed Golgi complex, abundant RER, small mitochondria, glycogen, and secretory granules of variable size.

3. **Function.** They synthesize and secrete **parathyroid hormone (PTH).** High blood calcium levels **inhibit** production of PTH.

C. **Oxyphil cells** are large, eosinophilic cells that are present singly or in small clusters within the parenchyma of the gland.

1. Oxyphil cells possess many large, elongated mitochondria, a poorly developed Golgi complex, and only a limited amount of RER.

2. Their function is not known.

D. **PTH** functions primarily to **increase blood calcium levels** by indirectly stimulating osteoclasts to resorb bone. With calcitonin, PTH provides a dual mechanism for regulating blood calcium levels. A near absence of PTH (hypoparathyroidism) may be caused by accidental surgical removal of the parathyroid glands, which leads to **tetany,** characterized by hyperexcitability and spastic skeletal muscle contractions throughout the body.

VI. Overview—Adrenal (Suprarenal) Glands. Adrenal glands lie embedded in fat at the superior pole of each kidney. They are derived from two embryonic sources: the ectodermal neural crest, which gives rise to the **adrenal medulla,** and the mesoderm, which gives rise to the **adrenal cortex.** The adrenal glands are invested by their own collagenous capsule.

A. The **adrenal cortex** (Table 13.2) contains parenchymal cells that synthesize and secrete, but **do not store,** various steroid hormones. It is divided into three concentric, histologically recognizable regions: the **zona glomerulosa, zona fasciculata, and zona reticularis.**

1. **Zona glomerulosa:**

a. Synthesizes and secretes **mineralocorticoids,** mostly **aldosterone** and some **deoxycorticosterone.** Hormone production is stimulated by angiotensin II and ACTH.

b. Is composed of small cells arranged in arch-like cords and clusters. These cells have a few small lipid droplets, an extensive network of smooth endoplasmic reticulum (SER), and mitochondria with **shelf-like cristae.**

2. **Zona fasciculata:**

a. Synthesizes and secretes **glucocorticoids,** namely **cortisol and corticosterone.** Hormone production is stimulated by ACTH (Figure 13.3).

b. Is composed of columns of cells and **sinusoidal capillaries** oriented perpendicularly to the capsule.

Table 13.2. Adrenal Gland Cells and Hormones

Cell	Hormone	Function
Adrenal cortex		
Zona glomerulosa	Mineralocorticoids (mostly aldosterone)	Regulate electrolyte and water balance via their effect on cells of renal tubules
Zona fasciculata	Glucocorticoids (cortisol and corticosterone)	Regulate carbohydrate metabolism by promoting gluconeogenesis, promote breakdown of proteins and fat, anti-inflammatory properties, suppress the immune response
Zona reticularis	Weak androgens (dehydro-epiandrosterone and androstenedione)	Promote masculine characteristics
Adrenal medulla		
Chromaffin cells	Epinephrine and norepinephrine	"Fight-or-flight" response Epinephrine increases heart rate and force of contraction, relaxes bronchiolar smooth muscle and promotes glycogenolysis and lipolysis Norepinephrine has little effect on cardiac output and is rarely used clinically

   c. Cells contain many lipid droplets and appear so vacuolated that they are called **spongiocytes.** These cells also possess spherical mitochondria with **tubular** and **vesicular cristae,** SER, RER, lysosomes, and **lipofuscin pigment granules.**

3. Zona reticularis:

   a. Synthesizes and secretes **weak androgens** (mostly **dehydroepiandrosterone** and some **androstenedione**) and perhaps small amounts of glucocorticoids. Hormone production is stimulated by ACTH.

   b. Is composed of cells, arranged in anastomosing cords. Many large **lipofuscin pigment granules** are common in these cells.

B. The **adrenal medulla** (see Table 13.2) is completely invested by the adrenal cortex. It contains two populations of parenchymal cells, called **chromaffin cells,** which synthesize, store, and secrete the catecholamines **epinephrine** and **norepinephrine.** It also contains scattered **sympathetic ganglion cells.**

   1. Chromaffin cells are large, polyhedral cells containing secretory granules that stain intensely with chromium salts (chromaffin reaction).

   a. Chromaffin cells are arranged in short, irregular cords surrounded by an extensive capillary network.

   b. They are innervated by **preganglionic sympathetic (cholinergic) fibers,** making these cells analogous in function to postganglionic sympathetic neurons.

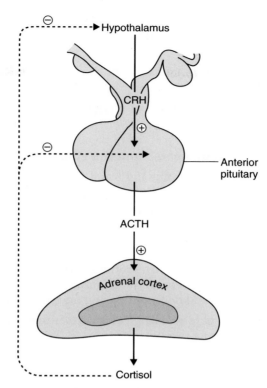

Figure 13.3. Regulation of glucocorticoid secretion by the adrenal cortex [via stimulation by corticotropin-releasing hormone (CRH) and adrenocorticotropic hormone (ACTH)] and the negative feedback of cortisol at both the pituitary and hypothalamic levels. (Adapted with permission from Junqueira LC, Carneiro J, Kelley RO: *Basic Histology*, 9th ed. Stamford, CT, Appleton & Lange, 1998, p 393.)

    c. They possess a well-developed Golgi complex, isolated regions of RER, and numerous mitochondria.

    d. They also contain large numbers of membrane-bound granules containing one of the catecholamines, ATP, enkephalins, and **chromogranins,** which may function as binding proteins for epinephrine and norepinephrine.

  2. Catecholamine release occurs in response to intense emotional stimuli and is mediated by the preganglionic sympathetic fibers that innervate the chromaffin cells.

C. Blood supply to the adrenal glands is derived from the superior, middle, and inferior adrenal arteries, which form three groups of vessels: to the capsule, to parenchymal cells of the cortex, and directly to the medulla.

  1. Cortical blood supply

    a. A **fenestrated** capillary network bathes cells of the zona glomerulosa.

    b. Straight, discontinuous, fenestrated capillaries supply the zona fasciculata and zona reticularis.

  2. Medullary blood supply

    a. Venous blood rich in hormones reaches the medulla via the discontinuous fenestrated capillaries that pass through the cortex.

     b. Arterial blood from direct branches of capsular arteries forms an extensive fenestrated capillary network among the chromaffin cells of the medulla.

     c. Medullary veins join to form the suprarenal vein, which exits the gland.

VII. Pineal Gland (Pineal Body; Epiphysis)

A. Overview

1. The pineal gland **projects from the roof of the diencephalon.**

2. Its secretions vary with the light and dark cycles of the day.

3. This gland has a capsule formed of the **pia mater,** from which septae (containing blood vessels and unmyelinated nerve fibers) extend to subdivide it into incomplete lobules.

4. It is composed primarily of pinealocytes and neuroglial cells.

5. It also contains calcified concretions **(brain sand)** in its interstitium. The function of these concretions is unknown, but they increase during short light cycles and decrease during periods of darkness.

B. Pinealocytes are pale-staining cells with numerous long processes that end in dilatations near capillaries.

1. Pinealocytes contain many secretory granules, microtubules, microfilaments, and unusual structures called **synaptic ribbons.**

2. These cells synthesize and secrete **serotonin** (during the day) and **melatonin** (at night). Melatonin is used to treat jet lag and seasonal affective disorder (SAD).

3. Pinealocytes may also produce **arginine vasotocin,** a peptide that appears to be an antagonist of LH and FSH.

C. Neuroglial (interstitial) cells resemble astrocytes, with elongated processes and a small, dense nucleus. They contain microtubules and many microfilaments and intermediate filaments.

VIII. Clinical Considerations

A. Pituitary adenomas are common tumors of the anterior pituitary.

1. They enlarge and often suppress secretions by the remaining pars distalis cells.

2. These tumors frequently destroy surrounding bone and neural tissues and are treated by surgical removal.

B. Graves disease is characterized by a diffuse enlargement of the thyroid gland and protrusion of the eyeballs (exophthalmic goiter).

1. This disease is associated with the presence of columnar-shaped thyroid follicular cells, excessive production of thyroid hormones, and decreased amounts of follicular colloid.

2. It is caused by the binding of autoimmune immunoglobulin G (IgG) antibodies to TSH receptors, which stimulates the thyroid follicular cells.

C. Simple goiter (enlargement of the thyroid gland) is caused by insufficient iodine (<10 micrograms/day) in the diet.

    1. It is usually not associated with either hyperthyroidism or hypothyroidism.

    2. Simple goiter is treated by administration of dietary iodine.

D. Hyperparathyroidism is overactivity of the parathyroid glands, resulting in excess secretion of PTH and subsequent bone resorption (see Chapter 7 II J 1).

    1. Hyperparathyroidism is associated with **high blood calcium levels,** which may lead to deposition of calcium salts in the kidneys and the walls of blood vessels.

    2. It may be caused by a benign tumor of the parathyroid glands.

E. Addison disease is characterized by secretion of inadequate amounts of adrenocortical hormones due to destruction of the adrenal cortex.

    1. Addison disease is most often caused by an autoimmune disease or can be a sequela of tuberculosis.

    2. This disease is life-threatening and requires steroid treatment.

F. Diabetes insipidus results from inadequate amounts of antidiuretic hormone (ADH); it is discussed in Chapter 18 VII C.

Review Test

Directions: Each of the numbered items or incomplete statements in this section is followed by answers or by completions of the statement. Select the ONE lettered answer or completion that is BEST in each case.

1. Protein hormones act initially on target cells by

(A) attaching to receptors on the nuclear membrane
(B) attaching to receptors in the nucleolus
(C) diffusing through the plasma membrane
(D) attaching to receptors on the plasma membrane

2. Which of the following statements concerning adrenal parenchymal cells is true?

(A) Those of the zona fasciculata produce androgens.
(B) Those of the adrenal medulla produce epinephrine and norepinephrine.
(C) Those of the zona glomerulosa produce glucocorticoids.
(D) Those of the cortex contain numerous secretory granules.

3. Characteristics of pinealocytes include which one of the following?

(A) They produce melatonin and serotonin.
(B) They resemble astrocytes.
(C) They contain calcified concretions of unknown function.
(D) They act as postganglionic sympathetic cells.

4. Prolactin is synthesized and secreted by which of the following cells?

(A) Acidophils in the pars distalis
(B) Basophils in the pars tuberalis
(C) Somatotrophs in the pars distalis
(D) Basophils in the pars intermedia

5. Adrenocorticotropic hormone (ACTH) is produced by which of the following cells?

(A) Chromophobes in the pars distalis
(B) Neurosecretory cells in the median eminence
(C) Basophils in the pars distalis
(D) Neurons of the paraventricular nucleus in the hypothalamus

6. The histological appearance of a thyroid gland being stimulated by thyroid-stimulating hormone (TSH) would show which of the following?

(A) Decreased numbers of follicular cells
(B) Increased numbers of parafollicular cells
(C) Columnar-shaped follicular cells
(D) An abundance of colloid in the lumen of the follicle
(E) Decreased numbers of parafollicular capillaries

7. A 40-year-old woman is diagnosed with Graves disease. Which of the following characteristics would be associated with her condition?

(A) Inadequate levels of iodine in her diet
(B) Weight gain
(C) Flattened thyroid follicular cells
(D) Excessive production of thyroid hormones
(E) Increased amounts of follicular colloid

Questions 8-10

The response options for the next three items are the same. Each item will state the number of options to select. Choose exactly this number.

(A) Calcitonin
(B) Parathyroid hormone
(C) Triiodothyronine (T_3)

For each of the following physiologic effects, select the hormone that exerts the effect.

8. Acts to stimulate calcium mobilization from bone (**SELECT 1 HORMONE**)

9. Lowers blood calcium levels by inhibiting bone resorption (**SELECT 1 HORMONE**)

10. Helps to regulate body temperature (**SELECT 1 HORMONE**)

Answers and Explanations

1-D. Protein hormones initiate their action by binding externally to transmembrane receptor proteins in the target-cell plasma membrane. Receptors for some hormones (e.g., thyroid-stimulating hormone, serotonin, epinephrine) are linked to G proteins; other receptors, including those for insulin and growth hormone, have protein kinase activity.

2-B. Chromaffin cells in the adrenal medulla synthesize and store epinephrine and norepinephrine in secretory granules, which also contain adenosine triphosphate (ATP), chromogranins, and enkephalins. The cortical parenchymal cells of the zona fasciculata produce glucocorticoids, and those of the zona glomerulosa produce mineralocorticoids. The cortical parenchymal cells do not store their secretory products and thus do not contain secretory granules.

3-A. Pinealocytes, the parenchymal cells of the pineal gland, produce melatonin and serotonin. The pineal gland also contains neuroglial cells that resemble astrocytes, and its interstitium has calcified concretions called brain sand.

4-A. Prolactin is produced by mammotrophs, one of the two types of acidophils located in the pars distalis of the pituitary gland. As their name implies, these cells produce a hormone that regulates the development of the mammary gland during pregnancy and lactation.

5-C. Adrenocorticotropic hormone (ACTH) is produced by corticotrophs, a type of basophil, present in the pars distalis of the pituitary gland.

6-C. Stimulation of the thyroid gland by thyroid-stimulating hormone (TSH) causes the follicular cells to become more active and columnar-shaped. They form apical pseudopods and engulf colloid, which is removed from the lumen of the follicle by endocytosis and broken down by controlled lysosomal hydrolysis to yield the thyroid hormones T_3 and T_4. Parafollicular cells and capillaries do not contain receptors for TSH.

7-D. Graves disease (exophthalmic goiter) results in an enlarged thyroid gland due to stimulation of the follicular cells by binding of autoimmune antibodies to thyroid-stimulating hormone (TSH) receptors. Follicular cells actively remove colloid from the lumen of the follicles. Heat intolerance and weight loss are common, but the disease is not caused by iodine deficiency.

8-B. Parathyroid hormone increases blood calcium levels by indirectly stimulating osteoclasts to mobilize calcium from bone.

9-A. Calcitonin lowers blood calcium levels and thus has an effect antagonistic to that of parathyroid hormone. It is produced by parafollicular cells of the thyroid gland.

10-C. Triiodothyronine and thyroxine both increase the basal metabolic rate, which affects heat production and body temperature. These thyroid hormones also have many other effects.

14

Skin

I. Overview—The Skin

A. The skin is the heaviest single organ and constitutes about 16% of the total body weight.

B. It is composed of two layers, the **epidermis** and **dermis,** which **interdigitate** to form an **irregular contour**.

C. A deeper superficial fascial layer, the **hypodermis,** lies under the skin. This layer, which is **not** considered part of the skin, consists of loose connective tissue and binds skin loosely to the subjacent tissue.

D. The skin contains several **appendages** (sweat glands, hair follicles, sebaceous glands, and nails). The skin and its appendages are called the **integument.**

E. **Function.** The skin protects the body against injury, desiccation, and infection; regulates body temperature; absorbs ultraviolet (UV) radiation, which is necessary for synthesis of vitamin D; and contains receptors for touch, temperature, and pain stimuli from the external environment.

II. Epidermis

A. Overview—Epidermis

1. The epidermis is the **superficial layer** of the skin. Primarily of **ectodermal origin,** it is classified as **stratified squamous keratinized epithelium.** The epidermis is composed predominantly of **keratinocytes** and three different types of nonkeratinocytes: **melanocytes, Langerhans cells,** and **Merkel cells.**

2. The epidermis is constantly being regenerated. **Regeneration,** which occurs approximately every 30 days, is carried out by the mitotic activity of keratinocytes, which typically divide at night.

3. The epidermis has deep invaginations **(interpapillary pegs)** that interdigitate with projections of the dermis **(dermal papillae)** resulting in a highly irregular interface. Where the epidermis overlies the dermal papillae, a series of **epidermal ridges** is produced. On the fingertips, these surface ridges are visible as **fingerprints,** whose configuration is genetically determined and thus unique to each individual.

B. Layers of the epidermis (Figure 14.1, Table 14.1)

1. The **stratum basale (germinativum)** is the **deepest layer** of the epidermis and is composed of keratinocytes that are cuboidal to columnar in shape. These **mitotically active** cells are attached directly to the basal

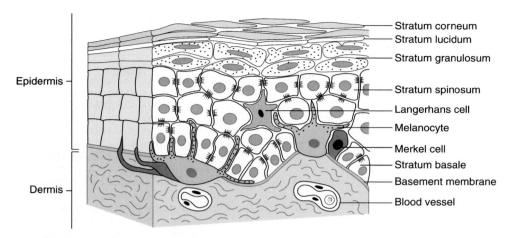

Figure 14.1. Diagram showing layers of epidermis. The stratum lucidum is present only in thick skin. Melanocytes lie between keratinocytes in the stratum basale. (Adapted with permission from Ham AH, Cormack DH: *Histology*, 8th ed. Philadelphia, JB Lippincott, 1979, p 625.)

lamina by **hemidesmosomes** (see Chapter 5 III B) and to each other by desmosomes. This layer also contains **melanocytes** and **Merkel cells.**

2. The **stratum spinosum** consists of a few layers of polyhedral keratinocytes **(prickle cells).** Their extensions, termed "intercellular bridges" by early histologists, are now known to terminate in **desmosomes** (see Chapter 5 II A 3). This layer also contains **Langerhans cells.**

   a. Keratinocytes in the deeper aspects of the stratum spinosum are also **mitotically active.**

   b. The **malpighian layer** (stratum malpighii) consists of the stratum spinosum and stratum basale. Nearly all the mitotic activity in the epidermis occurs in this region.

   c. In the superficial regions of the stratum spinosum are keratinocytes that contain **membrane-coating granules.** The contents of these granules are released into the intercellular spaces in the form of lipid-containing sheets that are **impermeable to water and many foreign substances.**

3. The **stratum granulosum** is the most superficial layer in which nuclei are still present. It comprises three to five layers of flattened keratinocytes that contain **keratohyalin granules, bundles of keratin filaments** (tonofilaments), and occasional **membrane-coating granules.**

   a. Keratohyalin granules (not membrane-bound) contain histidine- and cystine-rich proteins, which bind the keratin filaments together.

   b. The cytoplasmic aspect of the plasma membrane of keratinocytes in the stratum granulosum is reinforced by an electron-dense layer 10-12 nanometers (nm) thick.

4. The **stratum lucidum** is a clear, homogeneous layer just superficial to the stratum granulosum; it is often difficult to distinguish in histologic sections. It is found only in **palmar and plantar skin.** This layer consists of keratinocytes that **have neither nuclei** nor organelles but con-

Table 14.1. Histological Features of Skin

Divisions	Layers	Characteristics
Epidermis, classified as stratified squamous keratinized epithelium	Stratum corneum	The most superficial layer of epidermis Many flattened, dead "cells", called squames, packed with keratin filaments Surface cells are sloughed
	Stratum lucidum	Indistinct homogeneous layer of keratinocytes, present only in thick skin Cells lack nuclei and organelles Cytoplasm is packed with keratin filaments and eleidin
	Stratum granulosum	Flattened nucleated keratinocytes arranged in three to five layers Cells contain many coarse keratohyalin granules associated with tonofilaments Membrane-coating (waterproofing) granules occasionally present Present only in thick skin
	Stratum spinosum	Several layers of keratinocytes, called "prickle cells" because of their spiny shapes Desmosomes, associated with tonofilaments, interconnect cells between processes (intercellular bridges) Keratinocytes contain membrane-coating (waterproofing) granules Keratinocytes are mitotically active, especially in the deeper layers Langerhans cells also are located in this layer
	Stratum basale (stratum germinativum)	Deepest layer of epidermis, composed of a single layer of tall cuboidal keratinocytes Keratinocytes are mitotically active Melanocytes and Merkel cells also are present in this layer
Dermis, classified as dense irregular connective tissue	Papillary layer	Superficial thin layer of connective tissue that interdigitates with epidermal ridges of the epidermis Forms dermal papillae where Meissner corpuscles and capillary loops may be found Contains delicate collagen (type I and type III) fibers Contains anchoring fibrils (type VII collagen), microfibrils (fibrillin), and elastic fibers
	Reticular layer	Extensive part of the dermis, lying deep to the papillary layer Contains thick bundles of collagen (type I) fibers and elastic fibers Arteries, veins, and lymphatics are present Location of sweat glands and their ducts, pacinian corpuscles, and nerves In thin skin, contains hair follicles, sebaceous glands, and arrector pili muscles

tain keratin filaments and **eleidin,** a transformation product of keratohyalin.

5. The **stratum corneum** is the most **superficial layer** of the epidermis. It may consist of as many as 15 to 20 layers of flattened, nonnucleated, dead "cells" filled with **keratin.** These **nonviable, scale-like** structures are called **squames** (or horny cells), and have the shape of a 14-sided polygon. The outermost layer of squames is continuously shed by **desquamation.**

C. **Nonkeratinocytes in the epidermis**

1. **Melanocytes** (see Figure 14.1) are present in the **stratum basale** and originate from the neural crest.

   a. These cells synthesize a **dark brown pigment (melanin)** in oval-shaped organelles **(melanosomes).** Melanosomes contain **tyrosinase,** a UV-sensitive enzyme directly involved in melanin synthesis.

      (1) Melanosome content, size, rate of transfer, and aggregation pattern in keratinocytes vary with race.

      (2) Melanin protects against tissue damage caused by UV radiation.

   b. **Long, melanosome-containing processes** extend between the cells of the stratum spinosum. Melanin is transferred, via a unique mechanism known as **cytocrine secretion,** from these melanosome-filled tips into keratinocytes of the stratum spinosum.

2. **Langerhans cells** are **dendritic cells** (so named because of their long processes) that originate in the bone marrow. They are located primarily in the **stratum spinosum,** contain characteristic paddle-shaped **Birbeck granules,** and function as **antigen-presenting cells** in immune responses to contact antigens (contact allergies) and some skin grafts (see Chapter 12 III).

3. **Merkel cells** are present in small numbers in the **stratum basale,** near areas of well-vascularized, richly innervated connective tissue.

   a. They possess desmosomes and keratin filaments, suggesting an epithelial origin.

   b. Their pale cytoplasm contains **small, dense-cored granules** that are similar in appearance to those in some cells of the diffuse neuroendocrine system (DNES).

   c. They receive afferent nerve terminals and are believed to function as **sensory mechanoreceptors.**

D. **Thick and thin skin** are distinguished on the basis of the **thickness of the epidermis.**

1. **Thick skin** has an epidermis that is **400-600 micrometers (μm) thick.**

   a. It is characterized by a prominent stratum corneum, a well-developed stratum granulosum, and a distinct stratum lucidum.

   b. It lines the palms of the hands and the soles of the feet.

   c. Thick skin **lacks** hair follicles, sebaceous glands, and arrector pili muscles.

2. **Thin skin** has an epidermis that is **75-150 μm thick.**

    a. It has a less prominent stratum corneum than thick skin and generally lacks a stratum granulosum and stratum lucidum, although it contains individual cells that are similar to the cells of these layers.

    b. Thin skin is present over most of the body and contains hair follicles, sebaceous glands, and arrector pili muscles.

III. Dermis.
The dermis is the layer of the skin underlying the epidermis. It is of **mesodermal origin,** and is composed of dense, irregular connective tissue that contains many **type I collagen fibers** and networks of thick **elastic fibers.** Although it is divided into a **superficial** papillary layer and a **deeper,** more extensive reticular layer, no distinct boundary exists between these layers (see Table 14.1).

    A. The **dermal papillary layer** is uneven and forms **dermal papillae,** which interdigitate with the epidermal downgrowths (interpapillary pegs) and ridges of the overlying epidermis. This dermal layer is composed of thin, loosely arranged fibers and cells, and contains capillary loops and **Meissner corpuscles,** which are fine-touch receptors.

    B. The **dermal reticular layer** constitutes the major portion of the dermis. It is composed of dense **bundles of collagen fibers** and thick **elastic fibers** and may contain **pacinian corpuscles** (pressure receptors) and **Krause end-bulbs** (cold receptors) in its deeper aspects.

IV. Glands in the Skin

    A. Eccrine sweat glands (Figure 14.2) are **simple coiled tubular glands** consisting of a secretory unit and a single duct. These glands are present in skin throughout the body.

        1. The **secretory unit of eccrine sweat glands** is embedded in the dermis and is composed of three cell types.

        a. Dark cells line the lumen of the gland and contain many mucinogen-rich secretory granules.

        b. Clear cells underlie the dark cells, are rich in mitochondria and glycogen, and contain intercellular canaliculi that extend to the lumen of the gland. These cells secrete a watery, electrolyte-rich material.

        c. Myoepithelial cells lie scattered in an incomplete layer beneath the clear cells. They contract and aid in expressing the gland's secretions into the duct.

        2. The **duct of eccrine sweat glands** is narrow and lined by **stratified cuboidal epithelial cells,** which contain many keratin filaments and have a prominent terminal web. The cells forming the external (basal) layer of the duct have many mitochondria and a prominent nucleus.

        a. The duct leads from the secretory unit through the superficial portions of the dermis to penetrate an **interpapillary peg** of the epidermis and spiral through all of its layers to deliver sweat to the outside.

        b. As the secreted material passes through the duct, its cells reabsorb some electrolytes and excrete other substances (such as urea, lactic acid, ions, and certain drugs).

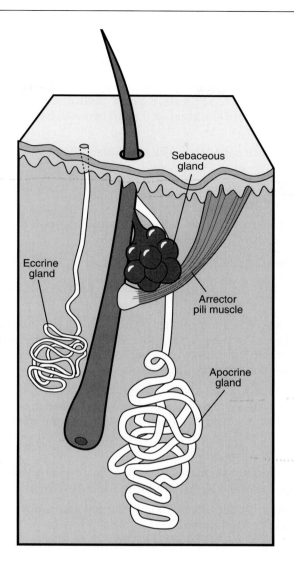

Figure 14.2. A hair follicle, arrector pili muscle, and sebaceous gland, as well as two coiled glands: an eccrine sweat gland and an apocrine gland. Note that the duct of the apocrine gland empties into the hair follicle, but the sweat gland duct empties onto the surface of the epidermis.

    B. Apocrine sweat glands include the **large, specialized sweat glands** located in various areas of the body (e.g., axilla, areola of the nipple, perianal region) and the **ceruminous (wax) glands** of the external auditory canal.

        1. These glands do not begin to function until puberty and are **responsive to hormonal influences.**

        2. Their large coiled secretory units are enveloped by scattered myoepithelial cells.

        3. These glands empty their viscous secretions into hair follicles at a location superficial to the entry of sebaceous gland ducts.

4. Although the term **apocrine** implies that a portion of the cytoplasm becomes part of the secretion, electron micrographs have shown that the **cytoplasm does not become part of the secretions** of apocrine sweat glands.

C. **Sebaceous glands** (see Figure 14.2) are **branched acinar glands** having a lobular appearance. The clustered acini of one sebaceous gland empty into a single short duct.

   1. The duct empties into the neck of a hair follicle.

   2. Sebaceous glands are embedded in the dermis over most of the body's surface but are absent from the palms and soles. They are most abundant on the face, forehead, and scalp.

   3. These **holocrine glands** release **sebum** (composed of an oily secretion and degenerating epithelial cells).

V. Hair Follicle and Arrector Pili Muscle

A. A **hair follicle** is an **invagination of the epidermis** extending deep into the dermis.

   1. The **hair shaft** is a long, slender filament that is located in the center of the follicle and extends above the surface of the epidermis. It consists of an inner **medulla, cortex,** and outer **cuticle of the hair.** At its deep end it is continuous with the **hair root.**

   2. The **hair bulb** is the terminal expanded region of the hair follicle in which the hair is rooted. It is deeply indented by a **dermal papilla,** which contains a capillary network necessary for sustaining the follicle. The hair bulb contains cells that form the internal root sheath and medulla of the hair shaft.

   3. The **internal root sheath** lies deep to the entrance of the sebaceous gland. It is composed of the **Henle layer,** the **Huxley layer,** and the **cuticle.**

   4. The **external root sheath** is a direct continuation of the stratum malpighii of the epidermis.

   5. The **glassy membrane** is a **noncellular layer** and represents a thickening of the basement membrane. It separates the hair follicle from the surrounding dermal sheath.

B. The **arrector pili muscle** attaches at an **oblique angle to the dermal sheath** surrounding a hair follicle.

   1. It extends superficially to underlie sebaceous glands, passing through the reticular layer of the dermis and **inserting into the papillary layer** of the dermis.

   2. The contraction of this **smooth muscle** elevates the hair and is responsible for formation of "goose bumps," caused by depressions of the skin where the muscle attaches to the papillary layer of the dermis.

VI. Nails. Nails are located on the distal phalanx of each finger and toe.

A. Nails are hard keratinized plates that rest on a bed of epidermis.

B. At the proximal end, each is covered by a fold of epidermis, called the **cuticle** or **eponychium,** which corresponds to the stratum corneum. The cuticle overlies the crescent-shaped, whitish **lunula.**

C. At the distal (free) edge, each is underlain by the **hyponychium,** which also is composed of stratum corneum.

D. Nails grow as the result of mitoses of cells in the matrix of the **nail root.**

VII. Clinical Considerations

A. Epidermolysis bullosa is a group of **hereditary** diseases of the skin characterized by **blister formation** following minor trauma. These diseases are caused by **defects in the keratinocyte intermediate filaments** that provide mechanical stability and in the **anchoring fibrils** that attach the epidermis to the dermis.

B. Warts (verrucae) are common **skin lesions** caused by a **virus.**

  1. They may occur anywhere on the skin (or oral mucosa), but are most frequent on the dorsal surfaces of the hands, often **close to the nails.**

  2. Histologic features of warts include marked epidermal hyperplasia, eosinophilic cytoplasmic inclusions, and deeply basophilic nuclei. By electron microscopy, many intranuclear viral particles can be observed in the keratinocytes.

C. Skin cancers commonly originate from cells in the epidermis. These cancers usually can be treated successfully if they are diagnosed early and surgically removed.

  1. Basal cell carcinoma arises from basal keratinocytes.

  2. Squamous cell carcinoma arises from cells of the stratum spinosum.

D. Malignant melanoma is a form of skin cancer that can be life-threatening.

  1. This form of cancer originates from **melanocytes** that divide, transform, and invade the dermis and then enter the lymphatic and circulatory systems, **metastasizing** to a wide variety of organs.

  2. Treatment involves **surgical removal** of the skin lesion and regional lymph nodes. Subsequent **chemotherapy** also is required because of the extensive metastases.

  3. Malignant melanoma represents about 3% of all cancers.

E. Keloids are swellings in the skin that result from increased collagen formation in hyperplastic scar tissue. They are most prevalent in African-Americans.

Review Test

Directions: Each of the numbered items or incomplete statements in this section is followed by answers or by completions of the statement. Select the ONE lettered answer or completion that is BEST in each case.

1. Intercellular bridges are characteristic of which of the following layers of the epidermis?

(A) Stratum granulosum
(B) Stratum lucidum
(C) Stratum corneum
(D) Stratum spinosum
(E) Stratum basale

2. Which of the following statements concerning the stratum granulosum is true?

(A) It contains melanosomes.
(B) It lies superficial to the stratum lucidum.
(C) It is the thickest layer of the epidermis in thick skin.
(D) It contains keratohyalin granules.

3. Which of the following statements about Langerhans cells is true?

(A) They commonly are found in the dermis.
(B) They function as sensory mechanoreceptors.
(C) They function as receptors for cold.
(D) They play an immunologic role in the skin.
(E) They are of epithelial origin.

4. Meissner corpuscles are present in which of the following regions of the skin?

(A) Dermal reticular layer
(B) Dermal papillary layer
(C) Hypodermis
(D) Stratum basale
(E) Epidermal ridges

5. Which of the following statements concerning thin skin is true?

(A) It does not contain sweat glands.
(B) It lacks a stratum corneum.
(C) It is less abundant than thick skin.
(D) It contains hair follicles.

6. Which of the following statements about eccrine sweat glands is true?

(A) They are absent in thick skin.
(B) They are holocrine glands.

(C) They have a narrow duct lined by a stratified cuboidal epithelium.
(D) They secrete an oily material called sebum.

7. Which of the following statements about hair follicles is true?

(A) They are always associated with an eccrine sweat gland.
(B) They are present in thin skin but not in thick skin.
(C) Their associated arrector pili muscle is composed of striated fibers.
(D) Their hair shaft inserts into the papillary layer of the epidermis.
(E) They do not extend into the dermis.

8. Which of the following statements concerning skin melanocytes is true?

(A) They synthesize a pigment that protects against damage caused by ultraviolet (UV) radiation.
(B) They are located only in the dermis.
(C) They produce keratinohyalin granules.
(D) They may give rise to basal cell carcinoma.

9. Which of the following statements concerning sebaceous glands is true?

(A) They do not begin to function until puberty.
(B) They employ the mechanism of holocrine secretion.
(C) They are present in thick skin.
(D) They secrete only in response to hormones.
(E) They produce a watery enzyme-rich secretion.

10. Which of the following is an appendage of skin?

(A) Meissner corpuscle
(B) Langerhans cell
(C) Krause end bulb
(D) Nail
(E) Pacinian corpuscle

Answers and Explanations

1-D. Observations with an electron microscope show that "intercellular bridges" are associated with desmosomes (maculae adherentes), linking the processes of adjacent cells in the stratum spinosum. Desmosomes also link cells within the other epidermal layers, but these cells do not form processes characteristic of bridges. The keratinocytes of the stratum basale also contain hemidesmosomes, which attach the cells to the underlying basal lamina.

2-D. The stratum granulosum contains a number of dense keratohyalin granules but not melanosomes. It lies just deep to the stratum lucidum and is a relatively thin layer in the epidermis of thick skin.

3-D. Langerhans cells in the epidermis function as antigen-presenting cells by trapping antigens that penetrate the epidermis and transporting them to regional lymph nodes, where they are presented to T lymphocytes. In this way, these cells assist in the immune defense of the body. They originate in the bone marrow and do not arise from epithelium.

4-B. Meissner corpuscles are encapsulated nerve endings present in dermal papillae, which are part of the papillary layer of the dermis. These corpuscles function as receptors for fine touch.

5-D. In contrast to thick skin, which lacks hair follicles, thin skin contains many of them.

6-C. Eccrine sweat glands are simple coiled tubular glands that have a duct lined by a stratified cuboidal epithelium. They are found in both thick and thin skin and are classified as merocrine glands, meaning they release only their secretory product, which does not include cells or portions of cells.

7-B. Hair follicles are present only in thin skin. They are associated with sebaceous glands and arrector pili smooth muscle bundles.

8-A. Melanocytes are present in the stratum basale of the epidermis. They synthesize melanin pigment and transfer it to keratinocytes to protect against damage caused by ultraviolet (UV) radiation. Melanocytes sometimes give rise to a form of skin cancer called malignant melanoma.

9-B. Sebaceous glands produce sebum, an oily material, and release it into the upper shaft of the hair follicle by a mechanism called holocrine secretion (which means the product and cellular debris are both released from the gland).

10-D. The nail is one appendage of the skin. Other skin appendages are hair follicles, sweat glands, and sebaceous glands.

15

Respiratory System

I. Overview—The Respiratory System

A. The respiratory system includes the **lungs** and a series of **airways** that connect the lungs to the external environment.

B. The respiratory system can be functionally divided into two major divisions: a **conducting portion,** consisting of airways that deliver air to the lungs, and a **respiratory portion,** consisting of structures within the lungs in which oxygen in the inspired air is exchanged for carbon dioxide in the blood.

C. The components of the respiratory system possess characteristic lining epithelia, supporting structures, glands, and other features, which are summarized in Table 15.1.

II. Conducting Portion of the Respiratory System. This portion of the respiratory system includes the nose, nasopharynx, larynx, trachea, bronchi, and bronchioles down to and including the terminal bronchioles. These structures **warm, moisten,** and **filter the air** before it reaches the sites where gas exchange occurs.

A. Nasal cavity

1. The **nares** are the nostrils; their outer portions are lined by **thin skin.** They open into the vestibule.

2. The **vestibule** is the first portion of the nasal cavity, where the epithelial lining becomes **nonkeratinized.** Posteriorly, the lining changes to **respiratory epithelium** (pseudostratified ciliated columnar epithelium with goblet cells).

a. The vestibule contains **vibrissae** (thick, short hairs), which filter large particles from the inspired air.

b. It has a lamina propria that is **vascular** (many venous plexuses) and contains **seromucous glands.**

3. Olfactory epithelium

a. Overview

(1) The olfactory epithelium is located in the roof of the nasal cavity, on either side of the nasal septum and on the superior nasal conchae.

(2) It is a tall **pseudostratified columnar epithelium** consisting of olfactory cells, supporting (sustentacular) cells, and basal cells.

(3) It has a lamina propria that contains many **veins** and **unmyelinated nerves** and houses **Bowman glands.**

b. Olfactory cells are **bipolar nerve cells** characterized by a bulbous apical projection **(olfactory vesicle)** from which several modified cilia extend.

   (1) Olfactory cilia (olfactory hairs):

   (a) Are very **long, nonmotile cilia** that extend over the surface of the olfactory epithelium. Their proximal third contains a typical **9 + 2 axoneme pattern,** but their distal two thirds are composed of 9 peripheral **singlet** microtubules surrounding a central pair of microtubules.

   (b) Act as receptors for odor

   (2) Supporting (sustentacular) cells:

   (a) Possess nuclei that are more apically located than those of the other two cell types

   (b) Have many **microvilli** and a prominent **terminal web** of filaments

   (3) Basal cells:

   (a) Rest on the basal lamina but do not extend to the surface

   (b) Form an incomplete layer of cells

   (c) Are believed to be **regenerative** for all three cell types

   (4) Bowman glands produce a **thin, watery secretion** that is released onto the olfactory epithelial surface via narrow ducts. Odorous substances dissolved in this watery material are detected by the olfactory cilia. The secretion also flushes the epithelial surface, preparing the receptors to receive new odorous stimuli.

B. Nasopharynx

1. The nasopharynx is the posterior continuation of the nasal cavities and becomes continuous with the oropharynx at the level of the soft palate.

2. It is lined by **respiratory epithelium;** the oropharynx and laryngopharynx are lined by **stratified squamous nonkeratinized epithelium.**

3. The lamina propria of the nasopharynx, located beneath the respiratory epithelium, contains **mucous** and **serous glands** as well as an abundance of lymphoid tissue, including the **pharyngeal tonsil.** When the pharyngeal tonsil is inflamed, it is called an **adenoid.**

C. Larynx

1. **Overview**

   a. The larynx connects the pharynx with the trachea.

   b. The wall of the larynx is supported by **hyaline cartilages** (thyroid, cricoid, and lower part of arytenoids) and **elastic cartilages** (epiglottis, corniculate, and tips of arytenoids).

   c. The wall also possesses **striated muscle,** connective tissue, and **glands.**

(Text continues on page 200)

Table 15.1. Comparison of Respiratory System Components

Division	Skeleton	Glands	Epithelium	Ciliated Cells	Goblet Cells	Special Features
Nasal cavity						
Vestibule	Hyaline cartilage	Sebaceous and sweat glands	Stratified squamous keratinized	No	No	Vibrissae
Respiratory	Bone and hyaline cartilage	Seromucous	Pseudostratified ciliated columnar	Yes	Yes	Large venous plexuses
Olfactory	Nasal conchae (bone)	Bowman glands	Pseudostratified ciliated columnar (tall)	Yes	No	Bipolar olfactory cells, sustentacular cells, basal cells, nerve fibers
Nasopharynx	Muscle	Seromucous	Pseudostratified ciliated columnar	Yes	Yes	Pharyngeal tonsil, entrance of eustachian tube
Larynx	Hyaline and elastic cartilage	Mucous and seromucous	Stratified squamous nonkeratinized and pseudostratified ciliated columnar	Yes	Yes	Vocal cords, striated muscle (vocalis), epiglottis
Trachea and primary bronchi	C-shaped hyaline cartilage "rings"	Mucous and seromucous	Pseudostratified ciliated columnar	Yes	Yes	Trachealis (smooth) muscle, elastic lamina, two mucous cell types, short cells, diffuse endocrine cells

Intrapulmonary bronchi	Plates of hyaline cartilage	Seromucous	Pseudostratified ciliated columnar	Yes	Yes	Two helically oriented ribbons of smooth muscle
Primary bronchioles	Smooth muscle	None	Simple ciliated columnar to simple cuboidal	Yes	Only in larger ones	Clara cells
Terminal bronchioles	Smooth muscle	None	Simple cuboidal	Some	None	Clara cells
Respiratory bronchioles	Some smooth muscle	None	Simple cuboidal except where interrupted by alveoli	Some	None	Occasional alveoli, Clara cells
Alveolar ducts	Smooth muscle at alveolar openings, some reticular fibers	None	Simple squamous	None	None	Linear structure formed of adjacent alveoli, type I and II pneumocytes, alveolar macrophages
Alveoli	Reticular fibers, elastic fibers at alveolar openings	None	Simple squamous	None	None	Type I and II pneumocytes, alveolar macrophages

Modified with permission from Gartner LP, Hiatt JL: *Color Atlas of Histology*, 2nd ed. Baltimore, Williams & Wilkins, 1994, p 240.

   2. The **vocal cords** consist of skeletal muscle (the **vocalis muscle**), the **vocal ligament** (formed by a band of elastic fibers), and a covering of **stratified squamous nonkeratinized epithelium.**

     a. Contraction of the laryngeal muscles changes the size of the opening between the vocal cords, which affects the pitch of the sounds caused by air passing through the larynx.

     b. Inferior to the vocal cords, the lining epithelium changes to **respiratory epithelium,** which lines air passages down through the trachea and intrapulmonary bronchi.

   3. Vestibular folds (false vocal cords) lie superior to the vocal cords.

     a. These folds of loose connective tissue contain glands, lymphoid aggregations, and fat cells.

     b. They are covered by **stratified squamous nonkeratinized epithelium.**

 D. Trachea and extrapulmonary (primary) bronchi

   1. Overview

     a. The walls of these structures are supported by **C-shaped hyaline cartilages** (C-rings), whose open ends face posteriorly. Smooth muscle (**trachealis muscle** in the trachea) extends between the open ends of these cartilages.

     b. Dense **fibroelastic** connective tissue is located between adjacent C-rings, which permits the elongation of the trachea during inhalation.

   2. Mucosa

     a. The **respiratory epithelium** in the trachea possesses the following cell types.

       (1) Ciliated cells:

         (a) Have **long, actively motile cilia,** which beat toward the mouth

         (b) Move inhaled particulate matter trapped in mucus toward the oropharynx, thus protecting the delicate lung tissue from damage

         (c) Also possess **microvilli**

       (2) Mature goblet cells are goblet-shaped and are filled with **large mucinogen droplets,** which are secreted to trap inhaled particles.

       (3) Small mucous granule cells:

         (a) Contain varying numbers of **small mucous granules**

         (b) Are sometimes called "brush" cells because of their many uniform **microvilli**

         (c) Actively **divide** and often replace recently desquamated cells

         (d) May represent goblet cells after they have secreted their mucinogen

       (4) Diffuse neuroendocrine cells (DNES cells):

(a) Are also known as **small granule cells,** amine precursor uptake and decarboxylation **(APUD) cells,** or **enteroendocrine cells**

(b) Contain many small granules concentrated in their **basal** cytoplasm

(c) Synthesize different **polypeptide hormones** and **serotonin,** which often exert a local effect on nearby cells and structures **(paracrine regulation).** The peptide hormones may also enter the bloodstream and have an **endocrine effect** on distant cells and structures.

(5) **Short (basal) cells:**

(a) Rest on the basal lamina but do not extend to the lumen; thus this epithelium is pseudostratified

(b) Are able to **divide** and replace the other cell types

b. The **basement membrane** is a very thick layer underlying the epithelium.

c. The **lamina propria** is a thin layer of connective tissue that lies beneath the basement membrane. It contains longitudinally oriented **elastic fibers** separating the lamina propria from the submucosa.

3. The **submucosa** is a connective tissue layer containing many **seromucous glands.**

4. The **adventitia** contains **C-shaped hyaline cartilages** and forms the outermost layer of the trachea.

E. Intrapulmonary bronchi (secondary bronchi)

1. Intrapulmonary bronchi arise from subdivisions of the primary bronchi.

2. They divide many times and give rise to **lobar** and segmental bronchi.

3. Their walls contain **irregular cartilage plates.**

4. They are lined by **respiratory epithelium.**

5. **Spiraling smooth muscle bundles** separate the lamina propria from the submucosa, which contains **seromucous glands.**

F. Primary and terminal bronchioles lack glands in their submucosa. Their walls contain **smooth muscle** rather than cartilage plates.

1. **Primary bronchioles**

a. Primary bronchioles have a diameter of 1 millimeter (mm) or less.

b. They are lined by epithelium that varies from **ciliated columnar with goblet cells** in the larger airways to **ciliated cuboidal with Clara cells** in the smaller passages.

c. They divide to form several terminal bronchioles after entering the **pulmonary lobules.**

2. **Terminal bronchioles** (Figure 15.1)

a. Terminal bronchioles are the **most distal part of the conducting portion** of the respiratory system.

b. They have a diameter of less than 0.5 mm.

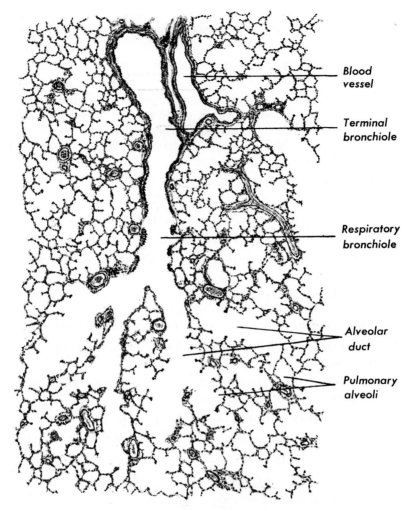

Figure 15.1. Drawing of a section of the lung. (Reprinted with permission from Kelly DE, Wood RL, Enders AC: *Bailey's Textbook of Microscopic Anatomy*, 18th ed. Baltimore, Williams & Wilkins, 1984, p 631.)

    c. They are lined by a **simple cuboidal epithelium** that contains mostly **Clara cells,** some ciliated cells, and no goblet cells.

    d. Function. Clara cells have the following functions:

        (1) Clara cells **divide,** and some of them differentiate to form ciliated cells.

        (2) They **secrete glycosaminoglycans.**

        (3) They **metabolize airborne toxins,** a process that is carried out by cytochrome P-450 enzymes present in their abundant smooth endoplasmic reticulum (SER).

III. Overview—Respiratory Portion of the Respiratory System. This portion of the respiratory system includes the respiratory bronchioles, alveolar ducts, alveolar sacs, and alveoli, all of which are located within the lung (see Figure 15.1). The **exchange of gases** takes place in this portion of the respiratory system.

A. Respiratory bronchioles

1. The respiratory bronchioles mark the transition from the conducting to the respiratory portion of the respiratory system.

2. They are lined by a **simple cuboidal epithelium** consisting of **Clara cells** and some **ciliated cells,** except where their walls are interrupted by **alveoli,** the sites where gas exchange occurs.

B. Alveolar ducts

1. Alveolar ducts are **linear passageways** continuous with the respiratory bronchioles.

2. Their walls consist of **adjacent alveoli,** which are separated from one another only by an **interalveolar septum.**

3. They are the most distal portion of the respiratory system to contain **smooth muscle,** which is present in their walls at the openings of adjacent alveoli.

4. They are lined by **type II pneumocytes** and the highly **attenuated simple squamous epithelium** of **type I pneumocytes**.

C. Alveolar sacs are expanded outpouchings of numerous alveoli located at the distal ends of alveolar ducts.

D. Alveoli

1. **Overview**

   a. Alveoli are pouch-like evaginations [about 200 micrometers (μm) in diameter] present in the walls of respiratory bronchioles, in alveolar ducts, and in alveolar sacs.

   b. They have thin walls, across which oxygen and carbon dioxide can diffuse between the air and the blood.

   c. They are separated from each other by **interalveolar septae,** which may contain one or more **alveolar pores** (pores of Kohn). These pores permit equalization of pressure between alveoli.

   d. They are rimmed by **elastic fibers** at their openings and are supported by many **reticular fibers** in their walls.

   e. They are lined by a **highly attenuated simple squamous epithelium** composed of type I and type II **pneumocytes.**

2. **Alveolar cells**

   a. **Type I pneumocytes (type I alveolar cells):**

      (1) Cover about 95% of the alveolar surface and form part of the blood-gas barrier where exchange of O_2 and CO_2 occurs

      (2) Have an extremely **thin cytoplasm** that may be less than 80 nanometers (nm) in thickness

      (3) Form **tight junctions** with adjacent cells

      (4) May have **phagocytic** capabilities

      (5) Are **not** able to divide

   b. Type II pneumocytes (type II alveolar cells; great alveolar cells; granular pneumocytes; septal cells):

     (1) Are **cuboidal** in shape and are most often **located near septal intersections**

     (2) Bulge into the alveolus and have a free surface that contains short **microvilli** around their peripheral borders

     (3) Are able to **divide** and **regenerate** both types of alveolar pneumocytes

     (4) Form **tight junctions** with adjacent cells

     (5) Synthesize **pulmonary surfactant,** which is stored in cytoplasmic **lamellar bodies**

       (a) Structure—pulmonary surfactant. Pulmonary surfactant consists of **phospholipids** and at least four different **proteins.** It forms **tubular myelin** (a network-like configuration) when it is first released from lamellar bodies; it then spreads to produce a **monomolecular film** over the alveolar surface, forming a **lower aqueous phase** and a **superficial lipid phase.**

       (b) Function—pulmonary surfactant. Pulmonary surfactant **reduces the surface tension** of the alveolar surface, permitting the alveoli to expand easily during inspiration and preventing alveolar collapse during expiration.

   c. Alveolar macrophages (alveolar phagocytes; dust cells):

     (1) Are the principal mononuclear **phagocytes** of the alveolar surface

     (2) Remove inhaled dust, bacteria, and other particulate matter trapped in the pulmonary surfactant, thus providing a vital line of defense in the lungs

     (3) Migrate to the bronchioles after becoming filled with debris. From there, they are carried via **ciliary action** to the upper airways, eventually reaching the oropharynx, where they are swallowed or expectorated

     (4) May also exit by migrating into the interstitium and leaving via lymphatic vessels

E. Interalveolar septum

  1. The interalveolar septum is the wall, or partition, between two adjacent alveoli.

  2. It is bounded on its outer surfaces by the extremely thin simple squamous epithelium lining the alveoli.

  3. It contains many **elastic and reticular fibers** in its thicker regions.

  4. It houses **continuous capillaries** in its central (interior) region.

  5. It accommodates the **blood-gas barrier,** which separates the alveolar airspace from the capillary lumen.

  a. Structure—blood-gas barrier

      (1) The **thinnest regions** of the barrier are 0.2 μm or less in thickness and consist of the following layers:

         (a) Type I pneumocytes and layer of surfactant lining the alveolar airspace

         (b) Fused basal laminae of type I pneumocytes and capillary endothelial cells

         (c) Endothelium of the continuous capillaries within the interalveolar septum

      (2) Thicker regions of the barrier measure about 0.5 μm across and have an **interstitial area** interposed between the two basal laminae, which are not fused.

    b. Function—blood-gas barrier. The blood-gas barrier permits the **diffusion of gases** between the alveolar airspace and the blood. **Oxygen** passes from the alveolus into the capillary, and **carbon dioxide** passes from the capillary blood into the alveolus.

IV. Lung Lobules

    A. Lung lobules vary greatly in size and shape, but each has an apex directed toward the pulmonary hilum and a wider base directed outward.

    B. Each lobule contains a **single primary bronchiole,** which enters at the apex and branches to form five to seven terminal bronchioles. The terminal bronchioles, in turn, divide, ultimately giving rise to alveoli at the base of the lobule.

V. Pulmonary Vascular Supply

    A. Pulmonary artery

      1. The pulmonary artery carries blood to the lungs to be **oxygenated.**

      2. It enters the root of each lung and extends branches along the divisions of the bronchial tree.

      3. It enters lung lobules, where **its branches follow the bronchioles.**

    B. Pulmonary veins

      1. In lung lobules, pulmonary veins run in the intersegmental connective tissue, **separated from the arteries.**

      2. After leaving the lobules, the pulmonary veins come close to divisions of the bronchial tree and **run parallel to branches of the pulmonary artery** as they accompany bronchi to the root of the lung.

    C. Bronchial arteries and veins

      1. Bronchial arteries and veins provide nutrients to and remove wastes from the nonrespiratory portions of the lung (bronchi, bronchioles, interstitium, and pleura).

      2. They follow the branching pattern of the bronchial tree and form anastomoses with the pulmonary vessels near capillary beds.

VI. Pulmonary Nerve Supply. The pulmonary nerve supply consists primarily of **autonomic fibers to the smooth muscle of bronchi and bronchioles.** Axons are also present in the thicker parts of the interalveolar septae.

    A. Parasympathetic stimulation causes **contraction** of pulmonary smooth muscle.

    B. Sympathetic stimulation causes **relaxation** of pulmonary smooth muscle and can be mimicked by certain drugs that cause dilation of bronchi and bronchioles.

VII. Clinical Considerations

    A. Hyaline membrane disease (respiratory distress syndrome)

        1. Hyaline membrane disease is frequently observed in **premature infants who lack adequate amounts of pulmonary surfactant.**

        2. It is characterized by **labored breathing,** which results from difficulty expanding the alveoli due to a high alveolar surface tension.

        3. If detected before birth, hyaline membrane disease can often be prevented by the administration of **glucocorticoids,** which induce synthesis of surfactant.

    B. Asthma

        1. Asthma is marked by widespread **constriction of smooth muscle in the bronchioles,** causing a decrease in their diameter.

        2. It is associated with **extremely difficult expiration** of air, **accumulation of mucus** in the passageways, and **infiltration of inflammatory cells.**

        3. It is treated with drugs that relax the bronchiolar smooth muscle cells and dilate the passageways.

    C. Emphysema

        1. Emphysema results from **destruction of alveolar walls** and formation of large cyst-like sacs, reducing the surface area available for gas exchange.

        2. It is marked by **decreased elasticity** of the lungs, which are unable to recoil adequately during expiration. In time, the lungs expand and enlarge the thoracic cavity ("barrel chest").

        3. Emphysema is associated with exposure to **cigarette smoke and other substances that inhibit α_1-antitrypsin,** a protein that normally protects the lungs from the action of **elastase** produced by alveolar macrophages.

        4. It can be a hereditary condition resulting from a defective α_1-antitrypsin. In such cases, gene therapy with recombinant α_1-antitrypsin is being used in an effort to correct the problem.

Review Test

Directions: Each of the numbered items or incomplete statements in this section is followed by answers or by completions of the statement. Select the ONE lettered answer or completion that is BEST in each case.

1. Characteristics of olfactory epithelium include which one of the following?

(A) It is located in the inferior region of the nasal cavity.
(B) It is classified as simple columnar.
(C) It has an underlying lamina propria containing mucous glands.
(D) It has modified cilia, which act as receptors for odor.

2. Which of the following statements concerning terminal bronchioles is true?

(A) They are part of the conducting portion of the respiratory system.
(B) They function in gas exchange.
(C) They do not contain ciliated cells.
(D) They have cartilage plates present in their walls.

3. The trachea possesses which one of the following components?

(A) Irregular cartilage plates in its wall
(B) Skeletal muscle in its wall
(C) An epithelium containing only two cell types
(D) A thick basement membrane underlying its epithelium

4. Which of the following statements concerning respiratory bronchioles is true?

(A) No gas exchange occurs in them.
(B) They do not have alveoli forming part of their wall.
(C) They contain goblet cells in their lining epithelium.
(D) They are included in the conducting portion of the respiratory system.
(E) Ciliated cells comprise a portion of their lining epithelium.

5. True statements about asthma include which one of the following?

(A) It is due to a loss of lung elasticity.
(B) It eventually causes the lungs to expand and leads to a "barrel chest".

(C) It is associated with difficulty expiring air from the lungs.
(D) It may be helped by gene therapy using recombinant α_1-antitrypsin.

6. Which of the following statements concerning alveolar macrophages is true?

(A) They secrete α_1-antitrypsin.
(B) They secrete elastase.
(C) They originate from blood neutrophils.
(D) They may play a role in causing hyaline membrane disease.

Questions 7-10

The response options for the next four items are the same. Each item will state the number of options to select. Choose exactly this number.

(A) Asthma
(B) Hyaline membrane disease
(C) Emphysema

For each of the following characteristics, select the disorder to which it applies.

7. May be treated with anti-elastase (α_1-antitrypsin) in some cases (**SELECT 1 DISORDER**)

8. Is caused by inadequate amounts of pulmonary surfactant (**SELECT 1 DISORDER**)

9. Is associated with a "barrel chest" (**SELECT 1 DISORDER**)

10. Is frequently treated with glucocorticoids (**SELECT 1 DISORDER**)

Answers and Explanations

1-D. The olfactory epithelium possesses nonmotile cilia, which act as receptors for odor. They are extensions of the bipolar nerve cells that form part of this tall pseudostratified epithelium located in the roof of the nasal cavity. Bowman glands, which lie in the lamina propria beneath this epithelium, produce a watery secretion, which moistens the olfactory surface.

2-A. Terminal bronchioles are the most distal components of the conducting portion of the respiratory system. They lack alveoli and thus do not function in gas exchange. They are lined by an epithelium composed of two cell types: secretory (Clara) cells and ciliated cells. Cartilage is not present in bronchioles.

3-D. The pseudostratified ciliated columnar epithelium lining the trachea rests on a thick basement membrane and contains five cell types. The trachea possesses C-shaped cartilages with smooth muscle (the trachealis) extending between their ends.

4-E. Respiratory bronchioles have alveoli interrupting their walls, so some gas exchange takes place at this level. Their remaining walls are lined by a simple cuboidal epithelium consisting of Clara cells and ciliated cells. Respiratory bronchioles are categorized as part of the respiratory portion of the system.

5-C. Asthma results from the constriction of smooth muscle in the bronchioles, which decreases their diameter and makes the expiration of air very difficult. Mucus accumulates in the airways, and inflammatory cells invade the bronchiolar walls.

6-B. Alveolar macrophages secrete elastase. Normally, α_1-antitrypsin (a serum protein) interacts with elastase, thereby protecting the lung against damage that may lead to emphysema. Alveolar macrophages, like all macrophages, arise from blood monocytes, and they are unrelated to the pathogenesis of hyaline membrane disease.

7-C. Hereditary forms of emphysema are now being treated with recombinant α_1-antitrypsin, which has anti-elastase activity.

8-B. Hyaline membrane disease is caused by inadequate amounts of pulmonary surfactant.

9-C. A loss of lung elasticity in emphysema makes it difficult for the lungs to recoil normally during expiration. The lungs and thoracic cavity enlarge, producing a "barrel chest."

10-B. Glucocorticoids, which stimulate synthesis of pulmonary surfactant, are often used to prevent or alleviate hyaline membrane disease.

16

Digestive System: Oral Cavity and Alimentary Tract

I. Overview—The Digestive System

A. The digestive system comprises the **oral region** and **alimentary canal** (**esophagus, stomach, small** and **large intestines**), and several **extrinsic glands**.

B. It consists of a hollow tube (highly modified in the oral cavity) of varying diameter, composed of a **mucosa, submucosa, muscularis externa,** and **serosa** (or **adventitia**).

C. Function. The digestive system secretes **enzymes** and **hormones** that function in ingestion, digestion, and absorption of nutrients as well as in the elimination of indigestible materials.

II. Oral Region.
The oral region includes the **lips, palate, teeth** and **associated structures,** and **tongue,** as well as the major salivary glands and lingual tonsils. It is covered in most places by a **stratified squamous epithelium** whose **epithelial ridges** interdigitate with tall **connective tissue papillae (connective tissue ridges)** of the subjacent connective tissue. The epithelial ridges and the connective tissue ridges are known collectively as the **rete apparatus**.

A. The **lips** are divided into an **external region,** a **vermilion zone,** and an **internal region.** The first two regions are covered by stratified squamous keratinized epithelium, whereas the internal region is lined by stratified squamous nonkeratinized epithelium.

1. A dense irregular connective tissue core envelops skeletal muscle.

2. Sebaceous glands, sweat glands, and **hair follicles** are present in the external region; **minor salivary glands** in the internal region; and occasional, nonfunctional sebaceous glands in the internal region and vermilion zone.

B. The **palate** is divided into an anterior **hard palate** (possessing a bony shelf in its core) and a posterior **soft palate** (possessing skeletal muscle in its core). The palate separates the nasal from the oral cavity. Therefore, the palate has a nasal aspect and an oral aspect. The entire palate is lined by **pseudostratified ciliated columnar epithelium** (respiratory epithelium) on its **nasal aspect.**

1. The **hard palate** is lined on its oral aspect by stratified squamous **parakeratinized** to stratified squamous **keratinized** epithelium. It contains adipose tissue (anteriorly) and minor mucous salivary glands (posteriorly) in the oral aspect of its mucosa.

2. The **soft palate** is lined on its oral aspect by stratified squamous **nonkeratinized** epithelium. It contains minor mucous salivary glands in the oral aspect of its mucosa.

C. Teeth

1. Overview—teeth

a. Teeth are composed of an internal soft tissue, the **pulp,** and three calcified tissues: **enamel** and **cementum,** which form the surface layer, and **dentin,** which lies between the surface layer and pulp. As in bone, **calcium hydroxyapatite** is the mineralized material in the calcified dental tissues.

b. Teeth contain an enamel-covered **crown,** a cementum-covered root, and a **cervix,** the region where the two surface materials meet.

2. Components—teeth

a. Enamel:

(1) Has a highly calcified matrix with an organic component that is composed mostly of the fibrous, keratin-like protein, **enamelin,** which is elaborated by **ameloblasts** during formation of the crown

(2) Is **acellular after tooth eruption** and therefore cannot repair itself

b. Dentin:

(1) Surrounds the central **pulp chamber** and **pulp (root) canal**

(2) Has a calcified matrix containing **type I collagen fibers**

(3) Is manufactured by **odontoblasts,** which persist and continue to elaborate dentin for the life of the tooth

c. Cementum:

(1) Has a **collagen-containing** calcified matrix, which is produced by cementoblasts

(2) Is **continuously elaborated** after tooth eruption, compensating for the decrease in tooth length resulting from abrasion of the enamel

d. Dental pulp:

(1) Is a gelatinous, richly vascularized connective tissue containing **odontoblasts** in its peripheral layer (closest to the dentin), fibroblasts and mesenchymal cells, and thin types I and III collagen fibers

(2) Contains **afferent nerve fibers.** All sensations from the pulp are interpreted as pain in the central nervous system.

3. Crown formation

a. The crown begins to form 6-7 weeks after conception as a horseshoe-shaped band, the **dental lamina,** which is derived from the oral **epithelium.** A dental lamina develops in each jaw and projects into the underlying **ectomesenchyme.**

    b. The crown forms **before** the root formation begins.

    c. The sequential stages of crown formation are **bud, cap, bell,** and **appositional stages.**

  4. Root formation follows completion of the crown and is accompanied by tooth eruption.

D. Dental supporting structures

  1. The **periodontal ligament** has type I collagen fibers that are arranged in five **principal fiber bundles,** which extend from cementum to bone, suspending the tooth in its **alveolus.**

  2. Gingivae (gums) are covered by stratified squamous **keratinized** (or **parakeratinized**) epithelium; their collagen fibers are also arranged in five **principal fiber bundles.**

  3. Alveolar bone consists of an inner layer **(cribriform plate)** and an outer layer **(cortical plate)** of compact bone with an intervening layer of cancellous bone **(spongiosa).**

E. Tongue

  1. Overview—tongue

    a. The tongue is divided into an anterior two thirds and a posterior one third by the **sulcus terminalis,** whose apex ends in the **foramen cecum.**

    b. Its dorsal surface is covered by stratified squamous **parakeratinized to keratinized** epithelium, whereas its ventral surface is covered by stratified squamous **nonkeratinized** epithelium. Both epithelial surfaces are underlain by a **lamina propria** and **submucosa** of dense irregular collagenous connective tissue.

    c. The tongue possesses a **core of skeletal muscle,** which forms the bulk of the tongue.

  2. Lingual papillae are located on the **dorsal surface of the anterior two thirds** of the tongue.

    a. Filiform papillae are short, narrow, **highly keratinized** structures lacking taste buds.

    b. Fungiform papillae are mushroom-shaped structures interspersed among the filiform papillae; they contain **occasional taste buds.**

    c. Foliate papillae are shallow, longitudinal furrows located on the lateral aspect of the posterior region of the anterior two thirds of the tongue. Their taste buds degenerate shortly after the second year of life.

    d. Circumvallate papillae are 10 to 15 large, circular papillae, each of which is surrounded by a moat-like furrow. They are located just anterior to the **sulcus terminalis** and possess taste buds.

      (1) Taste buds

        (a) Taste buds are **intraepithelial structures** located on the **lateral surfaces** of circumvallate papillae and the **walls** of the surrounding moat-like furrows.

(b) Function. Taste buds perceive salt, sour, bitter, and sweet taste sensations.

(2) Glands of von Ebner are minor salivary glands that deliver their **serous secretion** into the furrow surrounding each papilla, assisting the taste buds in perceiving stimuli.

3. The **muscular core of the tongue** is composed of bundles of **skeletal muscle fibers** arranged in three planes with **minor salivary glands** interspersed among them.

4. **Lingual tonsils** are located on the dorsal surface of the **posterior one third** of the tongue.

III. **Divisions of the Alimentary Canal.** Divisions of the alimentary canal are determined by the histophysiological variations of the layers (Table 16.1).

A. **Esophagus**

1. The esophagus is lined by a **stratified squamous nonkeratinized epithelium.**

2. The lamina propria contains mucus-secreting **esophageal cardiac glands,** and the submucosa contains mucus-secreting **esophageal glands proper.**

3. The **muscularis mucosae** varies in thickness and is composed of a single longitudinal layer of smooth muscle.

4. The upper third of the **muscularis externa** contains **skeletal muscle;** the middle third contains a combination of **smooth and skeletal muscle;** and the lower third contains **smooth muscle.**

5. The esophagus conveys a **bolus** of food from the pharynx into the stomach by **peristaltic activity of the muscularis externa.** Two physiologic **sphincters** (the pharyngoesophageal and the gastroesophageal) in the muscularis externa ensure that the bolus is transported in one direction only, toward the stomach.

B. **Stomach.** The **stomach acidifies** and converts the bolus into a thick, viscous fluid known as chyme. It also produces **digestive enzymes** and **hormones.**

1. **General structure—stomach**

a. The stomach exhibits longitudinal folds of the mucosa and submucosa (called **rugae**), which disappear in the distended stomach.

b. It has many **gastric pits** (foveolae), which are shallowest in the cardia and deepest in the pylorus.

(1) Gastric mucosa

(a) The **simple columnar** epithelium of the gastric mucosa is composed of mucinogen-producing **surface lining cells** (not goblet cells).

(b) The **lamina propria** is a loose connective tissue housing smooth muscle cells, lymphocytes, plasma cells, mast cells, and fibroblasts. It contains **gastric glands.**

(c) The **muscularis mucosae** is composed of a poorly defined inner circular layer, an outer longitudinal layer, and, occasionally, an outermost circular layer of smooth muscle.

(2) Gastric submucosa:

(a) Is composed of dense irregular **collagenous** connective tissue

(b) Contains fibroblasts, mast cells, and lymphoid elements embedded in the connective tissue

(c) Houses **Meissner (submucosal) plexus**

(d) Possesses arterial and venous plexuses that supply and drain the vessels of the mucosa, respectively

(3) Gastric muscularis externa:

(a) Is composed of **three layers** of smooth muscle: an incomplete inner oblique layer; a thick middle circular layer, which forms the **pyloric sphincter;** and an outer longitudinal layer. Auerbach myenteric plexus is located between the middle circular and outer longitudinal smooth muscle layers.

(b) Is responsible for mixing of gastric contents and emptying of the stomach

(c) Is affected by various characteristics of the chyme (e.g., lipid content, viscosity, osmolality, caloric density, and pH), which influence the **emptying rate** of the stomach and the rate of muscle contraction

(4) A serosa covers the external surface of the stomach.

2. **Gastric glands** are simple branched tubular glands located in the lamina propria of the **cardia, fundus,** and **pylorus.** Each gland consists of an **isthmus,** which connects the gland to the base of a gastric pit, a **neck,** and a **base.**

a. **Cells of the fundic glands**

(1) Parietal (oxyntic) cells

(a) These are pyramid-shaped cells concentrated in the upper half of the gland.

(b) They secrete **hydrochloric acid** (HCl) and **gastric intrinsic factor.** The latter is necessary for absorption of vitamin B_{12} in the ileum.

(c) They possess a unique intracellular **tubulovesicular system,** many mitochondria, and secretory **intracellular canaliculi** (deep invaginations of the apical plasma membrane) lined by **microvilli.**

(d) When stimulated to secrete HCl, the number and length of microvilli increase and the complexity of the tubulovesicular system decreases (suggesting that tubulovesicle membranes are incorporated into the intracellular canaliculi, thus lengthening the microvilli).

Table 16.1. Selected Histologic Features of the Alimentary Canal

Region	Epithelium	Lamina Propria	Muscularis Mucosae*	Submucosa	Muscularis Externa†
Esophagus	Stratified squamous	Esophageal cardiac glands	One layer: longitudinal	Collagenous CT Esophageal glands proper	Two layers: inner circular and outer longitudinal
Stomach	Simple columnar (no goblet cells)	Gastric glands	Two (sometimes three layers): inner circular, outer longitudinal, and outermost circular in places	Collagenous CT No glands	Three layers: inner oblique, middle circular, and outer longtiudinal
Small intestine	Simple columnar with goblet cells	Villi, crypts of Lieberkühn, Peyer patches in ileum (extend into submucosa), lymphoid nodules	Two layers: inner circular and outer longitudinal	Fibroelastic CT— Brunner glands in duodenum	Two layers: inner circular and outer longitudinal
Large intestine Cecum and colon	Simple columnar with goblet cells	Crypts of Lieberkühn (lack Paneth cells), lymphoid nodules	Two layers: inner circular and outer longitudinal	Fibroelastic CT No glands	Two layers: inner circular and outer longitudinal, which is modified to form teniae coli

	Epithelium	Lamina Propria	Muscularis Mucosae*	Submucosa	Muscularis Externa†
Rectum	Simple columnar with goblet cells	Crypts of Lieberkühn (fewer but deeper than in colon), lymphoid nodules	Two layers: inner circular and outer longitudinal	Fibroelastic CT No glands	Two layers: inner circular and outer longitudinal
Anal canal	Simple columnar cuboidal (proximally), stratified squamous nonkeratinized (distal to anal valves), stratified squamous keratinized (anus)	Sebaceous glands, circumanal glands, lymphoid nodules, rectal columns of Morgagni (involve entire mucosa), hair follicles (anus)	Two layers: inner circular and outer longitudinal	Fibroelastic CT with large veins, no glands	Two layers: inner circular (which forms internal anal sphincter) and outer longitudinal
Appendix	Simple columnar with goblet cells	Crypts of Lieberkühn (shallow), lymphoid nodules (large, numerous and may extend into the submucosa)	Two layers: inner circular and outer longitudinal	Fibroelastic CT, confluent lymphoid nodules, no glands, fat tissue (sometimes)	Two layers: inner circular and outer longitudinal

CT = connective tissue.
*The muscularis mucosae is composed entirely of smooth muscle throughout the alimentary canal.
†The muscularis externa is composed entirely of smooth muscle in all regions except the esophagus. The upper third of the esophageal muscularis externa is all skeletal muscle, the middle third is a mixture of skeletal and smooth muscle, and the lower third is all smooth muscle.

(2) Chief (zymogenic) cells:

(a) Are pyramid-shaped cells located in the lower half of the **gland**

(b) Secrete **pepsinogen** (a precursor of the enzyme pepsin) and the precursors of two other enzymes, **rennin** and **lipase**

(c) Display an abundance of basally located rough endoplasmic reticulum (RER), a supranuclear Golgi complex, and many apical zymogen (secretory) granules

(3) Mucous neck cells:

(a) Are located in the neck of the gland (and may be able to divide)

(b) Possess short microvilli, apical mucous granules, a prominent Golgi complex, numerous mitochondria, and some basally located RER

(4) Diffuse neuroendocrine cells (DNES cells):

(a) Are also referred to as **enteroendocrine cells** or as **APUD cells** (**a**mine **p**recursor **u**ptake and **d**ecarboxylation cells)

(b) Include more than a dozen different types of cells that house many small hormone-containing granules, usually concentrated in the **basal** cytoplasm. A given enteroendocrine cell is believed to secrete only one hormone (Table 16.2).

(c) Possess an abundance of mitochondria and RER and a moderately well-developed Golgi complex

(5) Regenerative cells are located primarily in the neck and isthmus; they replace all the epithelial cells of the gland, gastric pit, and luminal surface.

b. Cardiac and pyloric glands are different from fundic glands in that they are coiled tubular mucus-secreting glands and lack chief cells.

3. Gastric juice contains water, HCl, mucus, pepsin, lipase, rennin, and electrolytes. It is **very acidic** (pH 2.0) and facilitates the activation of pepsinogen to pepsin, which catalyzes the partial hydrolysis of proteins.

4. Regulation of gastric secretion is effected by neural activity (vagus nerve) and by several hormones.

a. Gastrin, released by enteroendocrine cells in the gastric and duodenal mucosa, together with histamine and acetylcholine, **stimulates HCl secretion.**

b. Somatostatin, produced by enteroendocrine cells of the pylorus and duodenum, inhibits the release of gastrin and thus **indirectly inhibits HCl secretion.**

c. Urogastrone, produced by Brunner glands of the duodenum, and **gastric inhibitory peptide,** produced by enteroendocrine cells in the small intestine, **directly inhibit HCl secretion.**

C. Small intestine

1. Overview—small intestine

a. The small intestine is approximately 7 meters long and has three regions: the **duodenum** (proximal), **jejunum** (medial), and **ileum** (distal).

Table 16.2. Selected Hormones Secreted by Cells of the Alimentary Canal*

Hormone	Cell	Site of Secretion	Physiologic Effect
Cholecystokinin (CCK)	I	Small intestine	Stimulates release of pancreatic enzymes and contraction of gall bladder (with release of bile)
Gastric inhibitory peptide (GIP)	K	Small intestine	Inhibits gastric hydrochloric acid (HCl) secretion
Gastrin	G	Pylorus and duodenum	Stimulates gastric secretion of HCl and pepsinogen
Glicentin	GL	Stomach through colon	Stimulates hepatic glycogenolysis
Glucagon	A	Stomach and duodenum	Stimulates hepatic glycogenolysis
Motilin	Mo	Small intestine	Increases gut motility
Neurotensin	N	Small intestine	Inhibits gut motility, stimulates blood flow to the ileum
Secretin	S	Small intestine	Stimulates bicarbonate secretion by pancreas and biliary tract
Serotonin and substance P	EC	Stomach through colon	Increases gut motility
Somatostatin	D	Pylorus and duodenum	Inhibits nearby enteroendocrine cells
Urogastrone†		Duodenum (Brunner glands)	Inhibits gastric HCl secretion, enhances epithelial cell division
Vasoactive intestinal peptide (VIP)	VIP	Stomach through colon	Increases gut motility, stimulates intestinal ion and water secretion

A = alpha cell-like cell; D = delta cell-like cell; EC = enterochromaffin-like cell; G = gastrin-producing cell; GL = glicentin-producing cell; Mo = motilin-producing cell; N = neurotensin-producing cell; S = secretin-producing cell.
*Some of these hormones also are secreted in other parts of the body and have additional physiologic effects.
†Not produced by diffuse neuroendocrine (DNES) cell.

    b. Function. The small intestine secretes several **hormones;** it continues and largely completes the **digestion** of foodstuffs and **absorbs** the resulting metabolites.

  2. Luminal surface modifications—small intestine. The luminal surface of the small intestine possesses plicae circulares, intestinal villi, and microvilli, which collectively **increase the luminal surface area by a factor of 400 to 600.**

    a. Plicae circulares (valves of Kerckring) are permanent spiral folds of the **mucosa** and **submucosa** that are present in the distal half of the duodenum, the entire jejunum, and proximal half of the ileum. Plicae circulares **increase the surface area two-fold to three-fold.**

    b. Intestinal villi (Figure 16.1) are permanent evaginations that possess, in their connective tissue core (lamina propria), numerous plasma cells and lymphocytes, fibroblasts, mast cells, smooth muscle cells, capillary loops, and a **single lacteal** (blind-ended lymphatic capillary). Villi **increase the surface area ten-fold.**

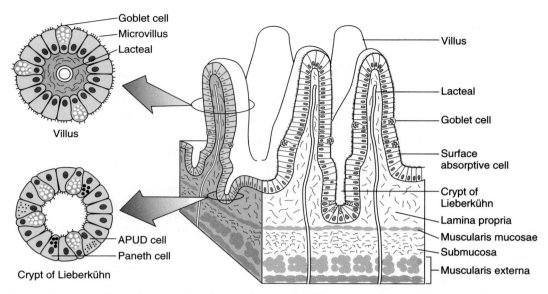

Figure 16.1. Three-dimensional drawing showing the spatial relationship of intestinal villi, crypts of Lieberkühn, and underlying muscularis mucosae, submucosa, and muscularis externa of the small intestine. Note that intestinal villi are evaginations of the epithelium and lamina propria. Each villus contains a single blind-ended lacteal and capillary loop. The submucosa, muscularis externa, and serosa are also depicted. APUD = amine precursor uptake and decarboxylation.

    c. Microvilli of the **apical** surface of the epithelial cells of each villus possess actin filaments that interact with myosin filaments in the terminal web. Microvilli **increase the surface area about twenty-fold.**

3. Mucosa of the small intestine

    a. The **epithelium** of the mucosa of the small intestine is **simple columnar,** composed of goblet cells, surface absorptive cells, and some DNES cells (see Figure 16.1).

        (1) Goblet cells:

            (a) Are **unicellular glands** that produce **mucinogen,** which accumulates in membrane-bounded granules, distending the **apical** region **(theca)** of the cell. After being released, mucinogen becomes hydrated and is thus converted to **mucin,** a thick, viscous substance that acts as a **protective coating** of the epithelial lining of the lumen.

            (b) Have their nucleus and other organelles in the basal region **(stem)** of the cell

            (c) Increase in number from the duodenum to the ileum

        (2) Surface absorptive cells:

            (a) Are tall columnar cells with numerous mitochondria, smooth and RER, and a Golgi complex

            (b) Possess a layer of closely packed microvilli **(striated border)** on their free apical surface

(c) Have a **glycocalyx,** which overlies the microvilli and binds various enzymes (disaccharidases and dipeptidases)

(d) Have well-developed **tight junctions and zonula adherens**

(3) **Diffuse neuroendocrine (DNES) cells** produce and secrete **gastrin, cholecystokinin, gastric inhibitory peptide,** and several other hormones (see Table 16.2).

b. Lamina propria:

(1) Occupies the cores of the villi and the interstices between the numerous **glands (crypts) of Lieberkühn**

(2) Consists of loose connective tissue with lymphoid cells, fibroblasts, mast cells, smooth muscle cells, nerve endings, and **lymphoid nodules**

(3) Also contains **lacteals** (blind-ended lymphatic vessels) and **capillary loops**

(a) **Crypts of Lieberkühn** are simple tubular glands that extend from the intervillous spaces to the muscularis mucosae of the intestine. They are composed of goblet cells (and oligomucous cells), columnar cells (similar to surface absorptive cells), enteroendocrine cells, regenerative cells, and Paneth cells.

(i) **Paneth cells,** located at the base of the crypts of Lieberkühn, are pyramid-shaped cells that secrete the antibacterial enzyme **lysozyme** stored in large, apical, membrane-bounded secretory granules. These cells also display extensive RER (basally), a large supranuclear Golgi complex, and many mitochondria.

(ii) **Regenerative cells,** located in the basal half of the crypts of Lieberkühn, are thin, tall, columnar **stem cells** that divide to replace themselves and the other types of epithelial cells.

(b) **Lymphoid nodules** are usually small and solitary in the lamina propria of the duodenum and jejunum. They increase in size and number in the ileum, where they form large contiguous aggregates, known as **Peyer patches,** which extend through the muscularis mucosae into the submucosa.

(i) **M (microfold) cells** are highly specialized, have an unusual shape, and lie in the epithelium over lymphoid nodules and Peyer patches. They are derived from undifferentiated cells of the crypts of Lieberkühn. They sample antigens as well as bacteria, viruses, and parasitic microorganisms. The endocytosed particles are conveyed, via transcytosis, to macrophages and lymphocytes that lie in the infoldings of the basal plasmalemma of M cells. These macrophages and the B and T lymphocytes are actually located in the lamina propria.

(ii) **Activated B lymphocytes** respond to antigenic challenge by forming more B cells, which enter the lymph and blood circulation, then home back to their original locations, where they populate the lamina propria and differentiate into immunoglobulin A (IgA)-producing plasma cells.

(iii) Plasma cells manufacture **IgA,** some of which is taken up by and bound to **secretory protein** within the epithelial cells and is transported across the intestinal epithelium **(transcytosis)** to the glycocalyx, where it remains as an immunologic defense against bacteria and antigens in the lumen. Much of the IgA enters blood vessels, and goes to the liver, where it is picked up by hepatocytes to be secreted as part of bile. From the liver, IgA enters the gall bladder to be released into the lumen of the duodenum.

 c. The **muscularis mucosae** is composed of an inner circular and an outer longitudinal layer of smooth muscle.

4. Submucosa of the small intestine:

 a. Consists of **fibroelastic** connective tissue containing blood and lymphatic vessels, nerve fibers, and **Meissner plexus**

 b. Also houses **Brunner glands,** which are present **only in the duodenum.** These glands produce an **alkaline fluid** and **urogastrone.** The former protects the duodenal epithelium from the acidic chyme; the latter is a polypeptide hormone that enhances epithelial cell division and inhibits gastric HCl production.

5. The **muscularis externa of the small intestine** is composed of **two** layers of smooth muscle: an inner circular and an outer longitudinal layer. The inner layer participates in the formation of the **ileocecal sphincter. Auerbach (myenteric) plexus** is housed between the two layers.

6. External layer of the small intestine

 a. Serosa covers all of the jejunum and ileum and part of the duodenum.

 b. Adventitia covers the remainder of the duodenum.

D. Large intestine

 1. Overview—large intestine

  a. The large intestine consists of the **cecum, colon** (ascending, transverse, descending, and sigmoid), **rectum, anal canal,** and **appendix.**

  b. The large intestine contains some digestive enzymes received from the small intestine.

  c. It houses bacteria that produce **vitamin B_{12} and vitamin K;** the former is necessary for hemopoiesis and the latter for coagulation.

  d. The large intestine produces **abundant mucus,** which lubricates its lining and facilitates the passage and elimination of feces.

  e. Function. The large intestine functions primarily in the **absorption of electrolytes, fluids, and gases.** Dead bacteria and indigestible remnants of the ingested material are compacted into **feces.**

2. Cecum and colon

  a. The **mucosa of the cecum and colon lacks villi** and possesses no specialized folds.

    (1) The **epithelium** of the mucosa of the cecum and colon is **simple columnar** with numerous goblet cells, surface absorptive cells, and occasional DNES cells.

    (2) The **lamina propria** is similar to that of the small intestine, possessing lymphoid nodules, blood and lymph vessels, and closely packed crypts of Lieberkühn, which lack Paneth cells.

    (3) The **muscularis mucosae** consists of an inner circular and outer longitudinal layer of smooth muscle cells.

  b. The **submucosa of the cecum and colon** is composed of **fibroelastic** connective tissue. It contains blood and lymphatic vessels, nerves, and Meissner (submucosal) plexus.

  c. The **muscularis externa of the cecum and colon** is composed of an inner circular and a modified outer longitudinal layer of smooth muscle. The outer layer is gathered into three flat, longitudinal ribbons of smooth muscle that form the **teniae coli.** When continuously contracted, the teniae coli form sacculations of the wall known as **haustra coli. Auerbach (myenteric) plexus** is housed between the two layers of smooth muscle.

  d. External layer of the cecum and colon

    (1) Adventitia covers the ascending and descending portions of the colon.

    (2) Serosa covers the cecum and the remainder of the colon. Fat-filled outpocketings of the serosa **(appendices epiploicae)** are characteristic of the transverse and sigmoid colon.

3. The **rectum** is similar to the colon but contains fewer and deeper crypts of Lieberkühn (see Table 16.1).

4. The **anal canal** is the constricted continuation of the rectum.

  a. The **anal mucosa** displays longitudinal folds called **anal columns** (or **rectal columns of Morgagni**), which join each other to form **anal valves.** The regions between adjacent valves are known as **anal sinuses.**

    (1) Epithelium of the anal canal:

      (a) Is simple columnar changing to **simple cuboidal** proximal to the anal valves

      (b) Is **stratified squamous nonkeratinized** distal to the anal valves

      (c) Changes to **stratified squamous keratinized** (epidermis) at the anus

    (2) The **lamina propria** is composed of **fibroelastic** connective tissue and contains sebaceous glands, circumanal glands, hair follicles, and large veins.

      (3) The **muscularis mucosae** consists of an inner circular and an outer longitudinal layer of smooth muscle, both of which terminate at the anal valves.

  b. The **anal submucosa** is composed of dense irregular fibroelastic connective tissue, which houses large veins.

  c. The **anal muscularis externa** is composed of an inner circular and an outer longitudinal layer of smooth muscle. The inner circular layer forms the **internal anal sphincter.**

  d. **Anal adventitia** attaches the anus to surrounding structures.

  e. The **external anal sphincter** is composed of **skeletal muscle** whose superficial and deep layers invest the anal canal. It exhibits **continuous tonus,** thus maintaining a closed anal orifice. The degree of tonus is under **voluntary control,** so the retention or evacuation of feces normally can be controlled at will.

5. Appendix

  a. Overview—appendix

    (1) The appendix is a short **diverticulum** arising from the blind terminus of the cecum.

    (2) It has a narrow, stellate, or irregularly shaped lumen, which often contains debris.

    (3) The wall is **thickened** due to the presence of large aggregates of **lymphoid nodules** in the mucosa and even in the submucosa (in middle-aged and older individuals).

  b. Mucosa of the appendix

    (1) The **epithelium** is **simple columnar** and contains surface columnar cells and goblet cells.

    (2) The **lamina propria** displays **numerous lymphoid nodules** (capped by M cells) and lymphoid cells. It does not form villi but possesses **shallow crypts of Lieberkühn** with some goblet cells, surface columnar cells, regenerative cells, occasional Paneth cells, and numerous enteroendocrine (DNES) cells (especially deep in the crypts).

    (3) The **muscularis mucosae** is composed of an inner circular and outer longitudinal layer of smooth muscle.

  c. The **submucosa of the appendix** is composed of **fibroelastic** connective tissue containing confluent lymphoid nodules and associated cell populations.

  d. The **muscularis externa of the appendix** is composed of an inner circular and an outer longitudinal layer of smooth muscle.

  e. The **serosa** completely surrounds the appendix.

IV. Digestion and Absorption

A. Carbohydrates

  1. Salivary and pancreatic amylases hydrolyze carbohydrates to **dis-**

accharides. This process begins in the oral cavity, continues in the stomach, and is completed in the small intestine.

2. **Disaccharidases** present in the glycocalyx of the brush border cleave disaccharides into monosaccharides.

3. **Monosaccharides are actively transported** into surface absorptive cells and then discharged into the lamina propria, where they enter the circulation.

B. Proteins

1. **Pepsin** in the lumen of the stomach partially hydrolyzes proteins, forming a mixture of high-molecular-weight **polypeptides.** Pepsin activity is greatest at low pH.

2. **Pancreatic proteases** within the lumen of the small intestine hydrolyze the polypeptides received from the stomach into dipeptides.

3. **Dipeptides** are cleaved into amino acids by dipeptidases present in the glycocalyx of the brush border. Amino acids are transported into surface absorptive cells. Amino acids are discharged into the lamina propria, where they enter the circulation.

C. Fats are degraded by **pancreatic lipase** into **monoglycerides, free fatty acids,** and **glycerol** in the lumen of the small intestine (Figure 16.2).

1. **Absorption of lipid digestion products** occurs primarily in the **duodenum** and **upper jejunum.**

   a. **Bile salts** act on the free fatty acids and monoglycerides, forming **water-soluble micelles.**

   b. **Micelles** and **glycerol** then enter the surface absorptive cells.

2. **Formation of chyle**

   a. **Triglycerides** are resynthesized from monoglycerides and free fatty acids within the smooth endoplasmic reticulum.

   b. **Chylomicrons** are formed in the Golgi complex by the complexing of the resynthesized triglycerides with proteins. Chylomicrons are transported to the lateral cell membrane and released by exocytosis; after crossing the basal lamina, they enter **lacteals** in the lamina propria to contribute to the formation of **chyle.**

   c. **Chyle** enters the submucosal lymphatic plexus by contraction of smooth muscle cells in the intestinal villi.

3. **Short-chain fatty acids** of less than 10-12 carbon atoms are not re-esterified but leave the surface absorptive cells directly and enter blood vessels of the lamina propria.

D. Water and electrolytes are absorbed by surface absorptive cells of both the small and large intestine, whereas **gases** are absorbed mostly in the large intestine.

V. Clinical Considerations

A. Disorders of the oral cavity

1. **Herpetic stomatitis** is caused by herpes simplex virus (HSV) type I. In the dormant state, this virus resides in the trigeminal ganglia.

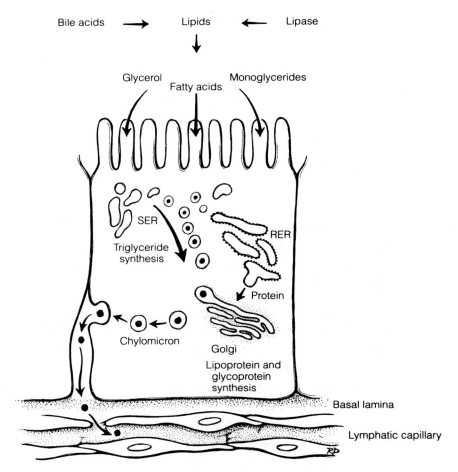

Figure 16.2. Diagram of the absorption of lipids by surface absorptive cells of the small intestine and the formation of chylomicrons. RER = rough endoplasmic reticulum; SER = smooth endoplasmic reticulum.

    a. HSV type 1 infection is very common; HSV is transmitted by kissing.

    b. This infection is characterized by painful fever blisters on the lips or in the vicinity of the nostrils. These blisters exude a clear fluid or are covered by a scab.

2. Cancers of the oral region most commonly affect the lips, tongue, and floor of the mouth. These cancers initially resemble leukoplakia and are asymptomatic. Survival rate is high if these cancers are recognized and treated in the early stages.

3. Malabsorption disorders may lead to malnutrition, resulting in wasting diseases, if major nutrients (carbohydrates, amino acids, ions) cannot be assimilated.

    a. Gluten enteropathy (nontropical sprue) results from the destructive effects of certain glutens (particularly of rye and wheat) on the intestinal villi, thus reducing the surface area available for absorption. It is treated by eliminating wheat and rye products from the diet.

b. Malabsorption of vitamin B$_{12}$ may cause pernicious anemia (see Chapter 10 VIII). It usually results from inadequate production of gastric intrinsic factor by the parietal cells of the gastric mucosa.

4. **Cholera-induced diarrhea** is caused by the action of cholera toxin, which blocks intestinal absorption of sodium ions and promotes excretion of water and electrolytes. It causes death shortly after onset unless the lost electrolytes and water are replaced.

5. **Colorectal carcinoma** is the second highest cause of cancer death in the United States. It most commonly affects individuals who are 55 years of age or older.

   a. It usually arises from adenomatous polyps and may be asymptomatic for many years. Rectal bleeding is frequently present.

   b. Colorectal carcinoma is probably diet-related. Diets high in fat and refined carbohydrates and low in fiber appear to be associated with colorectal carcinoma.

6. **Hemorrhoids** are very common in people over 50 years old. They present as rectal bleeding during defecation. Hemorrhoids are caused by the breakage of dilated, thin-walled vessels of venous plexuses either above (internal hemorrhoids) or below (external hemorrhoids) the anorectal line.

7. **Appendicitis** is usually associated with pain and/or discomfort in the lower right abdominal region, fever, nausea and vomiting, and an elevated white blood count.

Review Test

Directions: Each of the numbered items or incomplete statements in this section is followed by answers or by completions of the statement. Select the ONE lettered answer or completion that is BEST in each case.

1. The type of epithelium associated with the vermilion zone of the lips is

(A) stratified squamous nonkeratinized
(B) pseudostratified ciliated columnar
(C) stratified squamous keratinized
(D) stratified cuboidal

2. Which of the following cell types is present in the gastric glands of the pyloric stomach?

(A) Goblet cells
(B) Mucous neck cells
(C) Paneth cells
(D) Basal cells
(E) Chief cells

3. Secretin and cholecystokinin are produced and secreted by cells in the lining of the alimentary tract. Which of the following statements about these two substances is true?

(A) They are produced by diffuse neuroendocrine cells in the lining of the stomach and small intestine.
(B) They are digestive enzymes present within the lumen of the duodenum.
(C) They are produced by Paneth cells.
(D) They are hormones that have target cells in the pancreas and biliary tract.
(E) They are produced by Brunner glands and released into the lumina of the crypts of Lieberkühn.

4. If odontoblasts malfunction due to developmental anomalies, which of the following will be affected?

(A) Cementum
(B) Enamel
(C) Dentin
(D) Tooth crown only
(E) Tooth root only

5. Which of the following statements concerning the principal fiber bundles of the periodontal ligament is true?

(A) They are composed of elastin.
(B) They extend from the cementum to the enamel.
(C) They extend from the dentin to the cementum.

(D) They are composed of collagen.
(E) They extend from one tooth to the next.

6. A patient goes to the emergency room, and the physician notes one of the classic symptoms of appendicitis. Which of the following is that symptom?

(A) Apnea
(B) Vomiting of blood
(C) Depressed white cell count
(D) Rectal bleeding
(E) Abdominal pain

7. Passage of a bolus through the esophagus into the stomach is facilitated by which of the following?

(A) Peristaltic activity of the esophageal muscularis externa
(B) Peristaltic activity of the gastric muscularis mucosae
(C) Reflux through the pharyngoesophageal sphincter
(D) Smooth muscle in the esophageal muscularis mucosae
(E) Reflux through the gastroesophageal sphincter

8. The small intestine has three histologically distinct regions. Which of the following statements concerning the histologic differences in the three regions is true?

(A) Peyer patches are present only in the ileum.
(B) Goblet cells are present only in the epithelium of the duodenum.
(C) Brunner glands are located in the duodenum and jejunum but not the ileum.
(D) Lacteals are present only in the lamina propria of the ileum.
(E) The muscularis mucosae contains three layers of smooth muscle in the ileum and two layers in the duodenum and jejunum.

9. Which of the following materials can be absorbed directly by the surface lining cells of the stomach?

(A) Vitamin B_{12}
(B) Polysaccharides
(C) Chylomicrons
(D) Triglycerides
(E) Alcohol

Answers and Explanations

1-C. The external aspect and vermillion zone of the lips are covered by thin skin, which contains a stratified squamous keratinized epithelium. The internal aspect of the lips is lined by a wet mucosa containing a stratified squamous nonkeratinized epithelium.

2-B. Mucous neck cells are located in the neck of gastric glands in all parts of the stomach, whereas only fundic glands contain chief (zymogenic) cells.

3-D. Secretin and cholecystokinin are hormones produced by enteroendocrine cells in the small intestine. Secretin stimulates bicarbonate secretion in the pancreas and biliary tract. Cholecystokinin stimulates the release of pancreatic enzymes and contraction of the gall bladder.

4-C. Dentin is manufactured by odontoblasts.

5-D. The principal fiber bundles of the periodontal ligament are composed of collagen fibers. They suspend a tooth in its alveolus, extending from the cribriform plate of the alveolar bone to the cementum on the root of the tooth. The fibers that extend from one tooth to the next are the transseptal fibers of the gingivae.

6-E. Rectal bleeding or vomiting of blood often accompany gastrointestinal pathologies but not appendicitis. Elevated, not depressed, white cell count and abdominal pain are classic signs of appendicitis.

7-A. The smooth muscle of the muscularis mucosae plays no role in the movement of a bolus through the esophagus. This movement is accomplished by peristalsis of the esophageal muscularis externa, which contains both skeletal and smooth muscle. The sphincters at the proximal and distal ends of the esophagus permit movement of food in only one direction, toward the stomach.

8-A. The primary histologic differences in the three regions of the small intestine are the presence of Peyer patches in the lamina propria of the ileum and the presence of Brunner glands in the submucosa of the duodenum. The duodenum and jejunum lack Peyer patches, and the jejunum and ileum lack Brunner glands. Goblet cells are present throughout the small intestine.

9-E. Only a few simple substances, such as alcohol, can be absorbed by the epithelial lining of the stomach.

17

Digestive System: Glands

I. Overview—Extrinsic Glands of the Digestive System

A. The extrinsic glands of the digestive system include the **major salivary glands,** the **pancreas,** and the **liver** (with the associated **gallbladder**), all of which are located outside the wall of the digestive tract.

B. They produce enzymes, buffers, emulsifiers, and lubricants that are delivered to the lumen of the digestive tract via a system of ducts.

C. They also produce hormones, blood proteins, and other products.

II. Major Salivary Glands

A. Overview

1. The major salivary glands consist of three **paired exocrine** glands: the **parotid, submandibular,** and **sublingual.**

2. Function. They synthesize and secrete **salivary amylase, lysozyme, lactoferrin,** and a **secretory component,** which complexes with **immunoglobulin A (IgA)** (produced by plasma cells in the connective tissue), forming a complex that is resistant to enzymatic digestion in the saliva. They also release **kallikrein** into the connective tissue. This enzyme enters the bloodstream, where it converts kininogens into the vasodilator, **bradykinin.**

B. Structure. The major salivary glands are classified as **compound tubuloacinar** (tubuloalveolar) glands. They are further classified as **serous, mucous,** or **mixed** (serous and mucous), depending on the type of secretory acini they contain. These glands are surrounded by a capsule of dense irregular collagenous connective tissue with septa that subdivide each gland into lobes and lobules.

1. Salivary gland acini

a. Salivary gland acini consist of pyramid-shaped serous or mucous cells arranged around a central lumen that connects with an **intercalated duct.** Mucous acini may be overlain with a crescent-shaped collection of serous cells called **serous demilunes.**

b. They possess **myoepithelial cells** that share the basal lamina of the acinar cells.

c. They release a **primary secretion** that resembles extracellular fluid. This secretion is modified in the ducts to produce the **final secretion.**

d. Salivary glands are classified according to their types of salivary gland acini.

(1) **Parotid glands** consist of **serous** acini and are classified as serous.

(2) **Sublingual glands** consist mostly of **mucous** acini capped with **serous demilunes.** They are classified as mixed.

(3) **Submandibular glands** consist of both **serous** and **mucous** acini (some also have serous demilunes). They are classified as mixed.

2. **Salivary gland ducts**

a. **Intercalated ducts** originate in the acini and join to form striated ducts. They may deliver **bicarbonate ions** into the primary secretion.

b. **Striated (intralobular) ducts**

(1) Striated ducts are lined by **ion-transporting cells** that remove sodium and chloride ions from the luminal fluid (via a sodium pump) and actively pump potassium ions into it.

(2) In each lobule, they converge and become the **interlobular (excretory) ducts,** which run in the connective tissue septa. These ducts drain into the main duct of each gland, which empties into the oral cavity.

C. **Saliva**

1. Saliva is a **hypotonic** solution produced at the rate of about 1 liter (L)/day.

2. **Function**

a. Saliva **lubricates** and **cleanses** the oral cavity by means of its water and glycoprotein content.

b. It **controls bacterial flora** by the action of lysozyme, lactoferrin, and IgA, as well as by its cleansing action.

c. It initiates **digestion of carbohydrates** by the action of salivary amylase.

d. It acts as a solvent for substances that stimulate the taste buds.

e. It assists in the process of deglutition (swallowing).

III. Overview—Pancreas.

The pancreas has a slender connective tissue capsule. This gland produces **digestive enzymes** in its exocrine portion and several **hormones** in its endocrine portion **(islets of Langerhans).**

A. The **exocrine pancreas** is a **serous, compound tubuloacinar** gland.

1. **Pancreatic acinar cells**

a. Pancreatic acinar cells are pyramid-shaped serous cells arranged around a central lumen.

b. They possess a round, basally located nucleus, abundant rough endoplasmic reticulum (RER), an extensive Golgi complex, numerous mitochondria, and many free ribosomes.

c. **Zymogen (secretory) granules** are membrane-bound and densely packed in the **apical** region of pancreatic acinar cells.

    d. Their basal plasmalemma has receptors for cholecystokinin and acetylcholine.

  2. Pancreatic ducts

    a. The initial intra-acinar portion of the intercalated ducts is formed by **centroacinar cells,** which are low cuboidal in shape with a pale cytoplasm.

    b. From the initial portion, the intercalated ducts then converge into a small number of **intralobular ducts,** which in turn empty into large **interlobular ducts** that empty into the main (or accessory) pancreatic duct.

    c. The **main pancreatic duct** fuses with the common bile duct forming the ampulla of Vater, which delivers secretions of the exocrine pancreas and the contents of the gallbladder into the duodenum at the **major duodenal papilla.**

  3. Exocrine pancreatic secretions

    a. Enzyme-poor alkaline fluid

      (1) Enzyme-poor alkaline fluid is released in large quantities by **intercalated duct cells** stimulated by **secretin** (and possibly in conjunction with acetylcholine).

      (2) Function. It probably **neutralizes** the acidic chyme as it enters the duodenum.

    b. Digestive enzymes

      (1) Digestive enzymes are synthesized and stored in the pancreatic **acinar cells.** Their release is stimulated by **cholecystokinin,** previously known as **pancreozymin** (and possibly costimulated by acetylcholine released by postganglionic parasympathetic fibers).

      (2) Digestive enzymes are secreted as **enzymes** or **proenzymes** that must be activated in the intestinal lumen.

      (3) Enzymes include pancreatic amylase, pancreatic lipases, ribonuclease, and deoxyribonuclease; **proenzymes** include trypsin, chymotrypsin, carboxypeptidase, and elastase.

B. Islets of Langerhans (endocrine pancreas)

  1. Islets of Langerhans are richly vascularized spherical clusters [100-200 micrometers (μm) in diameter] of endocrine cells surrounded by a fine network of **reticular fibers.** They are scattered among the acini of the exocrine pancreas in an apparently random fashion.

  2. Islet cells (Table 17.1)

    a. Islet cells are of several types, which can be differentiated from each other only by immunocytochemistry or the use of special stains.

    b. They produce several polypeptide hormones, but **each cell type produces only one hormone.**

  3. Islet hormones

    a. Glucagon is produced by alpha cells and acts to **elevate the blood glucose level.**

Table 17.1. Comparison of Secretory Cells in Islets of Langerhans

Cell Type	Granule Characteristics	Relative Numbers	Location	Hormone Synthesized	Function
Alpha (A)	Spherical with a small halo between the membrane and electron-dense core	~20%	Positioned mostly at periphery of islets	Glucagon	Elevates blood glucose levels
Beta (B)	Small with an obvious halo between the membrane and irregular dense core	~70%	Concentrated in central region of islets but present throughout	Insulin	Decreases blood glucose levels
Delta (D)	Large and electron lucent	< 5%	Scattered throughout islets	Somatostatin	Inhibits hormone release by neighboring cells
Gastrin-producing cell (G)	Small	Rare	Scattered throughout islets	Gastrin	Stimulates hydrochloric acid (HCl) secretion
Pancreatic polypeptide-producing cell (PP)	Small	Rare	Scattered throughout islets	Pancreatic polypeptide	Inhibits release of exocrine pancreatic secretions

b. Insulin is produced by beta cells and acts to **decrease the blood glucose level.**

c. Somatostatin is produced by delta cells. It **inhibits** release of hormones by nearby secretory cells and **reduces** motility of the gastrointestinal tract and gall bladder by decreasing contraction of their smooth muscles.

d. Gastrin, produced by G cells, **stimulates** (in conjunction with histamine and acetylcholine) gastric hydrochloric acid (HCl) secretion.

e. Pancreatic polypeptide, produced by PP cells, **inhibits** release of exocrine pancreatic secretions.

IV. Liver

A. Overview

1. The liver is composed of a single type of parenchymal cell, the **hepatocyte.**

2. It is surrounded by the **Glisson capsule,** which is composed of dense irregular collagenous connective tissue. The capsule gives rise to septa that subdivide the liver into lobes and lobules.

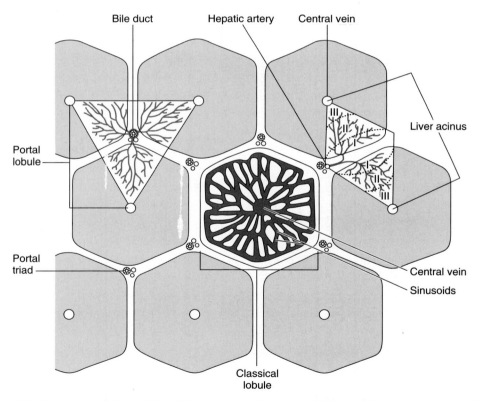

Figure 17.1. The defining characteristics of the classic liver lobule, portal lobule, and liver hepatic acinus of Rappaport. Observe the zonulation within the acinus of Rappaport. (Adapted with permission from Krause WJ, Cutts JH: *Concise Textbook of Histology*, 2nd ed. Baltimore, Williams & Wilkins, 1986, p 331.)

3. **Function.** The liver produces **bile** and plasma proteins and has a variety of other functions.

B. **Liver lobules** (Figure 17.1)

    1. The **classic liver lobule** is a hexagonal mass of tissue primarily composed of **plates of hepatocytes,** which radiate like spokes from the region of the **central vein** toward the periphery (Figure 17.2).

        a. **Portal areas (portal canals or portal triads)**

            (1) The portal areas are regions of the connective tissue between lobules that contain branches of the portal vein, hepatic artery, lymph vessel, and bile duct.

            (2) They are present at each corner of a classic liver lobule.

        b. **Liver sinusoids**

            (1) Liver sinusoids are sinusoidal capillaries that arise at the periphery of a lobule and run between adjacent plates of hepatocytes.

            (2) They receive blood from the vessels in the portal areas and deliver it to the central vein.

            (3) They are lined by **sinusoidal lining cells** (endothelial cells) that have large discontinuities between them, display **fenestrations,** and **lack basal laminae.**

            (4) They also contain **phagocytic cells (Kupffer cells)** derived from monocytes; these cells remove debris, old erythrocytes, and cellular fragments from the bloodstream.

        c. **Space of Disse**

            (1) The space of Disse is the **subendothelial space** located between hepatocytes and sinusoidal lining cells.

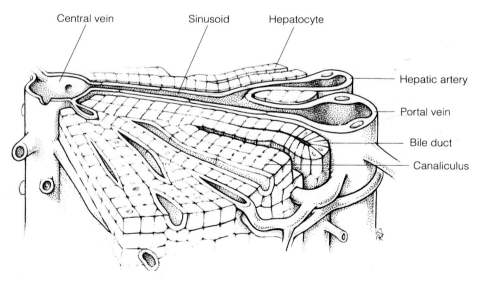

Figure 17.2. Three-dimensional representation of a portion of the classic liver lobule showing the area served by one portal triad.

(2) It contains the short microvilli of hepatocytes, reticular fibers (which maintain the architecture of the sinusoids), and occasional nonmyelinated nerve fibers.

(3) It also contains stellate-shaped **fat-storing cells (Ito cells, perisinusoidal stellate cells)**, which preferentially store vitamin A. However, when compromised, these cells can divide, change their phenotype, and begin to synthesize collagen, leading to fibrosis.

(4) **Function.** The space of Disse functions in the exchange of material between the bloodstream and hepatocytes. It is important to note that hepatocytes do not directly contact the bloodstream.

2. Portal lobule

a. The portal lobule is a **triangular region** that has three apices that are neighboring central veins, and a center that is located in a portal area (see Figure 17.1).

b. It contains portions of **three** adjacent classic liver lobules.

c. The portal lobule is defined in terms of **bile flow.** In this concept of liver lobulation, the bile duct is in the center of the lobule.

3. Hepatic acinus of Rappaport

a. The hepatic acinus of Rappaport is a **diamond-shaped region** encompassing triangular sections of **two** adjacent classic liver lobules (with apices that are the central veins) and is divided by the common distributing vessels (see Figure 17.1).

b. This concept of liver lobulation is defined in terms of **blood flow** from the distributing vessels in a single portal area. This concept was established to explain the histologic appearance of pathologic changes that occur in liver disease.

c. The hepatic acinus of Rappaport can be divided into **three zones** based on the proximity of the hepatocytes to the incoming blood.

C. Blood and bile flow (see Figure 17.2)

1. **Blood flow into the liver** is derived from two sources and is directed from the portal triads at the periphery of each classic liver lobule toward the central vein.

a. The **hepatic artery** brings oxygen-rich blood from the abdominal aorta and supplies 20%-30% of the liver's blood.

b. The **portal vein** brings nutrient-rich blood from the alimentary canal and spleen; it supplies 70%-80% of the liver's blood.

2. **Blood flow out of the liver** occurs via the **hepatic vein,** formed by the union of numerous **sublobular veins,** which collect blood from the central veins.

3. **Bile flow** is directed toward the periphery of the classic liver lobule (in the **opposite** direction of blood flow). Bile is carried in a system of ducts that culminate in the left and right **hepatic ducts,** which leave the liver and carry bile to the gallbladder.

a. **Bile canaliculi**

(1) The bile canaliculi are expanded intercellular spaces between adjacent hepatocytes.

(2) They receive the liver's **exocrine secretion** (bile) and carry it to the **canals of Hering** (bile ductules) located at the very periphery of classic liver lobules.

b. Bile ducts

(1) Bile ducts are located in the portal areas.

(2) They receive bile from the canals of Hering.

(3) They enlarge and fuse to form the hepatic ducts, which leave the liver at the porta hepatis.

D. Hepatocytes

1. Hepatocytes are large polyhedral cells (20-30 μm in diameter) that possess abundant RER and smooth endoplasmic reticulum (SER); numerous mitochondria, lysosomes, and peroxisomes; several Golgi complexes; and many lipid droplets and glycogen deposits.

2. They usually contain one round, centrally located nucleus; about 25% of the cells are binucleated. Occasionally, nuclei are polyploid.

3. **Hepatocyte surfaces**

a. **Hepatocyte surfaces facing the space of Disse** possess microvilli, which, by increasing the surface area, facilitate the transfer of materials (e.g., endocrine secretions) between the hepatocytes and the blood.

b. **Abutting surfaces of adjacent hepatocytes:**

(1) Frequently delineate bile canaliculi, small, tunnel-like expansions of the intercellular space. The **bile canaliculi** are sealed off from the remaining intercellular space by **occluding junctions** located on each side of each canaliculus.

(2) Possess microvilli that extend into the bile canaliculus

(3) Also have **gap junctions**

E. Hepatic functions

1. **Exocrine secretion** involves the production and release of 600-1200 milliliters of **bile** per day. Bile is a fluid composed of bilirubin glucuronide (bile pigment), bile acids (bile salts), cholesterol, lecithin, phospholipids, ions, IgA, and water. Hydrophobic bilirubin, a breakdown product of hemoglobin, is converted into water-soluble bilirubin glucuronide (a nontoxic compound) in the SER of the hepatocytes.

2. **Endocrine secretion** involves the production and release of several **plasma proteins** (e.g., prothrombin, fibrinogen, albumin, factor III, and lipoproteins) as well as urea. Hepatocytes can also manufacture and release nonessential amino acids.

3. **Metabolites** are **stored** in the form of **glycogen** (stored glucose) and **triglycerides** (stored lipid).

4. **Gluconeogenesis** is the conversion of amino acids and lipids into glucose, a complex process catalyzed by a series of enzymes.

5. **Detoxification** involves the inactivation of various drugs, noxious chemicals, and toxins by enzymes, such as the **microsomal mixed-function oxidase** system, that catalyze the oxidation, methylation, or conjugation of such substances. These reactions usually occur in the SER or, as in the case of alcohol, in peroxisomes.

6. **IgA transfer** involves the uptake of IgA across the space of Disse and its release into bile canaliculi. IgA is transported through the hepatic-biliary duct system to the intestine, where it serves an immunological protective function.

V. Gallbladder

A. The gallbladder communicates with the common hepatic duct via the **cystic duct,** which originates at the neck of the gallbladder.

B. It has a muscular wall whose contraction, stimulated by **cholecystokinin** (possibly, in conjunction with acetylcholine)**,** forces bile from its lumen into the duodenum. The wall has four layers:

1. The **mucosa** is composed of a **simple columnar epithelium** and a richly vascularized lamina propria. When the gallbladder is empty, the mucosa displays highly convoluted folds.

2. The **muscle layer** is composed of a thin layer of **smooth muscle cells,** oriented in an oblique fashion.

3. The **connective tissue layer** consists of dense irregular collagenous connective tissue and houses nerves and blood vessels.

4. The **serosa** covers most of the gallbladder, but adventitia is present where the organ is attached to the liver.

C. **Function.** The gallbladder concentrates, stores, and releases bile.

VI. Clinical Considerations

A. **Pancreatic disorders**

1. **Type I (insulin–dependent) diabetes mellitus (IDDM)**

   a. IDDM results from a **low level of plasma insulin.**

   b. It is characterized by **polyphagia** (insatiable hunger), **polydipsia** (unquenchable thirst), and **polyuria** (excessive urination).

   c. It usually has a **sudden onset** before age 20 and is distinguished by damage to and destruction of beta cells of the islets of Langerhans. Because of its early onset, IDDM is also known as **juvenile-onset diabetes mellitus.**

   d. It is treated with a combination of insulin therapy and diet.

2. **Type II (non–insulin-dependent) diabetes mellitus (NIDDM)**

   a. NIDDM does **not** result from low levels of plasma insulin and is **insulin resistant,** which is a major factor in its pathogenesis. The resistance to insulin is due to decreased binding of insulin to its plasmalemma receptors and to defects in postreceptor insulin action.

   b. It commonly occurs in overweight individuals over age 40.

   c. It is usually controlled by diet.

3. Pancreatic cancer is a malignant neoplasm, and most patients die within 6 months to 1 year after diagnosis. Most of the cases are **adenocarcinomas** in the head of the pancreas. Its incidence is 3 to 4 times greater in male than in female patients. Although its symptoms include anorexia, flatulence, fatty stool if the bile duct is obstructed, sudden loss of weight, weakness, back pain, and jaundice, exploratory biopsy is frequently required for a definitive diagnosis.

B. Liver diseases

1. Hepatitis is an inflammation of the liver, usually due to a viral infection but occasionally due to toxic materials.

a. Viral hepatitis A (infectious hepatitis) is caused by hepatitis A virus, which is frequently **transmitted by the fecal-oral route.** It has a **short incubation period** (2-6 weeks) and is usually nonfatal (but may cause jaundice).

b. Viral hepatitis B (serum hepatitis) is caused by hepatitis B virus, which is **transmitted by blood** and its derivatives.

(1) It has a **long incubation period** (6 weeks to 5 months).

(2) Its clinical symptoms are similar to those associated with viral hepatitis A, but with more serious consequences, including cirrhosis, jaundice, and death.

c. Viral hepatitis C is caused by hepatitis C virus and is responsible for the majority of transfusion-related cases of hepatitis. It is also associated with **hepatocellular carcinoma.**

2. Jaundice (icterus)

a. Jaundice is characterized by **excess bilirubin** in the blood and deposition of **bile pigment** in the skin and sclera of the eyes, resulting in a yellowish appearance.

b. It may be hereditary or caused by pathologic conditions such as excess destruction of red blood cells **(hemolytic jaundice),** liver dysfunction, and obstruction of the biliary passages **(obstructive jaundice).**

C. Gallstones (biliary calculi)

1. Gallstones are concretions, usually of fused crystals of **cholesterol,** that form in the gallbladder or bile duct.

2. Accumulation of gallstones may lead to the blockage of the cystic duct, which prevents emptying of the gallbladder.

3. Gallstones may need to be surgically removed if less invasive methods fail to dissolve or pulverize them.

Review Test

Directions: Each of the numbered items or incomplete statements in this section is followed by answers or by completions of the statement. Select the ONE lettered answer or completion that is BEST in each case.

1. An 18-year-old man presents subsequent to feeling faint. Symptoms include constant hunger, thirst, and excessive urination. The probable diagnosis is

(A) viral hepatitis A
(B) type I diabetes mellitus
(C) type II diabetes mellitus
(D) cirrhosis
(E) mumps

2. Which of the following statements concerning liver sinusoids is true?

(A) They are continuous with bile canaliculi.
(B) They are surrounded by a well-developed basal lamina.
(C) They are lined by nonfenestrated endothelial cells.
(D) They deliver blood to the central vein.
(E) They deliver blood to the portal vein.

3. A woman presents with yellow sclera and yellowish pallor. Blood test results indicate a low red blood cell count. The probable diagnosis is

(A) viral hepatitis A
(B) viral hepatitis B
(C) cirrhosis
(D) hemolytic jaundice
(E) type II diabetes mellitus

4. Which of the following statements concerning the gallbladder is true?

(A) It synthesizes bile.
(B) It is lined by a simple columnar epithelium.
(C) Bile leaves the gallbladder via the common bile duct.
(D) It has no muscle cells in the walls.
(E) It is affected by the hormone secretin.

5. A patient complains to her physician about sudden weight loss, a loss of appetite, weakness, and back pain. Because the patient's sclera and skin has a yellowish pallor, the doctor suspects which of the following?

(A) Type II diabetes
(B) Gallstones
(C) Pancreatic cancer
(D) Viral hepatitis A
(E) Viral hepatitis B

Questions 6-10

The response options for the next five items are the same. Each item will state the number of options to select. Choose exactly this number.

(A) Glucagon
(B) Lysozyme
(C) Insulin
(D) Plasma proteins
(E) Proteases

For each of the following cell types, select the appropriate secretory product.

6. Acinar cells of the exocrine pancreas (**SELECT 1 OPTION**)

7. Pancreatic alpha cells (**SELECT 1 OPTION**)

8. Pancreatic beta cells (**SELECT 1 OPTION**)

9. Submandibular acinar cells (**SELECT 1 OPTION**)

10. Hepatocytes (**SELECT 1 OPTION**)

Answers and Explanations

1-B. The three classic signs of type I diabetes (juvenile onset) are polyphagia (excessive eating), polyuria (excessive urination), and polydipsia (excessive drinking). The condition occurs in young individuals, usually before the age of 20.

2-D. Liver sinusoids are lined by fenestrated endothelial cells, lack a basal lamina, and deliver blood directly to the central vein.

3-D. Yellow skin color and yellow sclera are indicative of jaundice. Because the patient had a low red blood cell count, the most probable diagnosis is hemolytic jaundice.

4-B. The gallbladder is lined by a simple columnar epithelium.

5-C. Pancreatic cancer has all of the symptoms listed, namely sudden weight loss, a loss of appetite, weakness, back pain, and jaundice. Although jaundice is noted in hepatitis A and B and in the presence of obstructive jaundice caused by gallstones, sudden weight loss, loss of appetite, and weakness are not diagnostic of these diseases. Diabetes mellitus does not cause jaundice, loss of appetite, or weight loss.

6-E. Several proteases are synthesized by pancreatic acinar cells and are delivered via the pancreatic duct to the duodenum.

7-A. Glucagon is produced by alpha cells of the islets of Langerhans. They are the second most abundant secretory cells of the endocrine pancreas.

8-C. Insulin is produced by pancreatic beta cells, which are the most abundant cell type of the islets of Langerhans.

9-B. Lysozyme, an enzyme with antibacterial activity, is produced primarily by the submandibular salivary gland acinar cells.

10-D. Hepatocytes synthesize several plasma proteins, including fibrinogen, prothrombin, and albumin.

18

The Urinary System

I. Overview—The Urinary System

A. Structure. The urinary system is composed of the paired **kidneys** and **ureters** and the **bladder** and **urethra.**

B. Function. The urinary system produces and excretes **urine,** thereby clearing the blood of waste products. The kidneys also regulate the electrolyte levels in the extracellular fluid and synthesize renin and erythropoietin.

II. Kidneys

A. General structure

1. Kidneys are paired, bean-shaped organs, enveloped by a thin **capsule** of connective tissue.

2. Each kidney is divided into an outer **cortex** and an inner **medulla.**

3. Each kidney contains about 2 million **nephrons.** A nephron and the collecting tubule into which it drains form a **uriniferous tubule.**

B. The **renal hilum** is a concavity on the medial border of the kidney where arteries, veins, lymphatic vessels, and nerves are present, and where the renal pelvis is located.

C. The **renal pelvis** (Figure 18.1) is a funnel-shaped expansion of the upper end of the **ureter.** It is continuous with the **major renal calyces,** which in turn have several small branches, the **minor calyces.**

D. The **renal medulla** lies deep to the cortex but sends extensions **(medullary rays)** into the cortex.

1. **Renal (medullary) pyramids** are conical or pyramidal structures that compose the bulk of the renal medulla.

   a. Each kidney contains 10-18 renal pyramids.

   b. Each pyramid consists primarily of the thin limbs of **loops of Henle,** blood vessels, and **collecting tubules.**

2. The **renal papilla** is located at the **apex** of each renal pyramid. It has a perforated tip **(area cribrosa)** that projects into the lumen of a minor calyx.

E. The **renal cortex** is the superficial layer of the kidney located beneath the capsule. It consists primarily of **renal corpuscles** and **convoluted tubules.**

1. **Renal columns of Bertin** are extensions of cortical tissue between adjacent renal pyramids.

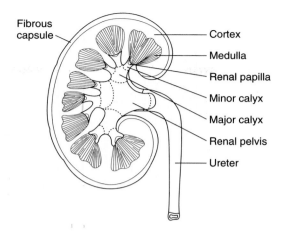

Figure 18.1. Diagram showing the internal structure of a bisected kidney.

      2. Medullary rays are **groups of straight tubules** that extend from the **base of each renal pyramid** into the cortex.

F. The **renal lobe** consists of a renal pyramid and its closely associated cortical tissue.

G. The **renal lobule** consists of a central medullary ray and the closely associated cortical tissue on either side of it, extending as far as an **interlobular artery.** Its **many nephrons** drain into the collecting tubules of the medullary ray.

H. The **renal interstitium** is the connective tissue compartment of the kidney. It consists primarily of **fibroblasts** and **mononuclear cells** (probably macrophages). In the medulla, it consists of two additional cell types:

      1. Pericytes, which are located along the blood vessels that supply the loops of Henle

      2. Interstitial cells, which have long **processes** that extend toward (and perhaps encircle) capillaries and tubules in the medulla. These cells manufacture **medullipin I,** a vasodepressor hormone that is converted to **medullipin II** in the liver. Medullipin II is a vasodilator that acts to reduce blood pressure.

III. Uriniferous Tubules (Table 18.1)

A. Nephrons. Nephrons consist of a **renal corpuscle, proximal convoluted tubule, loop of Henle,** and **distal convoluted tubule.**

      1. Classification. Nephrons can be classified as **cortical** or **juxtamedullary,** depending on the location of the renal corpuscle. Juxtamedullary nephrons possess longer loops of Henle than do cortical nephrons and are responsible for establishing the interstitial concentration gradient in the medulla.

      2. A **renal corpuscle** consists of the **glomerulus** and the **Bowman capsule** and is the structure in which the **filtration of blood** occurs.

      a. The Bowman capsule

(Text continues on page 244)

Table 18.1. Important Structural and Functional Characteristics of the Uriniferous Tubule

Region	Epithelium	Major Functions	Summary Comments
Renal corpuscle	Simple squamous epithelium lining Bowman capsule: podocytes (visceral layer) and outer (parietal layer)	Filtration of blood	Filtration barrier consists of fenestrated endothelial cells, fused basal laminae, filtration slits between podocyte secondary processes (pedicels)
Proximal convoluted tubule	Simple cuboidal epithelium with brush border, many compartmentalized mitochondria	Resorption of all glucose, amino acids, and filtered proteins and at least 80% of sodium and chloride and water	The activity of sodium pumps in basolateral membranes, transporting sodium ions out of tubule, reduces volume of ultrafiltrate and maintains its isotonicity with blood
Loop of Henle Descending thick limb	Lined by simple cuboidal epithelium with brush border	Same as for proximal convoluted tubule	Same as for proximal convoluted tubule
Loop of Henle Descending thin limb	Simple squamous epithelium	Permeable to water that enters interstitium, sodium and chloride enter ultrafiltrate	Ultrafiltrate becomes hypertonic with respect to blood; urea, from interstitium, also enters lumen of tubule
Loop of Henle Ascending thin limb	Simple squamous epithelium	Somewhat permeable to water that enters lumen, sodium and chloride exit ultrafiltrate	Ultrafiltrate remains hypertonic with respect to blood; urea, from interstitium, also enters lumen of tubule

Loop of Henle Ascending thick limb	Simple cuboidal epithelium, compartmentalized mitochondria	Impermeable to water; chloride is actively transported out of tubule into interstitium, and sodium follows	Ultrafiltrate becomes hypotonic with respect to blood; chloride pump in basolateral membranes is primarily responsible for establishing osmotic gradient in interstitium of outer medulla
Juxtaglomerular apparatus Macula densa	Simple cuboidal epithelium	Monitors level of sodium (or a decrease of fluid volume) in ultrafiltrate of distal tubule	Macula densa cells contact and communicate with JG cells in afferent arteriole via gap junctions
JG cells in afferent arteriole	Modified smooth muscle cells containing renin granules	Cells synthesize renin and release it into the bloodstream	The enzyme, renin, acts on plasma protein, triggering events that lead to formation of angiotensin II, and the release of aldosterone from adrenal glands
Distal convoluted tubule	Simple cuboidal cells, compartmentalized mitochondria	Cells respond to aldosterone by removing sodium from ultrafiltrate	Ultrafiltrate becomes more hypotonic in the presence of aldosterone; potassium, ammonium, and hydrogen ions enter the ultrafiltrate
Collecting tubules ducts	Simple cuboidal epithelium or simple columnar epithelium	In the absence of ADH, the tubule is impermeable to water, and therefore a hypotonic urine is excreted	In the presence of ADH, the tubule becomes permeable to water, which is removed from the filtrate, producing a hypertonic urine

ADH = antidiuretic hormone; JG = juxtaglomerular.

      (1) The **parietal layer** is the simple squamous epithelium that lines the outer wall of the Bowman capsule.

      (2) The **visceral layer** (glomerular epithelium) is the modified simple squamous epithelium, composed of **podocytes,** that lines the inner wall of the Bowman capsule and envelops the glomerular capillaries.

      (3) **The Bowman space** (also known as capsular space, or urinary space) is the narrow, chalice-shaped cavity between the visceral and parietal layers into which the ultrafiltrate passes.

      (4) The **vascular pole** is the site on the Bowman capsule where the afferent glomerular arteriole enters and the efferent glomerular arteriole leaves the glomerulus.

      (5) The **urinary pole** is the site on the Bowman capsule where the capsular space becomes continuous with the lumen of the proximal convoluted tubule.

  b. Podocytes are highly modified epithelial cells that form the **visceral layer** of the Bowman capsule. They have complex shapes and possess several **primary processes** that give rise to many secondary processes called **pedicels.**

    (1) **Pedicels**

      (a) Pedicels embrace the glomerular capillaries and interdigitate with pedicels arising from other primary processes.

      (b) Their surfaces facing the Bowman space are coated with **podocalyxin,** a protein that is believed to maintain their organization and shape.

    (2) **Filtration slits** are elongated spaces between adjacent pedicels. **Diaphragms,** composed of a layer of **filamentous material,** bridge each filtration slit.

  c. The **renal glomerulus** is the **tuft of capillaries** that extends into the Bowman capsule.

    (1) **Glomerular endothelial cells:**

      (a) Form the inner layer of the capillary walls

      (b) Have a thin cytoplasm that is thicker around the nucleus, where most organelles are located

      (c) Possess large **fenestrae** [60-90 nanometers (nm) in diameter], but **lack the thin diaphragms** that typically span the openings in other fenestrated capillaries

    (2) The **basal lamina** is located **between** the podocytes and the glomerular endothelial cells and is manufactured by **both** cell populations. It is unusually **thick** [0.15-0.5 micrometers (μm)] and contains three distinct **zones:**

      (a) The **lamina rara externa,** which is an electron-lucent zone adjacent to the podocyte epithelium

      (b) The **lamina densa,** which is a thicker, electron-dense intermediate zone of amorphous material

(c) The **lamina rara interna,** which is an electron-lucent zone adjacent to the capillary endothelium

(3) The **mesangium** is the interstitial tissue located between glomerular capillaries. It is composed of mesangial cells and an amorphous extracellular matrix elaborated by these cells.

  (a) **Mesangial cells:**

    (i) **Phagocytose** large protein molecules and debris, which may accumulate during filtration or in certain disease states

    (ii) Can also **contract,** thereby decreasing the surface area available for filtration

    (iii) Possess **receptors for angiotensin II** and **atrial natriuretic peptide**

  (b) The **mesangial matrix** helps support glomerular capillaries.

d. Renal filtration barrier

(1) **Structure.** The renal filtration barrier is composed of the **fenestrated endothelium** of the glomerular capillaries, the **basal lamina** (laminae rarae and lamina densa), and the **filtration slits** with diaphragms between pedicels.

(2) **Function.** The renal filtration barrier **permits passage** of water, ions, and small molecules from the bloodstream into the capsular space but **prevents passage** of large and/or most negatively charged proteins, thus forming an **ultrafiltrate of blood plasma** in the Bowman space.

  (a) The **laminae rarae** contain **heparan sulfate,** a polyanionic glycosaminoglycan that assists in **restricting the passage of negatively charged proteins** into the Bowman space.

  (b) The **lamina densa** contains **type IV collagen,** which acts as a **selective macromolecular filter** preventing passage of large protein molecules (molecular weight greater than 69,000 daltons) into the Bowman space.

3. Proximal convoluted tubule (Figure 18.2)

a. The proximal convoluted tubule is lined by a single layer of **irregularly shaped** (cuboidal to columnar) epithelial cells that have microvilli forming a prominent **brush border.** These cells exhibit the following structures:

(1) **Apical canaliculi, vesicles, and vacuoles** (endocytic complex), which function in **protein absorption**

(2) **Prominent interdigitations** along their lateral borders, which interlock adjacent cells with one another

(3) Numerous **mitochondria** compartmentalized in the basal region by extensive infoldings of the basal plasma membrane, which supply energy for the **active transport of sodium ions** out of the tubule

b. Function

(1) The proximal convoluted tubule drains the Bowman space at the urinary pole of the renal corpuscle.

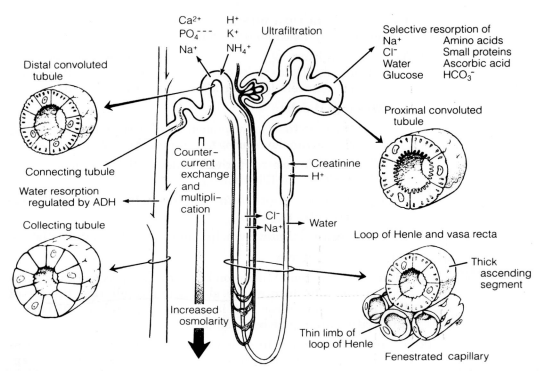

Figure 18.2. Schematic representation of a uriniferous tubule showing its major structural and functional features and its vascular associations. ADH = antidiuretic hormone. [Adapted with permission from Williams PL, Warwick R (eds): *Gray's Anatomy*, 36th British ed. London, Churchill Livingstone, 1980, p 1393.]

(2) It **resorbs** from the glomerular filtrate all of the glucose, amino acids, and small proteins and at least 80% of the sodium chloride and water.

(3) It **exchanges** hydrogen ions in the interstitium for bicarbonate ions in the filtrate.

(4) It **secretes** organic acids (e.g., creatinine) and bases and certain foreign substances into the filtrate.

4. Loop of Henle (see Figure 18.2)

a. Descending thick limb of the Henle loop

(1) The descending limb of the Henle loop is also known as the straight portion **(pars recta)** of the proximal tubule.

(2) It is lined by a simple **cuboidal** epithelium that has a prominent **brush border** and is similar to that lining the proximal convoluted tubule.

(3) Its function is to resorb, exchange, and secrete in a manner similar to that of the proximal convoluted tubule.

b. Thin limb of the Henle loop

(1) The thin limb of the Henle loop is composed of a descending segment, a loop, and an ascending segment, all of which are lined by simple **squamous** epithelial cells possessing a few short microvilli. The nuclei of these cells bulge into the lumen.

(2) In juxtamedullary nephrons, the thin limb can be divided into **three** distinct portions based on the shape of the epithelial cells, their organelle content, the depth of their tight junctions, and their permeability to water.

c. **Ascending thick limb of the Henle loop**

(1) The ascending thick limb of the Henle loop is also known as the straight portion **(pars recta)** of the distal tubule.

(2) It is lined by **cuboidal** epithelial cells that possess only a few microvilli, an apically located nucleus, and mitochondria compartmentalized within basal plasma membrane infoldings.

(3) It functions in establishing a gradient of osmolarity in the medulla (see Chapter 18 V).

(4) The ascending thick limb returns to the renal corpuscle of origin, where it is in close association with the afferent and efferent glomerular arterioles. In this region, the wall of the tubule is modified, forming the **macula densa,** which is part of the juxta-glomerular (JG) apparatus.

5. The **JG apparatus** is located at the **vascular pole** of the renal corpuscle.

a. **Components**

(1) **JG cells:**

(a) Are **modified smooth muscle cells** that exhibit some characteristics of protein-secreting cells

(b) Are located primarily in the wall of the **afferent arteriole,** but a few may also be present in the efferent arteriole

(c) Synthesize **renin** (a proteolytic enzyme) and store it in secretory granules

(2) **Macula densa cells:**

(a) Are tall, narrow, closely packed epithelial cells of the **distal tubule**

(b) Have elongated, closely packed nuclei that appear as a dense spot (macula densa) by light microscopy

(c) May **monitor the osmolarity and volume** of the fluid in the distal tubule and transmit this information to JG cells via the gap junctions existing between the two cell types

(3) **Extraglomerular mesangial cells:**

(a) Are also known as **polkissen** (pole cushion) or **lacis cells**

(b) Lie between the afferent and efferent glomerular arterioles

b. **Function.** The JG apparatus **maintains blood pressure** by the following mechanism:

(1) A **decrease in extracellular fluid volume** (perhaps detected by the macula densa) stimulates JG cells to release renin into the bloodstream.

(2) **Renin** acts on angiotensinogen in the plasma, converting it to **angiotensin I.** In capillaries of the lung (and elsewhere), an-

giotensin I is converted by angiotensin converting enzyme (ACE) to **angiotensin II,** a potent vasoconstrictor that stimulates release of **aldosterone** in the adrenal cortex.

(3) **Aldosterone** stimulates the epithelial cells of the distal convoluted tubule to remove sodium and chloride ions. **Water follows the ions,** thereby **increasing the fluid volume** in the extracellular compartment, which leads to an increase in blood pressure.

6. **Distal convoluted tubule** (see Figure 18.2)

   a. The distal convoluted tubule is continuous with the macula densa and is similar histologically to the ascending thick limb of the Henle loop.

   b. It is much shorter and has a wider lumen than the proximal convoluted tubule and **lacks a brush border.**

   c. **Function.** The distal convoluted tubule **resorbs sodium ions** from the filtrate and actively transports them into the renal interstitium; this process is stimulated by **aldosterone.** It also transfers potassium, ammonium, and hydrogen ions into the filtrate from the interstitium.

7. The **connecting** tubule is a short segment lying between the distal convoluted tubule and the collecting tubule into which it drains. It is lined by the following two types of epithelial cells.

   a. **Principal cells** have many infoldings of the basal plasma membrane. These cells **remove sodium ions** from the filtrate and **secrete potassium ions** into it.

   b. **Intercalated cells** have many apical vesicles and mitochondria. These cells **remove potassium ions** from the filtrate and **secrete hydrogen ions** into it.

B. **Collecting tubules** (see Figure 18.2) have a different embryologic origin than nephrons. They have segments in both the cortex and medulla and converge to form larger and larger tubules.

   1. **Cortical collecting tubules** are located primarily within medullary rays, although a few are interspersed among the convoluted tubules in the cortex **(cortical labyrinth).** They are lined by a simple epithelium containing two types of **cuboidal** cells.

      a. **Principal (light) cells** possess a round, centrally located nucleus and a single, central **cilium.**

      b. **Intercalated (dark) cells** are less numerous than principal cells and possess **microplicae** (folds) on their apical surface and numerous apical cytoplasmic **vesicles.**

   2. **Medullary collecting tubules.** In the **outer** medulla, medullary collecting tubules are similar in structure to cortical collecting tubules and contain both **principal** and **intercalated cells** in their lining epithelium. In the inner medulla, the collecting tubules are lined only by **principal cells.**

   3. **Papillary collecting tubules (ducts of Bellini)**

      a. Papillary collecting tubules are large collecting tubules (200-300 μm in diameter) formed from converging smaller tubules.

    b. They are lined by a simple epithelium composed of **columnar** cells that have a single central **cilium.**

    c. They empty at the **area cribrosa,** a region at the apex of each renal pyramid that has 10-25 openings through which the urine exits into a minor calyx.

IV. Renal Blood Circulation. The renal blood circulation is extensive, with total blood flow through both kidneys of about 1200 milliliters (mL)/minute. At this rate, all the circulating blood in the body passes through the kidneys every 4 to 5 minutes.

    A. Arterial supply to the kidney (Figure 18.3)

        1. Branches of the <u>renal artery</u> enter each kidney at the hilum and give rise to interlobar arteries.

        2. Interlobar arteries travel between the renal pyramids and divide into several **arcuate arteries,** which run along the corticomedullary junction in a direction parallel to the kidney's surface.

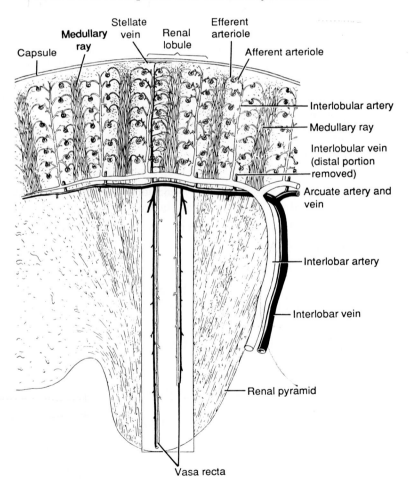

Figure 18.3. Blood circulation in the kidney. Arteries are shown in white and veins in black. Adjacent interlobular arteries, which extend outward from the arcuate artery, define the boundaries of a renal lobule. (Reprinted with permission from Junqueira LC, Carneiro J, Kelley RO: *Basic Histology,* 9th ed. Stamford, CT, Appleton & Lange, 1998, p 375.)

3. **Interlobular arteries**

   a. Interlobular arteries are smaller vessels that arise from the arcuate arteries.

   b. They enter the cortical tissue and travel outward **between adjacent medullary rays.** Adjacent interlobular arteries delimit a renal lobule.

   c. They give rise to **afferent (glomerular) arterioles** and also send branches to the interstitium just deep to the renal capsule.

4. **Afferent arterioles** are branches of the interlobular arteries and **supply the glomerular capillaries.**

5. **Efferent arterioles** arise from the glomerular capillaries and are associated with cortical and midcortical nephrons. They leave the glomerulus and give rise to an extensive **peritubular capillary network** that supplies the cortical labyrinth.

6. **Vasa recta**

   a. The vasa recta arise from the efferent arterioles supplying **juxtamedullary nephrons.**

   b. These long, thin vessels **(arteriolae rectae)** follow a straight path into the medulla and renal papilla, where they form capillaries, and then loop back and increase in diameter toward the corticomedullary boundary **(venulae rectae).**

   c. They are closely **associated with the Henle loops,** to which they supply nutrients and oxygen.

   d. These vessels play a critical role in countercurrent exchanges with the interstitium.

B. **Venous drainage of the kidney** (see Figure 18.3)

   1. **Stellate veins** are formed by convergence of **superficial cortical veins,** which drain the outermost layers of the cortex.

   2. **Deep cortical veins** drain the deeper regions of the cortex.

   3. **Interlobular veins**

      a. Interlobular veins receive both stellate and deep cortical veins.

      b. They join arcuate veins, which empty into interlobar veins. These then converge to form a branch of the renal vein, which exits the kidney at the hilum.

V. Regulation of Urine Concentration

A. **Overview**

   1. The regulation of urine concentration results in the excretion of large amounts of dilute **(hypotonic)** urine when water intake is high **(diuresis)** and of concentrated **(hypertonic)** urine when body water needs to be conserved **(antidiuresis).**

   2. This regulation depends on events that occur in the loops of Henle, vasa recta, and collecting tubules.

   3. It is affected by the presence or absence of **antidiuretic hormone (ADH),** which is secreted from the pars nervosa of the pituitary gland when water must be conserved.

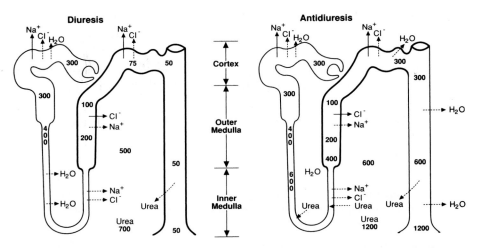

Figure 18.4. Summary of ion and water exchanges that occur in the uriniferous tubule in the absence (left) and presence (right) of antidiuretic hormone (ADH). The countercurrent multiplier system involving the loop of Henle produces an osmotic gradient in the medullary interstitium. Numbers refer to the local concentration in milliosmoles per liter (mosm/L). Segments of the tubule freely permeable to water are drawn with a thin line; impermeable segments are drawn with a thick line. In the distal convoluted tubule, some water follows sodium into the interstitium; sodium transport here is regulated by aldosterone. (Reprinted with permission from Weiss L: *Cell and Tissue Biology*, 6th ed. Baltimore, Urban & Schwarzenberg, 1988, p 840.)

B. The **countercurrent multiplier system** (Figure 18.4) depends on the **increasing osmotic concentration gradient** in the renal interstitium from the outer medulla to the renal papillae. It involves **ion and water exchanges** between the **renal interstitium** and the **filtrate in the loop of Henle.**

1. **In the descending limb of the loop of Henle**

   a. The **isotonic** filtrate coming from the proximal convoluted tubules loses water to the interstitium and gains sodium and chloride ions.

   b. The filtrate becomes **hypertonic.**

2. **In the ascending thick limb of the loop of Henle**

   a. No water is lost from the filtrate because this part of the nephron is **impermeable to water** in the presence or absence of ADH.

   b. Chloride ions are **actively transported** from the filtrate into the interstitium, and sodium ions follow.

   c. An **osmotic gradient** thus is established in the **interstitium** of the outer medulla.

   d. The filtrate becomes **hypotonic.**

3. **In the distal convoluted tubule,** active resorption of sodium ions from the filtrate may occur (in response to aldosterone), resulting in some water loss as well.

C. Role of collecting tubules

1. In the **absence of ADH,** the collecting tubules are **impermeable to water.** Thus the hypotonic filtrate coming from the ascending limb of the loop of Henle is not changed, and **hypotonic** urine is excreted.

2. In the **presence of ADH,** the collecting tubules become **permeable to water.** Thus the isotonic filtrate entering them from the distal convo-

luted tubule loses water, and a **hypertonic** (concentrated) urine is produced.

D. Countercurrent exchange system (Figure 18.5)

  1. The countercurrent exchange system involves passive ion and water exchanges between the renal interstitium and the blood in the vasa recta, the small straight vessels associated with the loops of Henle.

  2. This exchange acts to maintain the interstitial osmotic gradient created by changes taking place in the Henle loop.

E. Effect of urea (see Figure 18.4) is to aid in the production and maintenance of the interstitial osmotic gradient, mostly in the inner medulla.

  1. Urea concentrations in the filtrate progressively increase as water is lost from the medullary collecting tubules, causing the urea to diffuse out into the interstitium (thus contributing to the interstitial osmolarity).

  2. A high-protein diet increases urea levels in the filtrate and its subsequent entrapment in the interstitium, thus enhancing the kidney's ability to concentrate the urine.

VI. Excretory Passages

A. Overview (Table 18.2)

  1. The excretory passages include the minor and major **calyces** and the **renal pelvis,** located within each kidney, and the **ureters, urinary bladder,** and **urethra,** located outside the kidneys.

  2. These structures generally possess a three-layered wall composed of a **mucosa of transitional epithelium** (except in the urethra) lying on a lamina propria of connective tissue, a **muscularis (smooth muscle),** and an **adventitia.**

B. Ureter

  1. The ureter conveys urine from the renal pelvis of each kidney to the urinary bladder.

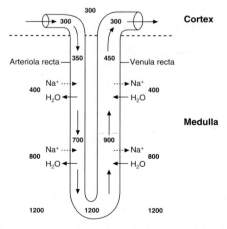

Figure 18.5. Summary of water and ion exchanges between the medullary interstitium and the blood in the vasa recta. These countercurrent exchanges are passive and do not disturb the osmotic gradient in the interstitial tissue. Numbers refer to the local osmolarity (mosm/L). (Adapted with permission from Junqueira LC, Carneiro J, Kelley RO: *Basic Histology,* 7th ed. Norwalk, CT, Appleton & Lange, 1992, p 390.)

Table 18.2. Features of Excretory Passages

Region	Epithelium	Lamina Propria	Muscularis	Comments
Calyces, minor and major	Transitional epithelium	Reticular and elastic fibers	A few inner longitudinal and outer circular smooth muscle fibers	Urine from collecting tubules (ducts of Bellini) empty into minor calyces
Renal pelvis	Transitional epithelium	Reticular and elastic fibers	Inner longitudinal, outer circular layer of smooth muscle	Expanded upper portion of ureter that receives urine from the major calyces
Ureters	Transitional epithelium lines a stellate-shaped lumen	Collagen and elastic fibers	Inner longitudinal, outer circular layer of smooth muscle; lower third has an additional outermost longitudinal layer	Peristaltic waves propel urine, so it enters bladder in spurts
Urinary bladder	Transitional epithelium: 5–6 cell layers in empty bladder; 3–4 cell layers in distended bladder	Fibroelastic connective tissue, rich in blood vessels	Three poorly defined layers of smooth muscle: inner longitudinal, middle circular, outer longitudinal	Plasmalemma of dome-shaped cells in epithelium have unique plaques and cells have elliptical vesicles, which underlie the remarkable (empty versus full) transition
	Trigone is a triangular region with apices that are the openings of the two ureters and the urethra			The trigone, unlike most of the bladder mucosa, always presents a smooth surface
Urethra Female	Transitional epithelium near bladder, remainder stratified squamous	Fibroelastic vascular connective tissue, mucus-secreting glands of Littre	Inner longitudinal, outer circular layer of smooth muscle; a sphincter of skeletal muscle surrounds urethra at urogenital diaphragm	Female urethra is conduit for urine The external sphincter of skeletal muscle permits voluntary control of micturition
Urethra Male prostatic	Transitional epithelium near bladder, pseudostratified or stratified columnar	Fibromuscular stroma of prostate gland, a few glands of Littre	Inner longitudinal, outer circular layer of smooth muscle	A conduit for urine and semen, receives secretions from prostate glands and from the paired ejaculatory ducts
Urethra Male membranous	Pseudostratified or stratified columnar	Fibroelastic stroma, a few glands of Littre	Striated muscle fibers of urogenital diaphragm form external sphincter	A conduit for urine and semen The external sphincter of skeletal muscle permits voluntary control of micturition
Urethra Male cavernous	Pseudostratified or stratified columnar, at fossa navicularis stratified squamous	Replaced by erectile tissue of corpus spongiosum, many glands of Littre	Replaced by sparse amounts of smooth muscle, and many elastic fibers, in septa lining vascular spaces in erectile tissue	A conduit for urine and semen Receives secretions of bulbourethral glands located in the urogenital diaphragm

2. It has a **transitional epithelium** that is thicker and contains more cell layers than that of the renal calyces.

3. It possesses a **two-layered muscularis** (an inner longitudinal and outer circular layer of smooth muscle) in its upper two thirds. The lowest one third possesses an additional outer longitudinal layer of smooth muscle.

4. It contracts its muscle layers, producing **peristaltic waves** that propel the urine so that it enters the bladder in spurts.

C. **Urinary bladder.** The **urinary bladder** possesses a **transitional epithelium** with a morphology that differs in the relaxed (empty) and distended states, a thin lamina propria of **fibroelastic** connective tissue, and a **three-layered muscularis.**

   1. **Epithelium of the relaxed bladder** is five to six cell layers thick and has **rounded** superficial **dome-shaped cells** that bulge into the lumen. These cells contain unique **plaques** (having a highly ordered substructure) in their thick luminal plasma membrane, and flattened, **elliptical vesicles** in their cytoplasm.

   2. **Epithelium of the distended bladder**

      a. The epithelium of the distended bladder is only three to four cell layers thick.

      b. It has **squamous** superficial cells.

      c. It is much thinner and has a larger luminal surface area than the relaxed bladder, which results from insertion of the elliptical vesicles into the luminal plasma membrane of the surface cells.

D. **Urethra**

   1. **Overview**

      a. The urethra conveys urine from the bladder to outside the body. In males, the urethra also carries semen during ejaculation.

      b. It has a **two-layered muscularis** consisting of an inner longitudinal and an outer circular layer of smooth muscle.

      c. It is surrounded at some point by an **external sphincter of skeletal muscle,** which permits its voluntary closure.

   2. **Male urethra**

      a. The male urethra is about 20 centimeters (cm) long and is divided into **prostatic, membranous,** and **cavernous** portions.

      b. It is lined by **transitional epithelium** in the prostatic portion and by **pseudostratified** or **stratified columnar epithelium** in the other two portions. The **fossa navicularis,** located at the distal end of the cavernous urethra, is lined by **stratified squamous epithelium.**

      c. It contains mucus-secreting **glands of Littre** in the lamina propria.

   3. **Female urethra**

      a. The female urethra is much shorter (4-5 cm long) than the male urethra.

    b. It is lined primarily by **stratified squamous epithelium,** although patches of pseudostratified columnar epithelium are present.

    c. It may contain **glands of Littre** in the lamina propria.

VII. Clinical Considerations

A. Glomerulonephritis

1. Glomerulonephritis is a type of nephritis characterized by **inflammation of the glomeruli.**

2. It is sometimes marked by proliferation of podocytes, endothelial cells, and mesangial cells in the glomerular tuft; infiltration of leukocytes also is common.

3. This disease often occurs **secondary to a streptococcal infection** elsewhere in the body, which is thought to result in deposition of immune complexes in the glomerular basal lamina. The immune complexes damage the glomerular basal lamina and markedly reduce its filtering ability.

4. It also may result from **immune** or **autoimmune disorders.**

5. It is associated with production of urine containing blood **(hematuria),** protein **(proteinuria),** or both; in severe cases, decreased urine output **(oliguria)** is common.

6. It occurs in acute, subacute, and chronic forms. The chronic form, in which the destruction of glomeruli continues, leads eventually to renal failure and death.

B. Chronic renal failure

1. Chronic renal failure can result from a variety of diseases (e.g., diabetes mellitus, hypertension, atherosclerosis) in which blood flow to the kidneys is reduced, causing a decrease in glomerular filtration and tubular ischemia.

2. It is associated with pathologic changes (hyalinization) in the glomeruli and atrophy of the tubules, which impair virtually all aspects of renal function.

3. It is marked by **acidosis** and **hyperkalemia** because the acid-base balance cannot be maintained, and by **uremia** because of the inability to eliminate metabolic wastes.

4. If untreated, chronic renal failure leads to neurologic problems, coma, and death.

C. Diabetes insipidus

1. Diabetes insipidus results from destruction of the supraoptic and paraventricular nuclei in the hypothalamus, which synthesize ADH (vasopressin) (see Figure 13.1).

2. It is associated with a **decreased ability of the kidney to concentrate urine** in the collecting tubules due to the reduced levels of ADH.

3. Signs and symptoms include dehydration, excessive thirst **(polydipsia),** and excretion of **high volumes of dilute urine.**

Review Test

Directions: Each of the numbered items or incomplete statements in this section is followed by answers or by completions of the statement. Select the ONE lettered answer or completion that is BEST in each case.

1. Which of the following statements concerning the structure of medullary rays is true?

(A) They contain arched collecting tubules.
(B) They contain proximal convoluted tubules.
(C) They do not extend into the renal cortex.
(D) They lie at the center of a renal lobule.
(E) They contain thin limbs of the loops of Henle.

2. Which one of the following structures is located in the renal cortex?

(A) Vasa recta
(B) Thin limbs of the loops of Henle
(C) Afferent arterioles
(D) Interlobar veins
(E) Area cribrosa

3. Which of the following structures is present in the male urethra but is not present in the female urethra?

(A) Stratified squamous epithelium
(B) Transitional epithelium
(C) Glands of Littre
(D) External sphincter of skeletal muscle
(E) Connective tissue layer underlying the epithelium

4. Which of the following statements concerning cortical collecting tubules is always true?

(A) They are lined by a simple epithelium containing two types of cells.
(B) They are also known as the ducts of Bellini.
(C) They empty on the area cribrosa.
(D) They are permeable to water.
(E) They are continuous with the ascending thick limb of the Henle loop.

5. A 35-year-old woman had surgery to remove a cerebral tumor. One month after the procedure, she reports being excessively thirsty and drinking several liters of water per day. She must also urinate so frequently that she avoids leaving the house. Laboratory tests indicate that her urine has a very low specific gravity. What is the most likely diagnosis of this woman's condition?

(A) Acute renal failure
(B) Glomerulonephritis
(C) Chronic renal failure
(D) Diabetes insipidus
(E) Urinary incontinence

6. The countercurrent multiplier system in the kidney involves the exchange of water and ions between the renal interstitium and

(A) the blood in the vasa recta
(B) the blood in the peritubular capillary network
(C) the filtrate in the proximal convoluted tubule
(D) the filtrate in the loop of Henle
(E) the filtrate in the medullary collecting tubule

Questions 7-10

The response options for the next four items are the same. Each item will state the number of options to select. Choose exactly this number.

(A) Isotonic
(B) Hypotonic
(C) Hypertonic

As the glomerular filtrate passes through the uriniferous tubule, ions and water are exchanged (actively or passively) with the renal interstitium. These exchanges result in the filtrate being isotonic, hypotonic, or hypertonic relative to blood plasma. For filtrate in each portion of the uriniferous tubule, select the relative tonicity that applies in a condition of antidiuresis.

7. Medullary collecting tubule (**SELECT 1 TONICITY**)

8. The Bowman (capsular) space (**SELECT 1 TONICITY**)

9. Distal portion of the ascending thick limb of the loop of Henle (**SELECT 1 TONICITY**)

10. Initial (thick portion) of the descending limb of the loop of Henle (**SELECT 1 TONICITY**)

256

Answers and Explanations

1-D. A medullary ray contains the straight portions of tubules projecting from the medulla into the cortex, giving the appearance of striations or rays. A renal lobule consists of a centrally located medullary ray and its closely associated cortical tissue.

2-C. Afferent arterioles, which arise from interlobular arteries and supply the glomerular capillaries, are located in the renal cortex.

3-B. Only the male urethra contains transitional epithelium (in the prostatic portion). Stratified squamous epithelium lines most of the female urethra and the distal end of the cavernous urethra in males. Mucus-secreting glands of Littre are always present in the male urethra and may be present in the female urethra.

4-A. Cortical collecting tubules are lined by a simple epithelium containing principal (light) cells and intercalated (dark) cells. They are permeable to water only in the presence of antidiuretic hormone; in the absence of this hormone, they are impermeable to water. The large papillary collecting tubules, called ducts of Bellini, empty on the area cribrosa at the apex of each renal papilla.

5-D. This woman is suffering from diabetes insipidus. Surgical removal of the cerebral tumor likely damaged her hypothalamus, which in turn greatly reduced or eliminated the production of ADH. Therefore, her kidney collecting tubules and distal tubules fail to resorb water, resulting in the production of vast quantities of dilute urine and causing excessive thirst.

6-D. The countercurrent multiplier system in the loop of Henle involves ion and water exchanges between the filtrate and the interstitium. It establishes an osmotic gradient in the interstitium of the medulla, which is greatest at the papilla.

7-C. The filtrate that enters the cortical collecting tubules is nearly isotonic. When antidiuretic hormone is present, water is removed from the filtrate in the collecting tubules, making the filtrate hypertonic by the time it reaches the medullary collecting tubules.

8-A. Filtration in the renal corpuscle yields an isotonic ultrafiltrate of blood plasma that enters the Bowman space.

9-B. The ascending thick limb of the loop of Henle is impermeable to water even in the presence of antidiuretic hormone, but actively transports chloride ions from the filtrate into the interstitium (sodium ions follow passively). As a result, the filtrate becomes hypotonic as it approaches the distal convoluted tubule.

10-A. As the filtrate passes through the proximal convoluted tubule, the loss of ions is offset by the loss of water. As a result, the filtrate is still isotonic as it enters the thick descending limb of the loop of Henle (also called the pars recta of the proximal tubule).

19

Female Reproductive System

I. Overview—Female Reproductive System

A. The female reproductive system consists of the paired **ovaries** and **oviducts;** the **uterus, vagina,** and **external genitalia;** and paired **mammary glands.**

B. It undergoes marked changes at the onset of puberty, which is initiated by **menarche.**

C. It exhibits monthly menstrual cycles and menses from puberty until the end of the reproductive years, which terminate at **menopause.**

II. Ovaries (Figure 19.1)

A. Overview

1. Ovaries are covered by a simple cuboidal epithelium called the **germinal epithelium.**

2. Ovaries possess a **tunica albuginea** composed of dense irregular collagenous connective tissue.

3. Each ovary is subdivided into a **cortex** and a **medulla,** which are not sharply delineated.

B. The **ovarian cortex** consists of **ovarian follicles** in various stages of development and a connective tissue **stroma** containing cells that respond in unique ways to hormonal stimuli.

1. **Ovarian follicles** (see Figure 19.1 and Table 19.1)

   a. **Primordial follicles** are composed of a **primary oocyte** enveloped by a single layer of **flat follicular cells.**

   (1) **Primary oocytes:**

   (a) Display a prominent, acentric, vesicular-appearing nucleus **(germinal vesicle)** with a single nucleolus

   (b) Possess many Golgi complexes, mitochondria, profiles of rough endoplasmic reticulum (RER), and well-developed annulate lamellae

   (c) Become **arrested in prophase of meiosis I** (by paracrine factors produced by the follicular cells) during fetal life and may remain in this stage for years

   (2) **Follicular cells:**

   (a) Are attached to one another by **desmosomes**

   (b) Are separated from the surrounding stroma by a basal lamina

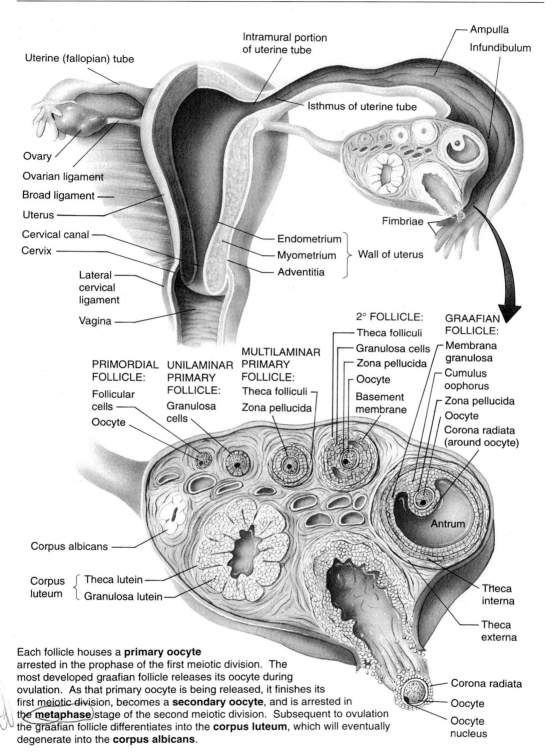

Uterine (fallopian) tube

Intramural portion
of uterine tube

Ampulla
Infundibulum

Isthmus of uterine tube

Ovary

Ovarian ligament

Broad ligament

Uterus

Cervical canal

Cervix

Lateral
cervical
ligament

Vagina

Endometrium
Myometrium
Adventitia

Wall of uterus

Fimbriae

PRIMORDIAL
FOLLICLE:

Follicular
cells

Oocyte

UNILAMINAR
PRIMARY
FOLLICLE:

Granulosa
cells

MULTILAMINAR
PRIMARY
FOLLICLE:

Theca folliculi

Zona pellucida

2° FOLLICLE:

Theca folliculi
Granulosa cells
Zona pellucida
Oocyte
Basement
membrane

GRAAFIAN
FOLLICLE:

Membrana
granulosa
Cumulus
oophorus
Zona pellucida
Oocyte
Corona radiata
(around oocyte)

Antrum

Corpus albicans

Corpus
luteum

Theca lutein
Granulosa lutein

Theca
interna

Theca
externa

Corona radiata

Oocyte

Oocyte
nucleus

Each follicle houses a **primary oocyte** arrested in the prophase of the first meiotic division. The most developed graafian follicle releases its oocyte during ovulation. As that primary oocyte is being released, it finishes its first meiotic division, becomes a **secondary oocyte**, and is arrested in the **metaphase** stage of the second meiotic division. Subsequent to ovulation the graafian follicle differentiates into the **corpus luteum**, which will eventually degenerate into the **corpus albicans**.

Figure 19.1. Structural features of the ovary. Note follicles and corpus luteum in different stages of development. (Reprinted with permission from Gartner LP, Hiatt JL: *Color Atlas of Histology*, 3rd ed. Baltimore, Lippincott Williams & Wilkins, 2000, p 342.)

Table 19.1. Stages in the Development of Ovarian Follicles

Stage	Zona Pellucida	Follicular Cell Layer (Granulosa)	Liquor Folliculi	Theca	Hormone Dependency
Primordial follicle	Not present	Single layer of flat cells	Not present	Not present	Dependent on local factors
Unilaminar primary follicle	Present	Single layer of cuboidal cells	Not present	Not present	Dependent on local factors
Multilaminar primary follicle	Present	Multiple layers of granulosa cells	Not present	Interna and externa are present	Dependent on local factors
Secondary follicle	Present	Spaces among granulosa cells	Accumulates in spaces among granulosa cells	Interna and externa are present	FSH dependent
Graafian follicle	Present	Forms membrana granulosa, cumulus oophorus	Fills the antrum	Interna and externa are present	FSH dependent until it becomes the dominant follicle

FSH = follicle-stimulating hormone.

b. Growing follicles

(1) **Primary follicles** are **not dependent on follicle-stimulating hormone (FSH)** for their development. They possess an amorphous layer (**zona pellucida**) surrounding and produced by the primary oocyte; a basal lamina is present outside the follicular cells.

    (a) **Unilaminar primary follicles:**

        (i) Develop from primordial follicles

        (ii) Are composed of a single layer of **cuboidal** follicular cells surrounding the primary oocyte

    (b) **Multilaminar primary follicles:**

        (i) Develop from unilaminar follicles by proliferation of follicular cells

        (ii) Consist of several layers of follicular cells; these follicular cells are now also known as **granulosa cells**

        (iii) Are circumscribed by two layers of stromal cells: an inner cellular layer (**theca interna**) and an outer fibrous layer (**theca externa**)

        (iv) Are separated from the theca interna by a basal lamina

(2) **Secondary (antral) follicles**

    (a) Secondary follicles are established when fluid (**liquor folliculi**) begins to accumulate in the intercellular spaces between granulosa cells. The fluid-filled spaces will begin to coalesce to form a single large cavity called an antrum.

    (b) Secondary follicles are **dependent on FSH,** which stimulates the granulosa cells to convert androgens (produced by the theca interna cells) into **estrogens,** and to manufacture plasmalemma **receptors for luteinizing hormone (LH).**

    (c) Narrow **processes** from the granulosa cells extend into the zona pellucida.

    (d) **Granulosa cells** contact each other via gap junctions and also form gap junctions with the cell membrane of the primary oocyte.

c. Graafian (mature) follicle

(1) The graafian follicle is the one follicle, selected from the secondary follicles, that **will ovulate.**

(2) It measures approximately 2.5 centimeters (cm) in diameter and is evident as a large bulge on the surface of the ovary.

(3) The primary oocyte is positioned off-center on a small mound of granulosa cells (**cumulus oophorus**) that projects into the antrum of the follicle. Granulosa cells surround the zona pellucida (and those contacting the zona pellucida are known as the **corona radiata**). Other granulosa cells line the antrum, forming the **membrana granulosa.**

(4) **Theca interna cells** manufacture **androgens**, which are transferred to granulosa cells, where they are converted into **estrogens.**

(5) The **theca externa** is mostly collagenous. It contains a few muscle cells and many blood vessels, which provide nourishment to the theca interna.

(6) **Ovulation**

(a) An LH surge from the pituitary gland triggers the **primary oocyte** to complete its first meiotic division just prior to ovulation, forming a **secondary oocyte** and the **first polar body.** The second meiotic division begins (due to the presence of local meiosis-inducing factors) but is blocked at metaphase.

(b) **Ovulation** also occurs in response to the LH surge. The secondary oocyte and its **corona radiata** cells leave the ruptured follicle at the ovarian surface to enter the fimbriated end of the oviduct.

2. The **corpus hemorrhagicus** is formed from the remnants of the graafian follicle.

3. **Corpus luteum**

a. **Overview**

(1) The corpus luteum is formed from the corpus hemorrhagicus.

(2) It is composed of **granulosa lutein cells** (modified granulosa cells) and **theca lutein cells** (modified theca interna cells).

(3) The formation of this richly vascularized **temporary endocrine gland** is dependent on LH.

b. **Granulosa lutein cells**

(1) Granulosa lutein cells are large [30 micrometers (μ) in diameter] pale cells that possess an abundance of smooth endoplasmic reticulum (SER), RER, many mitochondria, a well-developed Golgi complex, and lipid droplets.

(2) They are derived from cells of the membrana granulosa.

(3) **Function.** Granulosa lutein cells manufacture most of the body's **progesterone** and convert androgens formed by the theca lutein cells into **estrogens**.

c. **Theca lutein cells**

(1) These small (15 μ in diameter) cells are concentrated mainly along the periphery of the corpus luteum.

(2) They are derived from cells of the theca interna.

(3) **Function.** Theca lutein cells manufacture **progesterone** and **androgens** and small amounts of estrogen.

4. The **corpus albicans** is the remnant of the degenerated corpus luteum. It becomes a small scar on the surface of the ovary.

5. **Atretic follicles**

a. Atretic follicles are follicles (in various stages of maturation) that are undergoing degeneration.

    b. They are commonly present in the ovary; after a graafian follicle ovulates, the remaining secondary follicles degenerate.

    c. They often show pyknotic changes in the nuclei of the granulosa cells as well as other degenerative changes.

C. The **ovarian medulla** contains large blood vessels, lymphatic vessels, and nerves in a loose connective tissue stroma. They also possess a small number of **estrogen**-secreting **interstitial cells** and a few **androgen**-secreting **hilus cells**.

D. Hormonal regulation (Figure 19.2)

    1. Control of follicle maturation and ovulation

        a. The primary oocyte of unilaminar primary follicles secretes **activin**, which facilitates proliferation of granulosa cells.

        b. Gonadotropin-releasing hormone (GnRH) from the hypothalamus causes the release of FSH and LH from the pars distalis of the pituitary gland.

        c. FSH stimulates the growth and development of secondary (but not earlier stage) ovarian follicles and the appearance of LH receptors on the granulosa cell plasmalemma. The regulation of FSH and LH is influenced by the following:

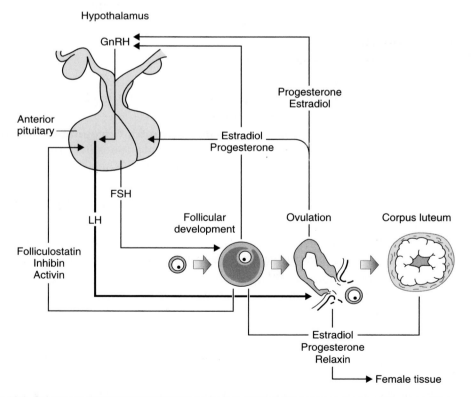

Figure 19.2. Schematic diagram illustrating the hormonal relationship between the hypophysis and the reproductive system. FSH = follicle-stimulating hormone; GnRH = gonadotropin-releasing hormone; LH = luteinizing hormone. (Adapted with permission from Gartner LP, Hiatt JL: *Color Textbook of Histology.* Philadelphia, WB Saunders, 1997, p 389.)

(1) Theca interna cells manufacture androgens, which are converted into estrogens by granulosa cells.

(2) Granulosa cells also secrete inhibin, follistatin, and activin, all of which (in addition to estrogen) regulate FSH secretion.

(3) By approximately the 14th day of the menstrual cycle, estrogen blood levels become sufficiently high to facilitate a sudden, brief surge of LH.

d. Surge of LH

(1) A surge of LH triggers the primary oocyte of the graafian follicle to complete meiosis I and to enter meiosis II, where it becomes arrested at metaphase.

(2) It initiates ovulation of the secondary oocyte from the graafian follicle.

(3) It promotes formation of the corpus luteum.

2. Fate of the corpus luteum

a. Luteal hormones

(1) **Progesterone,** the major hormone secreted by the corpus luteum, inhibits the release of LH (by suppressing the release of GnRH) but promotes development of the uterine endometrium.

(2) **Estrogen** inhibits the release of FSH (by suppressing the release of GnRH).

(3) **Relaxin** facilitates parturition.

b. In the event of pregnancy

(1) The syncytiotrophoblast of the developing placenta manufactures **human chorionic gonadotropin** (hCG) and **human chorionic somatomammotropin** (hCS).

(2) **hCG maintains the corpus luteum of pregnancy for about 3 months,** at which time the placenta takes over the production of progesterone, estrogen, and relaxin.

c. In the absence of pregnancy

(1) Neither LH nor hCG is present, and the **corpus luteum begins to atrophy.**

(2) Lack of estrogen and progesterone also triggers the release of FSH from the pituitary, thus re-initiating the menstrual cycle.

3. In each menstrual cycle up to 50 primordial follicles begin the process of maturation, less than five reach the graafian follicle stage, and usually only the dominant follicle undergoes ovulation. The dominant follicle is FSH-independent and produces a surge of inhibin that suppresses FSH production and leads to atrophy of the other maturing follicles.

III. Oviducts (Fallopian Tubes). The oviducts are subdivided into four regions: the **infundibulum,** which has a **fimbriated end;** the **ampulla,** which is the most common site of fertilization; the **isthmus;** and the **intramural portion,** which traverses the wall of the uterus. The wall of each oviduct consists of a **mucosa, muscularis,** and **serosa.**

A. The **mucosa** has extensive **longitudinal folds** in the infundibulum. The degree of folding progressively decreases in the remaining three regions of the oviduct.

    1. The **epithelium** is **simple columnar** and consists of peg cells and ciliated cells.

        a. Peg cells

            (1) The cells secrete a **nutrient-rich medium** that nourishes the spermatozoa (and preimplantation embryo).

            (2) Their cytoplasm contains abundant RER; a well-developed Golgi complex; and many apically located, electron-dense secretory granules.

        b. Ciliated cells

            (1) Ciliated cells possess many cilia, which beat mostly toward the lumen of the uterus.

            (2) Function. Ciliated cells may facilitate the transport of the developing embryo to the uterus.

    2. The **lamina propria** consists of loose connective tissue containing reticular fibers, fibroblasts, mast cells, and lymphoid cells.

B. Muscularis

    1. The muscularis is composed of an ill-defined inner circular and an outer longitudinal layer of smooth muscle.

    2. Function. By contracting rhythmically, the muscularis probably assists in moving the embryo toward the uterus.

C. The **serosa** covers the outer surface of the oviduct and is composed of a simple squamous epithelium overlying a thin connective tissue layer.

IV. Uterus. The uterus is subdivided into the **fundus,** body **(corpus),** and **cervix.**

A. The **uterine wall** consists of the **endometrium, myometrium,** and **adventitia** (or serosa).

    1. Endometrium

        a. Overview

            (1) The endometrium undergoes hormone-modulated cyclic alterations during different phases of the **menstrual cycle.**

            (2) It is lined by a **simple columnar** epithelium containing **secretory** and **ciliated cells.** The endometrium possesses **simple tubular glands.**

            (3) Its stroma resembles mesenchymal connective tissue, with **stellate-shaped cells** and an abundance of **reticular fibers.** Macrophages and leukocytes are also present.

        b. Layers of the endometrium

            (1) The **functional layer (functionalis)** is the thick superficial layer of the endometrium that is sloughed and re-established monthly due to hormonal changes during the menstrual cycle.

   (2) The **basal layer (basalis)** is the deeper layer of the endometrium that is preserved during menstruation. It has endometrial **glands,** which have basal cells that provide a source for **re-epithelialization** of the endometrium after the functional layer is sloughed.

   c. The **endometrial vascular supply** consists of two types of arteries derived from vessels in the stratum vasculare of the myometrium.

   (1) Coiled arteries extend into the functional layer and undergo pronounced changes during different stages of the menstrual cycle.

   (2) Straight arteries do not undergo cyclic changes and terminate in the basal layer.

2. Myometrium

   a. The myometrium is the thick smooth muscle tunic of the uterus.

   b. It is composed of inner and outer longitudinal layers and a thick middle circular layer. The circular layer is richly **vascularized** and is often referred to as the **stratum vasculare.**

   c. This layer thickens during pregnancy due to the hypertrophy and hyperplasia of individual smooth muscle cells.

   d. Near the end of pregnancy, the myometrium develops many gap junctions between its smooth muscle cells. These junctions coordinate contraction of the muscle cells during parturition.

   e. At parturition, the myometrium undergoes powerful contractions triggered by the hormone **oxytocin** and by **prostaglandins** (both of which are increased at term).

   f. After parturition the myometrium is reduced in size because smooth muscle cells, deprived of estrogen, undergo **apoptosis**.

3. External covering

   a. Serosa is present over surfaces of the uterus bulging into the peritoneal cavity.

   b. Adventitia is present along the retroperitoneal surfaces of the uterus.

B. The **menstrual cycle begins** on the day menstrual **bleeding appears.**

   1. Menstrual phase (days 1-4) is characterized by a **hemorrhagic discharge (menses)** of the functional layer of the endometrium.

   a. It is triggered by **spasms of contraction** and **relaxation** of the coiled arteries (caused by low levels of progesterone and estrogen). Long-term (2-3 days) **vasoconstriction** of these arteries causes ischemia and eventual necrosis.

   b. Vasoconstriction is followed by sudden, intermittent **vasodilation** of the coiled arteries, which ruptures their walls, flooding the stroma with blood, detaching the functional layer, and dislodging the necrotic tissue.

   c. Because the basal layer is supplied by short straight vessels that do not undergo prolonged vasoconstriction, it is not sloughed and does not become necrotic.

2. The **proliferative (follicular) phase (days 4-14)** follows the menstrual phase and involves **renewal of the entire functional layer,** including the repair of glands, connective tissue, and vascular elements (coiled arteries).

   a. The epithelium lining the luminal surface is renewed from mitotic activity of cells remaining in uterine glands of the basal layer of the endometrium.

   b. Glands are straight and lined by a simple columnar epithelium.

   c. Stromal cells divide, accumulate glycogen, and enlarge.

   d. Coiled arteries extend approximately two thirds of the way into the endometrium.

3. The **secretory (luteal) phase (days 15-28)** begins shortly after ovulation and is characterized by a **thickening of the endometrium,** resulting from edema and secretion by the endometrial glands.

   a. Glands become coiled; their lumens become filled with a secretion of glycoprotein material; and their cells accumulate large amounts of glycogen, basally.

   b. Coiled arteries become more highly coiled and longer, extending into the superficial aspects of the functional layer.

V. Cervix

A. The cervix does not participate in menstruation, but its **secretions change** during different stages of the menstrual cycle.

B. The cervical wall is composed mainly of dense collagenous connective tissue interspersed with numerous elastic fibers and a few smooth muscle cells.

C. The cervix has a **simple columnar (mucus-secreting) epithelium** except for the inferior portion (continuous with the lining of the vagina), which is covered by **a stratified squamous nonkeratinized epithelium.**

D. Branched **cervical glands** secrete a serous fluid near the time of ovulation that facilitates the entry of spermatozoa into the uterine lumen. During pregnancy, cervical glands produce a thick, viscous secretion that hinders the entry of spermatozoa (and microorganisms) into the uterus.

E. Prior to parturition, the cervix dilates and softens due to the lysis of collagen in response to the hormone **relaxin.**

VI. Fertilization and Implantation

A. Fertilization

1. Fertilization usually takes place within the **ampulla of the oviduct.**

2. It occurs when a spermatozoon penetrates the corona radiata, zona pellucida, and the plasma membrane of a **secondary oocyte.**

3. It triggers the resumption and completion of the second meiotic division, with the subsequent formation of an **ovum** and second polar body.

4. It is completed when the male haploid (n) pronucleus (derived from the spermatozoon) and the female haploid (n) pronucleus (derived from the ovum) fuse, forming a diploid (2n) cell known as a **zygote.**

B. Implantation

1. The **zygote** undergoes cell division (cleavage) and is transformed into a multicellular structure called a **morula,** which requires about 3 days to travel through the oviduct and enter the uterus.

2. The **conceptus** (the preimplantation embryo and its surrounding membranes) acquires a fluid-filled cavity and becomes known as a blastocyst.

3. The **blastocyst** implants into the endometrium of the uterus and becomes surrounded by an inner cellular layer, the **cytotrophoblast,** and an outer, multinucleated layer, the **syncytiotrophoblast.**

4. The **syncytiotrophoblast** further invades the endometrium in the wall of the uterus by the sixth day after fertilization. Formation of the placenta then begins.

VII. Placenta

A. The placenta is a **transient** structure, consisting of a **maternal portion** and a **fetal portion.**

B. Function

1. The placenta permits the exchange of various materials between the maternal and fetal circulatory systems. This exchange occurs **without** mixing of the two separate blood supplies.

2. It secretes **progesterone, human chorionic gonadotropin (hCG), chorionic thyrotropin,** and **human chorionic somatomammotropin** (**hCS**) (a lactogenic and growth promoting hormone).

3. It also produces **estrogen** with the assistance of the liver and adrenal cortex of the fetus.

4. Decidual cells of the stroma produce **prostaglandins** and **prolactin.**

VIII. Vagina

A. Overview

1. The vagina is a **fibromuscular** tube with a wall that is composed of three layers: an inner **mucosa,** a middle **muscularis,** and an external **adventitia.**

2. It is circumscribed by a **skeletal muscle** sphincter at its external orifice.

3. It lacks glands throughout its length and is lubricated by secretions from the cervix and by seepage of the extracellular fluid from the vascular supply of the lamina propria.

B. The **mucosa** is composed of a thick, **stratified squamous nonkeratinized epithelium** and a fibroelastic connective tissue, the **lamina propria.**

1. The **epithelium** contains **glycogen,** which is used by the vaginal bacterial flora to produce **lactic acid;** lactic acid lowers the pH during the follicular phase of the menstrual cycle and inhibits invasion by pathogens.

2. The **lamina propria** is a fibroelastic connective tissue that is **highly vascular** in its deeper aspect (which may be considered analogous to a submucosa).

C. The **muscularis** is composed of irregularly arranged layers of **smooth muscle** (thin inner circular layer and thicker outer longitudinal) interspersed with **elastic fibers.**

D. The **adventitia** is composed of **fibroelastic** connective tissue and fixes the vagina to the surrounding structures.

IX. External Genitalia (Vulva)

A. The **labia majora** are **fat-laden folds of skin; hair, sebaceous glands,** and **sweat glands** are present on their external surfaces.

B. **Labia minora**

1. The labia minora are folds of skin that possess a core of **highly vascular** connective tissue containing elastic fibers.

2. They lack hair follicles but contain numerous **sebaceous glands,** which open directly onto the epithelial surface.

C. The **vestibule** is the space between the two labia minora. **Glands of Bartholin (mucus-secreting glands)** and numerous small mucus-secreting glands located around the urethra and clitoris **(minor vestibular glands)** open into this space.

D. **Clitoris**

1. The clitoris is composed of two small, cylindrical **erectile bodies,** which terminate in the prepuce-covered **glans clitoridis.**

2. It contains many sensory nerve fibers and specialized nerve endings (e.g., Meissner corpuscles and pacinian corpuscles).

X. Mammary Glands.
Each mammary gland is composed of numerous **compound tubuloalveolar glands,** each with its own lactiferous sinus and a duct that opens at the apex of the nipple.

A. **Resting mammary glands** (in adult, nonpregnant women)

1. Resting mammary glands are composed of **lactiferous sinuses** and **ducts** lined in most areas by a stratified cuboidal epithelium that has a basal layer that consists of scattered **myoepithelial cells.**

2. A basal lamina separates the epithelial components from the underlying stroma.

B. **Active (lactating) mammary glands** are enlarged during pregnancy due to the development of **alveoli.**

1. **Alveolar cells** (secretory cells) (Figure 19.3)

a. Alveolar cells line the alveoli of active mammary glands and are surrounded by an incomplete layer of **myoepithelial cells.**

b. They are richly endowed with RER and contain several Golgi complexes, numerous mitochondria, lipid droplets, and vesicles containing milk protein (caseins) and lactose.

2. **Secretion by alveolar cells**

a. **Lipids** are released into the alveolar lumen (perhaps) via the **apocrine** mode of secretion.

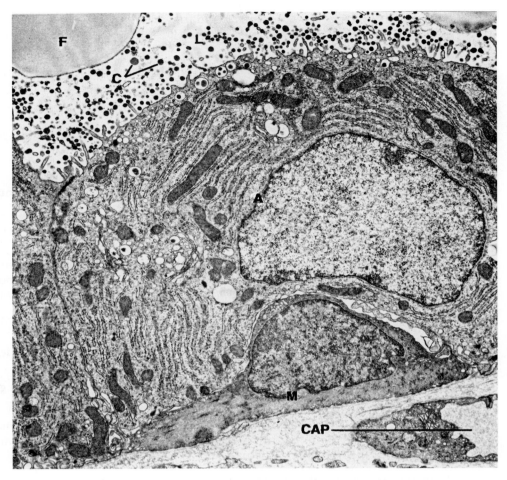

Figure 19.3. Transmission electron micrograph showing alveolar epithelial cell *(A)* from lactating mammary gland and an underlying myoepithelial cell *(M)*. CAP = capillary; L = lumen of alveolus containing milk; F = fat droplet; C = casein. (Reprinted with permission from Strum J: *A Study Atlas of Electron Micrographs*, 3rd ed. Baltimore, University of Maryland School of Medicine, 1992, p 105.)

    b. Proteins and sugars are released into the alveolar lumen via the **merocrine** mode of secretion (exocytosis).

C. Nipple

    1. The nipple is composed of dense collagenous connective tissue interlaced with smooth muscle fibers that act as a **sphincter.**

    2. It contains the openings of the lactiferous ducts.

    3. It is surrounded by pigmented skin **(areola)** that is more deeply pigmented during and subsequent to pregnancy and contains the **areolar glands (of Montgomery).**

D. Secretions of the mammary glands

    1. Colostrum (protein-rich, yellowish fluid)

       a. Colostrum is produced during the first few days after birth.

    b. It is rich in cells (lymphocytes, monocytes), lactalbumin, fat-soluble vitamins, and minerals and contains **immunoglobulin A (IgA).**

2. Milk

    a. Milk begins to be secreted by the third or fourth day after birth.

    b. It consists of proteins (caseins, IgA, lactalbumin), many lipid droplets, and lactose.

    c. It is released from the mammary glands, via the **milk ejection reflex,** in response to a variety of external stimuli related to suckling. The milk ejection reflex involves release of **oxytocin** (from axons in the pars nervosa of the pituitary gland), which induces contraction of the **myoepithelial cells,** forcing milk into the larger ducts and out of the breast.

XI. Clinical Considerations

A. Papanicolaou (Pap) smear

    1. In a Pap smear, epithelial cells are scraped from the lining of the cervix (or vagina) and are examined to detect cervical cancer.

    2. A Pap smear shows **variation in cell populations** with stages of the menstrual cycle.

B. Carcinoma of the cervix

    1. Carcinoma of the cervix originates from stratified squamous nonkeratinized epithelial cells.

    2. It may be contained within the epithelium and not invade the underlying stroma **(carcinoma in-situ),** or it may penetrate the basal lamina and metastasize to other parts of the body **(invasive carcinoma).**

    3. It occurs at a relatively high frequency, but may be cured by surgery if discovered early (by Pap smears) before it becomes invasive.

C. Endometriosis

    1. Endometriosis is a condition in which uterine endometrial tissue exists in the pelvic peritoneal cavity.

    2. It is associated with hormone-induced changes occurring in the ectopic endometrium during the menstrual cycle. As the endometrium is shed, bleeding occurs in the peritoneal cavity, causing severe pain and the formation of cysts and adhesions.

    3. It may lead to **sterility** because the ovaries and oviducts become deformed and embedded in scar tissue.

D. Ectopic (Tubal) pregnancy

    1. Ectopic pregnancy is the **implantation** of the early **embryo** in an abnormal site (e.g., **wall of the oviduct**).

    2. It can be **fatal** without immediate medical intervention.

E. Teratomas

Teratomas are germ cell tumors that are classified into three groups: mature, monodermal, and immature.

1. **Mature teratomas** are benign (although occasionally they may become malignant) and are usually present in young women. These are cysts with walls that frequently contain hair and other epidermal structures such as sebaceous glands, as well as bone, tooth, and cartilage fragments.

2. **Monodermal teratomas** are rare tumors that are also known as specialized teratomas. The two most frequent types of these tumors are struma ovarii and ovarian carcinoid. **Struma ovarii** is an ovarian tumor composed of well-developed thyroid follicles that produce thyroid hormone and may be responsible for hyperthyroidism. **Ovarian carcinoid** is a tumor that usually produces serotonin (5-OH-tryptamine).

3. **Immature teratomas** are fast-growing malignant tumors with a histology that resembles that of fetal rather than mature tissues. They are usually present in adolescents and very young women.

F. **Breast cancer**

1. Breast cancer may originate from the epithelium lining the ducts **(ductal carcinoma)** or the terminal ductules **(lobular carcinoma).**

2. If breast cancer is not treated early, the tumor cells **metastasize,** via lymphatic vessels, to the axillary nodes near the affected breast and later, via the bloodstream, to the lungs, bone, and brain.

3. In the United States, 180,000 new cases of breast cancer are diagnosed annually, and every year 43,000 women die of this disease.

4. **Early detection** by self-examination, mammography, or ultrasound has led to a reduction in the mortality rate associated with breast cancer.

5. Deficiency or mutation in the gene **BRCA1** has been shown to decrease the stability of or elevate the incidence of the mutation rate of tumor suppressor genes such as *p53*. It appears that **mutations in the BRCA1 gene** result in incapacitation of the checkpoint at G2-M of the cell cycle and, concurrently, the number of centrosomes of these cells is increased. Therefore, these mutated cells have the capability to proliferate unchecked.

Review Test

Directions: Each of the numbered items or incomplete statements in this section is followed by answers or by completions of the statement. Select the ONE lettered answer or completion that is BEST in each case.

1. Which of the following statements concerning secondary ovarian follicles is true?

(A) They lack liquor folliculi.
(B) They contain a secondary oocyte.
(C) Their continued maturation requires follicle-stimulating hormone.
(D) They lack a theca externa.
(E) They have a single layer of cuboidal follicular cells surrounding the oocyte.

2. Colostrum contains which of the following antibodies?

(A) Immunoglobulin A (IgA)
(B) IgD
(C) IgE
(D) IgG
(E) IgM

3. Which of the following statements concerning the corpus luteum is true?

(A) It produces luteinizing hormone.
(B) It produces follicle-stimulating hormone.
(C) It derives its granulosa luteal cells from the theca externa.
(D) It becomes the corpus albicans.
(E) It is derived from atretic follicles.

4. Luteinizing hormone exerts which one of the following physiologic effects?

(A) It triggers completion of the second meiotic division by secondary oocytes.
(B) It triggers ovulation.
(C) It suppresses release of estrogens.
(D) It induces primary follicles to become secondary follicles.

5. The basal layer of the uterine endometrium

(A) becomes sloughed during menstruation
(B) has no glands
(C) is supplied by coiled arteries
(D) is supplied by straight arteries

6. One of the recognized phases of the menstrual cycle is termed the

(A) gestational phase
(B) active phase
(C) follicular phase
(D) resting phase

7. During the proliferative phase of the menstrual cycle, the functional layer of the endometrium undergoes which of the following changes?

(A) Blood vessels become ischemic.
(B) The epithelium is renewed.
(C) The stroma swells due to edema.
(D) Glands become coiled.

8. Which of the following statements concerning the vaginal mucosa is true?

(A) It is lined by stratified columnar epithelium.
(B) It is lined by stratified squamous keratinized epithelium.
(C) It possesses no elastic fibers.
(D) It is lubricated by glands located in the cervix.
(E) Its cells secrete lactic acid.

9. Which of the following statements concerning the oviduct is true?

(A) It is lined by a simple cuboidal epithelium.
(B) Its epithelium contains peg cells.
(C) It functions in nourishing trilaminar germ discs.
(D) Fertilization most often occurs in its fimbriated portion.
(E) Its epithelium contains goblet cells.

10. Which one of the following teratomas is a malignant, fast-growing tumor?

(A) Mature teratoma
(B) Monodermal teratoma
(C) Immature teratoma
(D) Struma ovarii
(E) Ovarian carcinoid

Answers and Explanations

1-C. Secondary follicles are dependent on follicle-stimulating hormone for their continued development. They are established when liquor folliculi (an ultrafiltrate of plasma and granulosa cell secretions) begins to accumulate among the granulosa cells. Secondary follicles contain a primary oocyte blocked in the prophase of meiosis I.

2-A. Immunoglobulin A (IgA) antibodies are present in colostrum and milk. IgG antibodies are acquired by the fetus by placental transfer from the mother.

3-D. A corpus albicans is formed from a corpus luteum that has ceased to function. Luteinizing hormone and follicle-stimulating hormone are both produced in the anterior pituitary gland. Granulosa luteal cells are derived from the granulosa cells of an ovulated graafian follicle.

4-B. A sudden surge of luteinizing hormone near the middle of the menstrual cycle triggers ovulation.

5-D. The basal layer of the uterine endometrium is supplied by the straight arteries and contains the deeper portions of the uterine glands. Cells from these glands re-epithelialize the endometrial surface after the functional layer (supplied by the coiled arteries) has been sloughed.

6-C. The recognized phases of the menstrual cycle are the follicular (proliferative), secretory (luteal), and menstrual phases. The mammary glands are characterized by active (lactating) and resting phases. The term "gestational phase" refers to the period of pregnancy.

7-B. During the proliferative phase of the menstrual cycle, the entire functional layer of the endometrium is renewed, including the epithelium lining the surface and glands. Edema in the stroma and coiled glands are characteristic of the secretory phase of the cycle, and ischemia is responsible for the menstrual phase.

8-D. The vagina lacks glands throughout its length and is lubricated by secretions from cervical glands. It is lined by a stratified squamous nonkeratinized epithelium with cells that release glycogen, which is used by the normal bacterial flora of the vagina to manufacture lactic acid.

9-B. The oviduct is lined by a simple columnar epithelium that is composed of ciliated cells and peg cells but no goblet cells. Fertilization most often occurs in the ampulla of the oviduct, not in the infundibulum where fimbria are located. Under normal circumstances the trilaminar germ disc stage occurs after the blastocyst is implanted in the wall of the uterus and is not present in the oviduct.

10-C. Immature teratomas are fast-growing malignant tumors. The other teratomas listed are usually benign, but even those that do become malignant are slow growing.

20

Male Reproductive System

I. Overview—Male Reproductive System

A. The male reproductive system consists of the **testes, genital ducts,** accessory genital glands (**seminal vesicles, prostate gland,** and **bulbourethral glands**), and the **penis.**

B. Function. The male reproductive system produces **spermatozoa** (sperm), **testosterone,** and **seminal fluid.** Seminal fluid transports and nourishes the sperm as they pass through the excretory ducts. The penis delivers sperm to the exterior and also serves as the conduit for excretion of urine from the body.

II. Testes.
Testes develop in the abdominal cavity and later descend into the scrotum, where they are suspended at the ends of the **spermatic cords.** They are the sites of **spermatogenesis** and production of the male **sex hormones,** primarily **testosterone.**

A. Testicular tunicae

1. The **tunica vaginalis** is a **serous sac,** derived from the peritoneum, which partially covers the anterior and lateral surfaces of each testis.

2. Tunica albuginea

a. The tunica albuginea is the thick, fibrous connective tissue capsule of the testis.

b. It is lined by a highly vascular layer of loose connective tissue, the **tunica vasculosa.**

c. It is thickened posteriorly to form the **mediastinum testis,** from which incomplete connective tissue septae arise to divide the organ into approximately 250 compartments (lobuli testis).

B. Lobuli testis

1. The lobuli testis are pyramid-shaped intercommunicating compartments that are separated by incomplete septa.

2. Each contains 1-4 **seminiferous tubules.** These tubules are embedded in a meshwork of loose connective tissue containing blood and lymphatic vessels, nerves, and interstitial cells of Leydig.

C. Interstitial cells of Leydig

1. Interstitial cells of Leydig are round to polygonal cells located in the interstitial regions between seminiferous tubules.

2. They possess a large central nucleus, numerous mitochondria, a well-developed Golgi complex, and many lipid droplets. The lipid droplets contain cholesterol esters, which are precursors of testosterone.

3. They are richly supplied with capillaries and lymphatic vessels.

4. Function. Interstitial cells of Leydig are **endocrine cells** that produce and secrete **testosterone.** Secretion is stimulated by **luteinizing hormone** (LH; interstitial cell-stimulating hormone) produced in the pituitary gland. These cells mature and begin to secrete during puberty.

D. Seminiferous tubules

1. Overview

    a. Seminiferous tubules are 30-70 centimeters (cm) long with a diameter of 150-250 micrometers (μm).

    b. They are enveloped by a fibrous connective tissue tunic composed of several layers of fibroblasts and extensive capillary beds.

    c. They form tortuous pathways through the testicular lobules and then narrow into short, straight segments, the **tubuli recti,** which connect with the **rete testis.**

    d. They are lined by a thick complex epithelium (**seminiferous,** or **germinal epithelium**). This epithelium consists of 4 to 8 cell layers and contains **spermatogenic cells,** from which the germ cells eventually develop (spermatogenesis), and **Sertoli cells,** which have several functions.

2. Sertoli cells (Figure 20.1)

    a. Structure

        (1) Sertoli cells have a pale, oval nucleus, which displays frequent indentations; are highly infolded; and possess a large nucleolus.

        (2) They have a well-developed smooth endoplasmic reticulum (SER), some rough endoplasmic reticulum (RER), an abundance of mitochondria and lysosomes, and an extensive Golgi complex.

        (3) **Receptors for follicle-stimulating hormone (FSH)** are present on their plasma membranes.

        (4) They form **zonulae occludens** (tight junctions) with adjacent Sertoli cells near their bases, thus dividing the lumen of the seminiferous tubule into a **basal** and an **adluminal compartment.** These junctions are responsible for the **blood-testis barrier,** which protects developing sperm cells from autoimmune reactions.

    b. Function

        (1) Sertoli cells support, protect, and nourish the spermatogenic cells.

        (2) They phagocytize excess cytoplasm discarded by maturing spermatids.

        (3) They secrete a fluid into the lumen that transports spermatozoa through the seminiferous tubules to the genital ducts.

        (4) They synthesize **androgen-binding protein (ABP)** under the influence of FSH. **ABP** assists in maintaining the necessary concentration of testosterone in the seminiferous tubule so that spermatogenesis can progress.

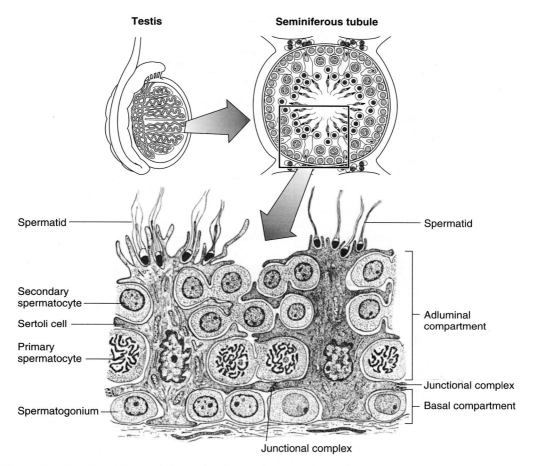

Testis   **Seminiferous tubule**

Spermatid

Spermatid

Secondary spermatocyte

Sertoli cell

Primary spermatocyte

Adluminal compartment

Junctional complex

Basal compartment

Spermatogonium

Junctional complex

Figure 20.1. Drawing of the seminiferous (germinal) epithelium. Note the intercellular bridges between spermatocytes and the junctional complexes near the bases of adjacent Sertoli cells. These junctional complexes of the Sertoli cells divide the epithelium into an adluminal and a basal compartment. (Reprinted with permission from Krause WJ, Cutts JH: *Concise Text of Histology*, 2nd ed. Baltimore, Williams & Wilkins, 1986, p 414.)

    (5) They secrete **inhibin,** a hormone that inhibits the synthesis and release of FSH by the anterior pituitary.

    (6) They establish a blood-testis barrier.

    (7) They synthesize and release **antimüllerian hormone,** which determines maleness.

  3. Spermatogenesis

    a. Spermatogenesis is the entire process of spermatozoon formation. It is divided into three phases:

      (1) Spermatocytogenesis—process of differentiation of spermatogonia into primary spermatocytes

      (2) Meiosis—reduction division to reduce the diploid chromosomal compliment of primary spermatocytes to form haploid spermatids

      (3) Spermiogenesis—process of transformation of spermatids into spermatozoa

   b. Spermatogenesis does **not** occur simultaneously or synchronously in all seminiferous tubules, but rather in wave-like sequences of maturation, referred to as **cycles of the seminiferous epithelium.**

   c. During spermatogenesis, daughter cells remain connected to each other via **intercellular bridges.** The resultant **syncytium** may be responsible for the **synchronous development** of germ cells along **any one** seminiferous tubule.

4. Spermatogenic cells (see Figure 20.1)

   a. Spermatogonia are **diploid** germ cells located adjacent to the basal lamina of the seminiferous epithelium. At puberty, they become influenced by testosterone to enter the cell cycle.

      (1) Pale type A spermatogonia possess a pale-staining nucleus, spherical mitochondria, a small Golgi complex, and abundant free ribosomes. They are **mitotically active** (starting at puberty) and give rise either to more cells of the same type (to maintain the supply) or to type B spermatogonia.

      (2) Dark type A spermatogonia represent mitotically **inactive** (reserve) cells (in the G_0 phase of the cell cycle) with dark nuclei; they have the potential to resume mitosis and produce pale type A cells.

      (3) Type B spermatogonia undergo mitosis and give rise to primary spermatocytes.

   b. Spermatocytes

      (1) Primary spermatocytes are large **diploid** cells with 4CDNA content. They undergo the **first meiotic division** (reductional division) to form secondary spermatocytes (see Chapter 2).

      (2) Secondary spermatocytes are **haploid** cells with 2CDNA that quickly undergo the **second meiotic division** (equatorial division), without an intervening S phase, to form spermatids.

   c. Spermatids

      (1) Spermatids are small **haploid** cells containing only **1CDNA.**

      (2) They are located near the lumen of the seminiferous tubule.

      (3) Their nuclei often display regions of condensed chromatin.

      (4) They possess a pair of centrioles, mitochondria, free ribosomes, SER, and a well-developed Golgi complex.

5. Spermiogenesis is a unique process of **cytodifferentiation** whereby **spermatids** shed much of their cytoplasm and become **transformed into spermatozoa,** which are released into the lumen of the seminiferous tubule. It is divided into four phases.

   a. Golgi phase

      (1) The Golgi phase is characterized by the formation of an **acrosomal granule,** enclosed within an **acrosomal vesicle,** which becomes attached to the anterior end of the nuclear envelope of a spermatid.

      (2) In this phase, centrioles migrate away from the nucleus to form the **flagellar axoneme.** The centrioles then migrate back toward the nucleus to assist in forming the **connecting piece** associated with the tail.

b. The **cap phase** is characterized by expansion of the acrosomal vesicle over much of the nucleus, forming the **acrosomal cap.**

c. **Acrosomal phase**

(1) The **nucleus** becomes condensed, flattened, and located in the head region.

(2) **Mitochondria** aggregate around the proximal portion of the flagellum, which develops into the middle piece of the tail.

(3) The **spermatid** elongates; this process is aided by a temporary cylinder of microtubules called the **manchette.**

(4) By the end of the acrosomal phase, the spermatid is oriented with its acrosome pointing toward the base of the seminiferous tubule.

d. **Maturation phase**

(1) The maturation phase is characterized by the loss of excess cytoplasm and intercellular bridges connecting spermatids into a syncytium. The discarded cytoplasm is phagocytized by Sertoli cells.

(2) This phase is completed when the **nonmotile** spermatozoa are released (tail first) into the lumen of the seminiferous tubule. Spermatozoa remain immotile until they leave the epididymis. They become **capacitated** (capable of fertilizing) in the female reproductive system.

6. **Spermatozoon**

a. **Head of the spermatozoon**

(1) The spermatozoon head is flattened and houses a dense, homogeneous nucleus containing 23 chromosomes.

(2) It also possesses the acrosome, which contains **hydrolytic enzymes** (e.g., acid phosphatase, neuraminidase, hyaluronidase, and proteases) that assist the sperm in penetrating the corona radiata and zona pellucida of the oocyte. Release of these enzymes is termed the **acrosomal reaction.**

b. **Tail of the spermatozoon**

(1) The **neck** houses the centrioles and the **connecting piece,** which is attached to the nine **outer dense fibers** of the remainder of the tail.

(2) The **middle piece** extends from the neck to the **annulus** and contains the axoneme, nine outer dense fibers, and a spirally arranged **sheath of mitochondria.**

(3) The **principal piece** extends from the annulus to the end piece and contains the axoneme with its surrounding dense fibers, which in turn are encircled by a **fibrous sheath** that has circumferentially oriented ribs.

(4) The **end piece** consists of the axoneme and the surrounding plasma membrane.

E. Regulation of spermatogenesis

1. **Critical testicular temperature** is 35°C for spermatogenesis to occur.

2. **Hormonal interactions** (Figure 20.2)

   a. **Certain neurons in the hypothalamus** produce gonadotropin-releasing hormone (GnRH), which initiates the release of LH and FSH from the adenohypophysis.

   b. **Stimulation of testicular hormone production** is effected by these two pituitary gonadotropins.

      (1) **LH** stimulates the interstitial cells of Leydig to secrete **testosterone.**

      (2) **FSH** promotes the synthesis of **ABP** by Sertoli cells.

   c. **Testosterone** is necessary for the normal development not only of male germ cells but also of secondary sex characteristics.

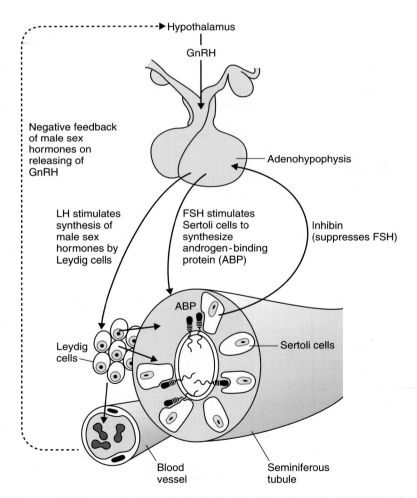

Figure 20.2. Diagram of the hormonal control of testicular function. Note the feedback inhibition of the pituitary by testicular hormones. ABP = androgen-binding protein; FSH = follicle-stimulating hormone; LH = luteinizing hormone; GnRH = gonadotropin-releasing hormone. (Adapted with permission from Fawcett DW: *Bloom and Fawcett's A Textbook of Histology*, 10th ed. New York, Chapman and Hall, 1975, p 839.)

d. ABP binds testosterone and maintains it at a high concentration in the seminiferous tubule; it can also bind estrogens, thus inhibiting spermatogenesis.

e. Inhibition of FSH and LH release

    (1) Excess levels of testosterone inhibit LH release.

    (2) Inhibin, a hormone secreted by Sertoli cells, inhibits FSH release.

III. Genital Ducts (Table 20.1)

A. Intratesticular ducts

    1. Tubuli recti are short, straight tubules lined by a **simple cuboidal epithelium** with **microvilli** and a single **flagellum.**

    2. The **rete testis** is a labyrinthine plexus of anastomosing channels lined by a **simple cuboidal epithelium;** many of the cells possess a single luminal **flagellum.**

Table 20.1. Histology and Functions of the Male Genital Ducts

Duct	Epithelium	Connective Tissue	Muscle Layers	Function
Tubuli recti	Sertoli cells in proximal half and simple cuboidal epithelium in distal half	Loose connective tissue	No smooth muscle	Conduct spermatozoa from the seminiferous tubules to the rete testis
Rete testis	Simple cuboidal epithelium	Vascular connective tissue	No smooth muscle	Conduct spermatozoa from the tubuli recti to the ductuli efferentes
Ductuli efferentes	Regions of ciliated columnar cells alternating with nonciliated cuboidal cells	Thin loose connective tissue	Thin layer of circularly arranged smooth muscle cells	Conduct spermatozoa from the rete testis to the epididymis
Epididymis	Pseudostratified epithelium composed of tall principal cells (with stereocilia) and short basal cells	Thin loose connective tissue	Layer of circularly arranged smooth muscle cells	Conduct spermatozoa from the ductuli efferentes to the ductus deferens
Ductus (vas) deferens	Pseudostratified stereociliated columnar epithelium	Loose fibroelastic connective tissue	Thick three-layered smooth muscle coats: inner and outer longitudinal, middle circular	Deliver spermatozoa from tail of epididymis to ejaculatory duct
Ejaculatory duct	Simple columnar epithelium	Subepithelial connective tissue is thrown into folds, giving the lumen an irregular appearance	No smooth muscle	Deliver spermatozoa and seminal fluid to prostatic urethra at colliculus seminalis

B. Extratesticular ducts

1. Ductuli efferentes

a. The ductuli efferentes are a collection of 10 to 20 tubules leading from the rete testis to the ductus epididymidis.

b. They have a thin circular layer of **smooth muscle** beneath the basal lamina of the epithelium.

c. They are lined by a **simple epithelium** composed of **alternating clusters** of **nonciliated cuboidal cells** and **ciliated columnar cells.**

d. Function. The ductuli efferentes reabsorb fluid from the semen.

2. Ductus epididymidis

a. The ductus epididymidis, together with the ductuli efferentes, constitutes the **epididymis.**

b. It is surrounded by **circular layers of smooth muscle** that undergo **peristaltic contractions,** which assist in conveying sperm toward the ductus deferens.

c. It is lined by a **pseudostratified columnar epithelium,** which is supported by a basal lamina and contains the following two cell types:

  (1) Basal cells, which are round and appear undifferentiated, and apparently serve as precursors of the principal cells

  (2) Principal cells, which are columnar in shape and possess nonmotile **stereocilia** (long, irregular microvilli) on their luminal surfaces

    (a) These cells have a large Golgi complex, RER, lysosomes, and many apically located pinocytic and coated vesicles; the latter suggest that these cells function in **fluid resorption.**

    (b) They secrete **glycerophosphocholine,** which probably inhibits **capacitation** (the process whereby a sperm becomes capable of fertilizing an oocyte).

3. Vas (ductus) deferens

a. The vas deferens has a **thick muscular wall** with inner and outer layers of longitudinally oriented smooth muscle, which are separated from one another by a middle circular layer.

b. It has a narrow, irregular lumen lined by a **pseudostratified columnar epithelium** similar to that of the ductus epididymidis.

4. Ejaculatory duct

a. The ejaculatory duct is the straight continuation of the ductus deferens beyond where it receives the duct of the seminal vesicle.

b. It lacks a muscular wall.

c. It enters the prostate gland and terminates in a slit on the **colliculus seminalis** in the prostatic urethra.

IV. Accessory Genital Glands

A. Seminal vesicles

1. **Epithelium**

   a. The epithelium of seminal vesicles is **pseudostratified columnar,** with a height that varies with testosterone levels; it lines the **extensively folded mucosa.**

   b. It contains many **yellow lipochrome pigment granules** and secretory granules, a large Golgi complex, many mitochondria, and an abundant RER.

2. The **lamina propria** consists of **fibroelastic** connective tissue surrounded by an inner circular and outer longitudinal layer of smooth muscle.

3. The **adventitia** is composed of **fibroelastic** connective tissue.

4. The seminal vesicles **secrete** a yellow, viscous fluid containing substances that **activate sperm** (e.g., fructose); this fluid constitutes about 70% of the human ejaculate.

B. Prostate gland

1. **Overview**

   a. The prostate gland surrounds the urethra as it exits the urinary bladder.

   b. It consists of 30 to 50 discrete **branched tubuloalveolar glands,** which empty their contents, via excretory ducts, into the prostatic urethra. These glands are arranged in three concentric layers (**mucosal, submucosal,** and **main**) around the urethra.

   c. It is covered by a **fibroelastic capsule** that contains smooth muscle. Septa from the capsule penetrate the gland and divide it into lobes.

2. **Epithelium**

   a. The epithelium of the prostate gland is **simple** or **pseudostratified columnar** and lines the individual glands that comprise the prostate.

   b. It is composed of cells that contain abundant RER, a well-developed Golgi complex, numerous lysosomes, and many secretory granules.

3. **Corpora amylacia** are **concretions,** composed of glycoprotein, which may become calcified; their numbers increase with age.

4. The **prostate secretes** a whitish, thin fluid containing proteolytic enzymes, citric acid, acid phosphatase, and lipids. Its synthesis and release are regulated by dihydrotestosterone.

C. Bulbourethral (Cowper) glands

1. Bulbourethral glands are located adjacent to the membranous urethra.

2. They empty their secretion into the lumen of the membranous urethra to lubricate it.

3. They are lined by a **simple cuboidal** or **columnar epithelium.**

4. They are surrounded by a **fibroelastic capsule** containing smooth and skeletal muscle.

V. Penis

A. Corpora cavernosa

1. The corpora cavernosa are **paired** masses of erectile tissue that contain **irregular vascular spaces** lined by a continuous layer of endothelial cells. These spaces are separated from each other by trabeculae of connective tissue and smooth muscle.

2. The vascular spaces decrease in size toward the periphery of the corpora cavernosa.

3. During erection, the vascular spaces become engorged with blood due to **parasympathetic impulses,** which constrict arteriovenous shunts and dilate the helicine arteries.

B. Corpus spongiosum

1. The corpus spongiosum is a single mass of **erectile tissue** that contains vascular spaces of uniform size.

2. It possesses **trabeculae** that contain more elastic fibers and less smooth muscle than those of the corpora cavernosa.

C. Connective tissue and skin

1. The **tunica albuginea** is a thick fibrous connective sheath that surrounds the paired corpora cavernosa and the corpus spongiosum. The arrangement of dense collagen bundles permits extension of the penis during erection.

2. **Glans penis**

   a. The glans penis is the dilated distal end of the corpus spongiosum.

   b. It contains dense connective tissue and longitudinal muscle fibers.

   c. It is covered by retractable skin, the **prepuce,** which is lined by stratified squamous lightly keratinized epithelium.

3. **Glands of Littre** are mucus-secreting glands present throughout the length of the penile urethra.

VI. Clinical Considerations

A. Cryptorchidism

1. Cryptorchidism is a developmental defect characterized by **failure of the testes to descend** into the scrotum.

2. This condition results in sterility because the temperature of the undescended testes (i.e., normal body temperature) inhibits spermatogenesis; however, it does not affect testosterone production.

3. It is **associated with** a much higher incidence of **testicular malignancy** than in normally descended testes.

4. It can be **surgically corrected,** usually between 5 and 7 years of age. After corrective surgery, however, affected individuals may have abnormal sperm.

B. Benign prostatic hypertrophy

1. Benign prostatic hypertrophy most commonly involves only the **mucosal glands,** which become enlarged.

2. It is frequently associated with an inability to begin and cease urination because the urethra is partially strangulated by the enlarged prostate.

3. It leads to **nocturia** (urination at night) and sensory urgency (the desire to urinate without having to void).

4. This disease is common in older men, occurring in about 50% of men over 50 years of age and in 95% of those over 70.

C. Adenocarcinoma of the prostate gland

1. Adenocarcinoma of the prostate gland may be diagnosed by palpation through the rectum.

2. This cancer commonly **metastasizes to bone** via the circulatory system.

3. It occurs in about one third of men over the age of 75 and is the second most common form of cancer in men.

4. It is associated with an elevated **prostatic specific antigen (PSA)** level in blood. Men with elevated PSA values are candidates for more aggressive diagnostic tests for prostatic cancer.

Review Test

Directions: Each of the numbered items or incomplete statements in this section is followed by answers or by completions of the statement. Select the ONE lettered answer or completion that is BEST in each case.

1. Which one of the following functions is attributed to the Sertoli cells?

(A) Secretion of follicle-stimulating hormone
(B) Secretion of testosterone
(C) Secretion of androgen-binding protein
(D) Secretion of luteinizing hormone
(E) Secretion of interstitial cell stimulating hormone

2. Which of the following statements concerning the cells of Leydig is true?

(A) They become functional at puberty.
(B) They are located within the seminiferous tubules.
(C) They are stimulated by follicle-stimulating hormone.
(D) They secrete much of the fluid portion of semen.
(E) They respond to inhibin.

3. Type A spermatogonia are germ cells that

(A) develop from secondary spermatocytes
(B) undergo meiotic activity subsequent to sexual maturity
(C) develop through meiotic divisions
(D) give rise to primary spermatocytes
(E) may be dark or pale

4. Which of the following statements is related to the ductus epididymidis?

(A) It begins at the rete testis.
(B) It is lined by a pseudostratified columnar epithelium.
(C) It secretes a large volume of fluid into its lumen.
(D) It possesses motile cilia.
(E) It capacitates spermatozoa.

5. Testosterone is produced by which of the following?

(A) Interstitial cells of Leydig
(B) Sertoli cells
(C) Spermatogonia
(D) Spermatids
(E) Spermatocytes

6. Spermatozoa are conveyed from the seminiferous tubules to the rete testis via the

(A) ductus epididymidis
(B) tubuli recti
(C) ductuli efferentes
(D) ductus deferens

7. Manchette formation occurs during which of the following phases of spermiogenesis?

(A) Meiotic phase
(B) Maturation phase
(C) Golgi phase
(D) Cap phase
(E) Acrosomal phase

8. The structural feature that best distinguishes the ductus deferens from the other genital ducts is its

(A) smooth-bore lumen
(B) thick muscular wall containing three muscle layers
(C) lining of transitional epithelium
(D) flattened mucosa
(E) nonmotile stereocilia

9. Inhibin, a hormone that inhibits synthesis and release of follicle-stimulating hormone, is secreted by which of the following structures?

(A) Prostate gland
(B) Sertoli cells
(C) Seminal vesicles
(D) Bulbourethral glands
(E) Interstitial cells of Leydig

10. A 55-year-old man presents with urinary complications. He complains of difficulty urinating and reduced urinary flow. He also has an elevated prostate-specific antigen (PSA) level, along with palpable hard nodules on the prostate. The possible diagnosis is

(A) benign prostatic hyperplasia
(B) adenocarcinoma of the prostate
(C) prostatic concretions
(D) initiation of impotence

Answers and Explanations

1-C. Sertoli cells have many functions, including the synthesis of androgen-binding protein. Testosterone is secreted by the interstitial cells of Leydig, whereas follicle-stimulating hormone and luteinizing hormone (also known as interstitial cell-stimulating hormone) are secreted by the pituitary gland.

2-A. Interstitial cells of Leydig become functional at puberty due to the action of luteinizing hormone produced in the pituitary gland.

3-E. Type A spermatogonia, which may be pale or dark, are primitive germ cells. Pale type A spermatogonia become mitotically active at puberty and give rise to type B spermatogonia, which undergo mitoses, giving rise to primary spermatocytes.

4-B. The ductus epididymidis begins at the termination of the ductuli efferentes and is lined by a pseudostratified columnar epithelium, which has principal cells that possess nonmotile stereocilia. These cells are involved in fluid resorption, and secrete glycerophosphocholine, which inhibits capacitation.

5-A. The hormone testosterone is produced by the interstitial cells of Leydig.

6-B. The seminiferous tubules are connected to the rete testis by the tubuli recti.

7-E. Spermiogenesis is the process of cytodifferentiation by which spermatids are transformed into spermatozoa. It does not involve cell division. Manchette formation occurs during the acrosomal phase of spermiogenesis.

8-B. The ductus (vas) deferens possesses three layers of smooth muscle in its wall, whereas the other genital ducts do not. Like the ductus epididymidis, the ductus deferens is lined by a pseudostratified columnar epithelium with principal cells that possess nonmotile stereocilia.

9-B. Inhibin is secreted by Sertoli cells.

10-B. Although some of the symptoms are characteristic of benign prostatic hypertrophy, rectal palpation indicating hard nodules, along with an elevated prostate-specific antigen (PSA) level, indicates probable prostatic adenocarcinoma.

21

Special Senses

I. Overview—Special Sense Receptors

A. Special sense receptors are responsible for the five special senses of **taste, smell, seeing, hearing,** and **feeling** (touch, pressure, temperature, pain, and proprioception).

B. **Function.** Special sense receptors **transduce stimuli from the environment into electrical impulses.**

II. Specialized Diffuse Receptors (Figure 21.1)

A. **Overview**

1. Specialized diffuse receptors are **dendritic nerve endings** located in the skin, fascia, muscles, joints, and tendons.

2. They respond to stimuli related to **touch, pressure, temperature, pain,** and **proprioception.**

3. These receptors are specialized to receive only **one** type of sensory stimulus, although they will respond to other types of stimuli if they are intense enough.

4. They are divided morphologically into **free nerve terminals** and **encapsulated nerve endings,** which are ensheathed in a connective tissue capsule.

B. **Touch and pressure receptors**

1. **Pacinian corpuscles**

   a. Pacinian corpuscles are large, ellipsoid **encapsulated** receptors located in the dermis and hypodermis and in the connective tissue of the mesenteries and joints.

   b. They are especially abundant in the digits and breasts.

   c. They are composed of a **multilayered capsule** consisting of fibroblasts and collagen and bathed in tissue fluid. The capsule surrounds an **inner unmyelinated nerve terminal.**

   d. They resemble a sliced onion in histologic section.

   e. **Function.** Pacinian corpuscles perceive **pressure, touch,** and **vibration.**

2. **Ruffini endings**

   a. Ruffini endings are **encapsulated** receptors located in the dermis and joints.

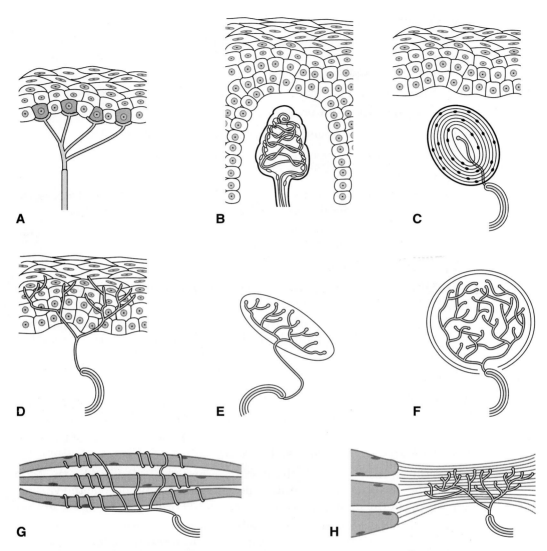

Figure 21.1. Diagram illustrating various types of specialized receptors. *(A)* Merkel disk. *(B)* Meissner corpuscle. *(C)* Pacinian corpuscle. *(D)* Free nerve endings, nociceptors, and thermoreceptors. *(E)* Ruffini corpuscle. *(F)* Krause end bulb (cold receptor). *(G)* Neuromuscular spindle. *(H)* Golgi tendon organ.

    b. They are composed of groups of **branched terminals** from myelinated nerve fibers and are surrounded by a thin connective tissue capsule.

    c. Function. Ruffini endings function in **pressure** and **touch** reception.

  3. Meissner corpuscles

    a. Meissner corpuscles are ellipsoid, **encapsulated** receptors located in the dermal papillae of thick skin, eyelids, lips, and nipples.

    b. They have a connective tissue capsule that envelops the nerve terminal and its associated Schwann cell.

    c. Function. Meissner corpuscles function in **fine-touch** perception.

   4. Free nerve endings

    a. Free nerve endings are **unencapsulated, unmyelinated** terminations located in the skin in longitudinal and circular arrays around most of the hair follicles.

    b. Function. Free nerve endings function in **touch** perception.

C. Temperature and pain receptors

  1. Cold receptors respond to temperatures **below 25°C-30°C.**

  2. Heat receptors respond to temperatures above **40°C-42°C.**

  3. Nociceptors

    a. Nociceptors are sensitive to **pain stimuli** from mechanical stress, extremes in temperature, or the presence of certain cytokines.

    b. They are delicate, myelinated fibers that lose their myelin before entering the epidermis.

D. Proprioceptive receptors (see Chapter 8)

  1. Golgi tendon organs are encapsulated mechanoreceptors sensitive to stretch and tension in tendons.

  2. Muscle spindle receptors are 3-12 small encapsulated intrafusal muscle fibers called **flower spray endings** and **annulospiral endings** that sense differences in muscle length and tension.

III. Eye (Figure 21.2)

A. Overview

  1. The eye is the photosensitive organ responsible for vision.

  2. It is composed of three layers: the **tunica fibrosa** (outer layer), **tunica vasculosa** (middle layer), and **retina.**

  3. It receives **light** through the **cornea.** The light is focused by the **lens** onto the **retina,** which contains specialized cells that encode the various patterns of the image for transmission to the brain via the **optic nerve.**

  4. The eye possesses **intrinsic muscles** that adjust the aperture of the iris and alter the lens diameter, permitting **accommodation** for close vision.

  5. It possesses **extrinsic muscles,** attached to the external aspect of the orb (eyeball), which move the eyes in a coordinated manner to access the desired visual fields.

  6. The orb is moistened on its anterior surface with **lacrimal fluid** (tears) secreted by the **lacrimal gland.**

  7. It is covered by the upper and lower eyelids, which protect the anterior surface.

B. Tunica fibrosa

  1. The **sclera** is the **opaque,** relatively avascular, fibrous connective tissue layer that covers the posterior five sixths of the orb, which receives insertions of the extrinsic ocular muscles.

  2. The **cornea** is the **transparent,** highly innervated, **avascular** anterior one sixth of the tunica. It joins the sclera in a region called the **limbus,** which is highly vascularized, and is composed of five layers.

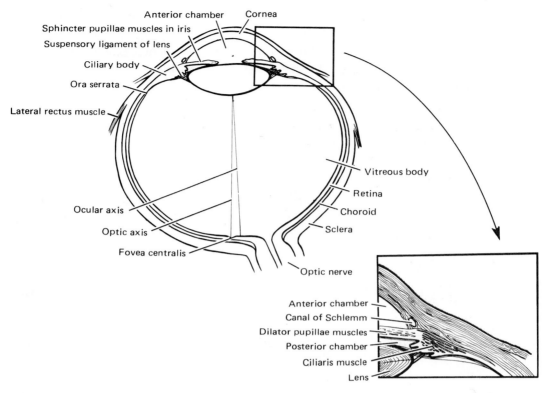

Figure 21.2. Diagram of the internal anatomy of the orb. (Adapted with permission from Hiatt JL, Gartner LP: *Textbook of Head and Neck Anatomy*, 2nd ed. Baltimore, Williams & Wilkins, 1987, p 178.)

a. Corneal epithelium

    (1) The corneal epithelium lines the **anterior aspect** of the cornea.

    (2) It is a **stratified squamous nonkeratinized epithelium.**

    (3) It possesses **microvilli** in its superficial layer; the microvilli trap moisture, protecting the cornea from dehydration.

b. Bowman membrane is a homogeneous **noncellular** layer that functions to provide form, stability, and strength to the cornea.

c. Corneal stroma

    (1) The corneal stroma is the thickest corneal layer.

    (2) It has channels, located in the region of the limbus, that are lined by **endothelium,** forming the **canal of Schlemm.** This canal drains fluid from the anterior chamber of the eye into the venous system.

d. Descemet membrane is a thick [5-10 micrometers (μm)] basal lamina separating the stroma from the endothelium lining the cornea.

e. Corneal endothelium

    (1) The corneal endothelium lines the **posterior aspect** of the cornea.

(2) It is a **simple squamous epithelium** with cells that exhibit numerous **pinocytic vesicles.**

(3) It **resorbs fluid** from the stroma, thus contributing to the transparency of the cornea, contributory to light refraction.

C. **Tunica vasculosa (uvea)** is composed of three parts:

1. **Choroid**

   a. The choroid is the **highly vascular, pigmented layer** of the eye located on the posterior wall of the orb; its loose connective tissue contains many **melanocytes.**

   b. It is loosely attached to the tunica fibrosa.

   c. It has a deep **choriocapillary layer** and **Bruch membrane** (basement membrane), which extends from the **optic disk** to the **ora serrata.**

2. **Ciliary body**

   a. The ciliary body is the wedge-shaped **anterior expansion of the choroid.**

   b. It completely encircles the lens and separates the ora serrata from the iris.

   c. It is lined on its inner surface by two layers of cells: an **outer, pigmented columnar epithelium** rich in melanin and an **inner, nonpigmented simple columnar** epithelium.

      (1) **Ciliary processes:**

         (a) Are radially arranged extensions (about 70) of the ciliary body

         (b) Have a connective tissue core containing many **fenestrated capillaries**

         (c) Are covered by two epithelial layers. The **unpigmented inner layer** transports components from the plasma filtrate in the posterior chamber and thus forms the **aqueous humor,** which flows to the anterior chamber via the pupillary aperture.

         (d) Possess **suspensory ligaments** (zonulae) that arise from the processes and insert into the capsule of the lens, serving to anchor it in place

      (2) **Ciliary muscle:**

         (a) Is attached to the sclera and ciliary body in such a manner that its contractions stretch the ciliary body and release tension on the suspensory ligament and lens. Contraction permits the lens to become more convex, allowing the eye to focus on nearby objects **(accommodation).** With advancing age, the lens loses its elasticity, thereby gradually losing the ability to accommodate.

         (b) It is innervated via **parasympathetic fibers** of the oculomotor nerve (CN III).

3. **Iris**

   a. **Overview**

      (1) The iris is the most anterior extension of the choroid, separating the anterior and posterior chambers of the orb.

(2) It incompletely covers the anterior surface of the lens, forming an aperture called the **pupil** that is continually adjusted by intrinsic pupillary muscles.

(3) It is covered by an incomplete layer of pigmented cells and fibroblasts on its anterior surface.

(4) It has a wall composed of loose, vascular connective tissue containing melanocytes and fibroblasts.

(5) It is covered on its deep surface by a two-layered epithelium with pigmented cells that blocks light from entering the interior of the eye except via the pupil.

b. Eye color is blue if only a few melanocytes are present. Increasing amounts of pigment impart darker colors to the eye.

c. Dilator pupillae muscle

(1) Dilator pupillae muscle is a **smooth muscle** with fibers that radiate from the periphery of the iris toward the pupil.

(2) It contracts upon stimulation by **sympathetic** nerve fibers, **dilating the pupil.**

d. Sphincter pupillae muscle

(1) Sphincter pupillae muscle is **smooth muscle** arranged in concentric rings around the pupillary orifice.

(2) It contracts upon stimulation by **parasympathetic** nerve fibers, **constricting the pupil.**

D. Refractive media of the eye

1. Aqueous humor

a. The aqueous humor is a **plasma-like fluid,** located in the anterior compartment of the eye, that is **formed by epithelial cells lining the ciliary processes.**

b. It is secreted into the posterior chamber of the eye and then flows to the anterior chamber; from there it enters the venous system via the canal of Schlemm (see Figure 21.2).

2. The **lens** is a biconvex, **transparent,** flexible structure composed of the lens capsule, subcapsular epithelium, and lens fibers.

a. The **lens capsule** is a thick basal lamina that envelops the entire lens epithelium.

b. The **subcapsular epithelium** (located on the anterior lens surface only) is composed of a single layer of cuboidal cells that communicate with each other via **gap junctions** and interdigitate with lens fibers.

c. Lens fibers represent highly differentiated, elongated cells that, when mature, **lack** both nuclei and organelles. Lens fibers are filled with a group of proteins called **crystallins.**

d. The **suspensory ligament** stretches between the lens and the ciliary body, keeping tension on the lens and **enabling it to focus on distant objects.**

3. The **vitreous body** is a **refractile gel** composed mainly of water, collagen, and hyaluronic acid. This gel fills the interior of the globe posterior to the lens.

E. Retina (Figure 21.3)

1. **Overview**

    a. The retina is the innermost of the three tunics of the eye and is responsible for **photoreception.**

    b. It has a shallow depression in its posterior wall that contains only cones; this avascular region, called the **fovea centralis,** exhibits the greatest visual acuity.

    c. It displays **10 distinct layers;** they are discussed in order below from the outermost to the innermost.

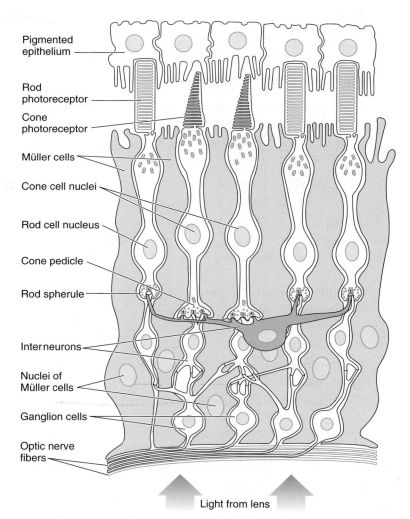

Figure 21.3. Schematic diagram of the layers of the retina. (Adapted with permission from Gartner LP, Hiatt JL: *Color Textbook of Histology*. Philadelphia, WB Saunders, 1997, p 429.)

2. The **retinal pigment epithelium** is a layer of **columnar cells** firmly attached to the **Bruch membrane.**

   a. **Structure**

      (1) Retinal pigment epithelial cells have **junctional complexes** and **basal invaginations** that contain mitochondria, suggesting the involvement of these cells in ion transport.

      (2) They contain smooth endoplasmic reticulum (SER) and many **melanin granules** located apically in cellular processes.

      (3) They extend **pigment-filled microvillus processes** that invest the tips of the rods and cones.

   b. **Function**

      (1) Retinal pigment epithelial cells **esterify vitamin A** (used in the formation of visual pigment by rods and cones).

      (2) They **phagocytize** the shed tips of the outer segments of rods.

      (3) They **synthesize melanin,** which absorbs light after the rods and cones have been stimulated.

3. The **photoreceptor layer** consists of **neurons (photoreceptor cells)** referred to as **rods** or **cones.** Their dendrites interdigitate with cells of the pigmented epithelium, and their bases form synapses with cells of the bipolar layer.

   a. **Rods (sensitive to light of low intensity)**

      (1) Overview

         (a) Rods have **outer** and **inner segments,** a **nuclear region,** and a **synaptic region.**

         (b) They may synapse with bipolar cells, giving rise to **summation.**

         (c) They contain an **incomplete cilium** terminating in a basal body within the inner segment.

         (d) They face the back of the eye; therefore, light must pass through all the other retinal layers before reaching the photosensitive region.

      (2) The **outer segments of rods:**

         (a) Consist mainly of hundreds of **flattened membranous disks** that contain **rhodopsin**

         (b) Eventually shed their disks, which are subsequently phagocytized by the pigment epithelial cells

      (3) The **inner segments of rods** possess mitochondria, glycogen, polyribosomes, and proteins, which migrate to the outer segments to become incorporated into the membranous disks.

      (4) Photoreception by rods is initiated by the interaction of **light** with **rhodopsin,** which is composed of the integral membrane protein **opsin** bound to **retinal,** the aldehyde form of vitamin A.

(a) The retinal moiety of rhodopsin **absorbs light** in the visible range.

(b) Retinal dissociates from opsin. This reaction, referred to as **bleaching,** eventually closes the Na+ channels, thus permitting diffusion of bound **Ca²⁺ ions** into the cytoplasm of the outer segment of a rod cell.

(c) Excess **Ca²⁺ ions** act to **hyperpolarize** the cell because Na⁺ ions are prevented from entering the cell.

(d) Ionic alterations in the rod generate electrical activity, which is relayed to other rods via gap junctions.

(e) Dissociated retinal and opsin **re-assemble** by an active process in which Müller and pigment epithelial cells also participate.

(f) Ca²⁺ ions are recaptured by the membranous disks, leading to reopening of the Na⁺ channels and **re-establishment** of the normal resting membrane potential.

b. Cones (sensitive to light of high intensity)

(1) Cones are **much less numerous than rods,** but produce **greater visual acuity** than do rods.

(2) They are generally similar in structure to rods and mediate photoreception in the same way, with the following exceptions:

(a) The membranous disks in the outer segments of cones are invaginations of the plasma membrane, whereas in rods they are not.

(b) The proteins synthesized in the inner segments of cones are passed to the entire outer segment, whereas in rods they are added only to newly forming disks.

(c) Cones possess **iodopsin** in their disks. This photopigment varies in amount in different cones, making them differentially sensitive to red, green, or blue light.

(d) Each cone synapses with a **single** bipolar neuron, whereas each rod may synapse with several bipolar neurons.

4. External limiting membrane

a. The external limiting membrane is not a true membrane but an area where **zonulae adherens** (belt desmosomes) are located between the photoreceptor cells and the retinal glial cells (Müller cells).

b. This membrane also contains microvilli that project from the Müller cells.

5. The **outer nuclear layer** consists primarily of the **nuclei of the rods and cones.**

6. Outer plexiform layer

a. The outer plexiform layer contains **axodendritic synapses** between the axons of photoreceptor cells and the dendrites of bipolar and horizontal cells.

    b. It displays **synaptic ribbons** within the rod and cone cells at synaptic sites.

  7. The **inner nuclear layer** contains the **cell bodies of bipolar neurons,** horizontal cells, and amacrine cells and the nuclei of Müller cells.

  8. Inner plexiform layer

    a. The inner plexiform layer contains **axodendritic synapses** between the axons of bipolar cells and the dendrites of ganglion cells.

    b. The processes of amacrine cells are located in this layer.

  9. The **ganglion cell layer** contains the **somata of ganglion cells,** which form the final link in the retina's neural chain.

    a. Structure—ganglion cells

      (1) Ganglion cells are typical neurons that project their axons to a specific region of the retina called the **optic disk.**

      (2) These cells are **midget, diffuse,** and **stratified ganglion cells.**

    b. Function—ganglion cells. Ganglion cells are activated by **hyperpolarization of rods and cones** and generate an **action potential,** which is transmitted to horizontal and amacrine cells and carried to the visual relay system in the brain.

 10. The **optic nerve fiber layer** consists primarily of the **unmyelinated axons** of ganglion cells, which form the fibers of the **optic nerve.** As each fiber pierces the sclera, it acquires a myelin sheath.

 11. The **inner limiting membrane** consists of the terminations of Müller cell processes and their basement membranes.

F. Accessory structures of the eye

  1. Conjunctiva (transparent mucous membrane)

    a. The conjunctiva lines the eyelids and is reflected onto the anterior portion of the orb up to the cornea, where it becomes continuous with the corneal epithelium.

    b. It is a **stratified columnar epithelium** possessing many **goblet cells.**

    c. It is separated by a basal lamina from an underlying lamina propria of loose connective tissue.

  2. Eyelids

    a. The eyelids are lined internally by conjunctiva and externally by skin that is elastic and covers a supportive framework of **tarsal plates.**

    b. They contain highly modified **sebaceous glands** (meibomian glands), **modified sebaceous glands** (glands of Zeis), and **sweat glands** (glands of Moll).

  3. Lacrimal apparatus

    a. Lacrimal gland

      (1) The lacrimal gland is a compound **tubuloalveolar gland** with secretory units that are surrounded by an incomplete layer of **myoepithelial cells.**

(2) It secretes **tears.** Tears drain via 6-12 ducts into the conjunctival **fornix,** from which the tears flow over the cornea and conjunctiva, keeping them moist. Tears (which contain **lysozyme,** an antibacterial enzyme) then enter the lacrimal puncta and the lacrimal canaliculi.

b. Lacrimal canaliculi are lined by a **stratified squamous epithelium** and unite to form a common canaliculus, which empties into the lacrimal sac.

c. The **lacrimal sac** is lined by a **pseudostratified ciliated columnar epithelium.**

d. The **nasolacrimal duct** is the inferior continuation of the lacrimal sac and also is lined by a **pseudostratified ciliated columnar epithelium.** The duct empties into the floor of the nasal cavity.

IV. Ear (Vestibulocochlear Apparatus). The ear consists of three parts: the **external ear,** which receives sound waves; the **middle ear,** through which sound waves are transmitted; and the **internal ear,** where sound waves are transduced into nerve impulses. The vestibular organ, responsible for equilibrium, is also located in the inner ear.

A. External ear (Figure 21.4)

1. The **auricle (pinna)** is composed of irregular plates of **elastic cartilage** covered by **thin skin.**

2. The **external auditory meatus** is lined by **skin** containing hair follicles, sebaceous glands, and **ceruminous glands** (modified sweat glands that produce **earwax (cerumen).**

3. **Tympanic membrane (eardrum)**

   a. The tympanic membrane is covered by **skin** on its external surface and by a **simple cuboidal epithelium** on its inner surface.

   b. It possesses **fibroelastic connective tissue** interposed between its two epithelial coverings.

   c. Function. The tympanic membrane **transmits sound vibrations** that enter the ear to the ossicles in the middle ear.

B. Middle ear (tympanic cavity)

1. The tympanic cavity houses the **ossicles (malleus, incus, and stapes),** which **transmit movements of the tympanic membrane to the oval window** (a membrane-covered opening in the bony wall).

2. It is connected to the pharynx via the **auditory tube (eustachian tube).**

3. It is lined by a **simple squamous epithelium,** which changes to **pseudostratified ciliated columnar epithelium** near its opening to the auditory tube.

4. It has a **lamina propria** composed of dense connective tissue tightly adherent to the bony wall.

C. Internal ear

1. The **bony labyrinth,** which is filled with **perilymph,** houses the membranous labyrinth.

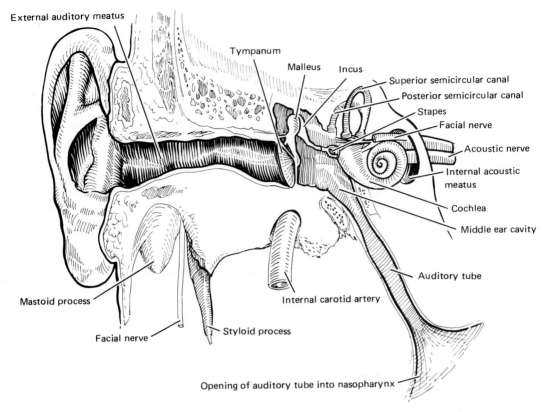

Figure 21.4. Diagram of the external, middle, and internal ears. (Reprinted with permission from Hiatt JL, Gartner LP: *Textbook of Head and Neck Anatomy*, 2nd ed. Baltimore, Williams & Wilkins, 1987, p 309.)

   a. Semicircular canals house the **semicircular ducts** of the membranous labyrinth.

   b. The **vestibule** houses the **saccule** and **utricle.**

   c. Cochlea

      (1) The cochlea winds 2 1/2 times around a bony core (the **modiolus**), which contains blood vessels and the spiral ganglion.

      (2) It is subdivided into three spaces: the **scala vestibuli** and **scala tympani,** which are both filled with perilymph, and the **scala media,** or cochlear duct, which is filled with endolymph.

  2. The **membranous labyrinth** is filled with **endolymph** and possesses various **sensory structures,** which represent **specializations of the epithelium.**

   a. Saccule and utricle (within the vestibule)

      (1) Overview

         (a) The saccule and utricle are sac-like bodies composed of a thin sheath of connective tissue lined by **simple squamous epithelium.**

(b) Each gives rise to a duct; the two ducts join, forming the **endolymphatic sac.**

(c) They possess small, specialized regions, called **maculae,** which contain type I and type II neuroepithelial hair cells, supporting cells, and a gelatinous layer **(otolithic membrane).**

(2) Vestibular hair cells

(a) These **neuroepithelial cells** contain many mitochondria and a well-developed Golgi complex.

(b) They possess 50-100 elongated, **rigid stereocilia** (sensory microvilli) arranged in rows and a single cilium **(kinocilium).** These cilia extend from the apical surface of the hair cells into the otolithic membrane. They function in the **detection of linear acceleration.**

(c) Types of vestibular hair cells include:

(i) Type I hair cells (bulbar in shape), which are almost completely surrounded by a **cup-shaped afferent nerve ending**

(ii) Type II hair cells (columnar in shape), which make contact with **small afferent terminals** containing synaptic vesicles

(3) Supporting cells are generally columnar and possess a round, basally located nucleus, many microtubules, and an extensive terminal web.

(4) The **otolithic membrane** is a **gelatinous layer** of glycoprotein that contains small calcified particles called **otoliths,** or **otoconia.**

b. Semicircular ducts are continuous with and arise from the utricle. The three semicircular ducts are positioned perpendicular to each other so that they can **detect angular acceleration** of the head in three-dimensional space.

(1) The **ampullae** are dilated regions of the semicircular ducts located near their junctions with the utricle.

(2) Cristae ampullares:

(a) Are specialized **sensory regions** within the ampullae of the semicircular ducts

(b) Are similar to maculae but have a thicker, cone-shaped glycoprotein layer **(cupula),** which does not contain otoliths

c. The **endolymphatic duct** ends in the expanded endolymphatic sac.

d. Endolymphatic sac

(1) The endolymphatic sac has an epithelial lining containing **electron-dense** columnar cells, which have an irregularly shaped nucleus, and **electron-lucent** columnar cells, which possess long microvilli, many pinocytic vesicles, and vacuoles.

(2) It contains **phagocytic cells** (macrophages, neutrophils) in its lumen.

(3) **Function.** The endolymphatic sac may function in **resorption of endolymph.**

e. **Cochlear duct** (Figure 21.5)

(1) **Overview**

(a) The cochlear duct is a specialized diverticulum of the saccule that contains the **spiral organ of Corti.**

(b) It is bordered above by the **scala vestibuli** and below by the **scala tympani** of the bony cochlea. These scalae, which contain perilymph, communicate with each other at the **helicotrema,** located at the apex of the cochlea.

(2) The **vestibular (Reissner) membrane:**

(a) Is composed of **two layers** of flattened squamous epithelium separated by an intervening basement membrane

(b) Helps maintain the high ionic gradients between the perilymph in the scala vestibuli and the endolymph in the cochlear duct

(3) The **stria vascularis:**

(a) Is a **vascularized** pseudostratified epithelium that lines the lateral aspect of the cochlear duct

(b) Is composed of basal, intermediate, and marginal cells

(c) May **secrete endolymph**

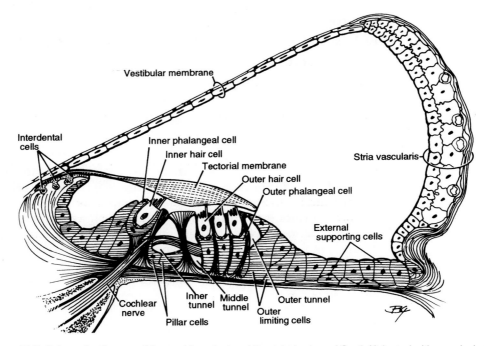

Figure 21.5. Schematic diagram of the cochlear duct and the spiral organ of Corti. (Adapted with permission from Dellmann H-D: *Textbook of Veterinary Histology,* 4th ed. Baltimore, Williams & Wilkins, 1993, p 332.)

(4) Spiral prominence

    (a) The spiral prominence is an epithelium-covered protuberance that extends the length of the cochlear duct. This epithelium is continuous with that of the stria vascularis and is reflected onto the basilar membrane, where it follows an indentation to form the **external spiral sulcus.**

    (b) Its cells become cuboidal and continue onto the basilar membrane, where they are known as the **cells of Claudius,** which overlie the polyhedral-shaped **cells of Boettcher.**

(5) The **basilar membrane:**

    (a) Is a thick layer of amorphous material containing **keratin-like fibers**

    (b) Extends from the spiral ligament to the tympanic lip of the limbus spiralis

    (c) Has two zones: the medial **zona arcuata** and the lateral **zona pectinata**

(6) The **tectorial membrane:**

    (a) Makes contact with the processes of the hair cells

    (b) Is secreted by the interdental cells of the spiral sulcus

(7) The **spiral organ of Corti:**

    (a) Lies upon both parts of the basilar membrane

    (b) Displays the **inner tunnel of Corti** and the **outer tunnel (space of Nuel),** which communicate with each other via intercellular spaces

    (c) Is composed of **hair cells** and various **supporting cells**

(8) Hair cells of the organ of Corti:

    (a) Are **neuroepithelial cells** containing a round, basally located nucleus, which is surrounded by many mitochondria

    (b) Possess many long, **stiff stereocilia** (arranged in a W-shaped formation) on their free surfaces, as well as a **basal body** (but no kinocilium)

    (c) Are divided into two types: **inner hair cells,** which are organized in a **single row** along the entire length of the cochlear duct and receive numerous afferent synaptic terminals on their basal surface, and **outer hair cells,** which are organized in **three or more rows** and are ensconced within a cup-shaped afferent nerve ending, where synaptic contacts are made

    (d) Function in the **reception of sound** and can respond to different sound frequencies

(9) Inner and outer pillar cells of the organ of Corti:

    (a) Are intimately associated with each other; both types rest on the basilar membrane

    (b) Enclose and support the inner tunnel of Corti, which lies between the inner and outer pillar cells

(c) Possess a wide base and have elongated processes, which contain microtubules, intermediate filaments, and actin filaments

(10) **Inner and outer phalangeal cells** of the organ of Corti:

(a) Are intimately associated with the inner and outer hair cells, respectively

(b) Support the slender nerve fibers that form synapses with the hair cells

(11) **Cells of Hensen and border cells of the organ of Corti** delineate the inner and outer borders of the spiral organ of Corti.

3. Auditory function of the inner ear

a. Movement of the stapes at the oval window causes disturbances in the perilymph, which cause deflection of the **basilar membrane.** Oscillations set in motion at the oval window are dissipated at the secondary tympanic membrane covering the round window of the cochlea. (At a very loud concert, the amount of energy absorbed is very high, and it may take up to 72 hours for it to be dissipated. A residual humming noise may be heard for 2 to 3 days.)

b. Large areas of the basilar membrane vibrate at many frequencies. However, optimal vibrations are detected at only specific areas. Sound waves of low frequency are detected farther away from the oval window.

c. The **pillar cells** attached to the basilar membrane move laterally in response to this deflection, in turn causing a lateral shearing of the stereocilia of the sensory hair cells of the organ of Corti against the tectorial membrane.

d. Movement of the stereocilia is transduced into electrical impulses that travel via the **cochlear nerve** to the brain.

4. Vestibular function of the inner ear

a. A change in the position of the head causes a flow of the endolymph in the semicircular ducts **(circular movement)** or in the saccules and utricles **(linear movement).**

b. Movement of the endolymph in the semicircular ducts displaces the cupula overlying the **cristae ampullares,** causing bending of the stereocilia of the sensory hair cells.

c. Movement of the endolymph in the saccules and utricles displaces the **otoliths.** This deformation is transmitted to the **maculae** via the overlying gelatinous layer, causing bending of the stereocilia of the sensory hair cells.

d. In both cases, **movement of the stereocilia** is transduced into **electrical impulses,** which are transmitted to the brain via **vestibular nerve fibers.**

V. Clinical Considerations

A. Conjunctivitis

1. Conjunctivitis is an inflammation of the conjunctiva producing red sclera and a discharge.

2. It may be caused by bacteria, viruses, parasites, and allergens.

3. Some forms are contagious and may cause blindness if left untreated.

B. Glaucoma

1. Glaucoma is a condition of abnormally **high intraocular pressure.** It is caused by obstructions that prevent drainage of aqueous humor from the eye.

2. **Chronic glaucoma,** the most common form of glaucoma, may be associated with few symptoms except for a gradual loss of peripheral vision. However, it usually can be controlled with eye drops.

C. Cataract

1. A cataract is an **opacity of the lens** resulting from the accumulation of pigment or other substances in the lens fibers.

2. This condition is often associated with **aging.**

3. **If untreated,** it leads to a **gradual loss of vision.**

D. Detachment of the retina

1. Retinal detachment occurs when the neural and pigmented retinae become separated from each other.

2. This condition can be treated successfully by **surgery.**

E. Conductive hearing loss

1. Conductive hearing loss results from a defect in the conduction of sound waves in the external or middle ear.

2. It may be caused by **otitis media,** a common inflammation of the middle ear; **obstruction** by a foreign body; or **otosclerosis** of the middle ear.

F. Nerve deafness

1. Nerve deafness results from a **lesion** in any of the nerves transmitting impulses from the spiral organ of Corti to the brain.

2. It may be caused by disease, exposure to drugs, or prolonged exposure to loud noises.

Review Test

Directions: Each of the numbered items or incomplete statements in this section is followed by answers or by completions of the statement. Select the ONE lettered answer or completion that is BEST in each case.

1. Which of the following specialized receptors exhibits a large ovoid capsule consisting of many concentric lamellae each separated by a space containing tissue fluid?

(A) Cold receptors
(B) Pacinian corpuscles
(C) Ruffini endings
(D) Meissner corpuscles

2. Aqueous humor is drained from the eye by passing

(A) into the ciliary processes
(B) from the anterior chamber into the posterior chamber
(C) through the canal of Schlemm
(D) into the vitreous body

3. Which of the following statements is characteristic of the choroid?

(A) It is avascular.
(B) It is the posterior portion of the tunica fibrosa.
(C) It is tightly attached to the sclera.
(D) It contains many melanocytes.

4. Communication of the scala vestibuli and scala tympani occurs at the

(A) round window
(B) oval window
(C) helicotrema
(D) endolymphatic sac

5. The bony ossicles of the middle ear cavity are arranged in a series bridging the tympanic cavity beginning at the tympanic membrane and ending at the

(A) endolymphatic duct
(B) round window
(C) helicotrema
(D) oval window

6. Which of the following cells in the inner ear are involved in detecting movements of the head?

(A) Hair cells in the maculae
(B) Outer pillar cells
(C) Inner pillar cells
(D) Cells of Hensen
(E) Hair cells in the organ of Corti

7. Rods and cones form synapses with which of the following cells?

(A) Amacrine
(B) Bipolar *and Horizontal cells*
(C) Ganglion
(D) Müller

8. Which of the following statements is characteristic of the cornea?

(A) It represents the anterior portion of the tunica vasculosa.
(B) It is the anterior transparent portion of the tunica fibrosa.
(C) It forms the anterior boundary of the posterior chamber of the eye.
(D) It is devoid of nerve endings.

9. A patient exhibiting a high intraocular pressure in both eyes is symptomatic for

(A) cataract
(B) detached retina
(C) glaucoma
(D) conjunctivitis

Answers and Explanations

1-B. Pacinian corpuscles are usually macroscopic in size. Their capsules are composed of several lamellae containing fibroblasts and collagen fibers. The lamellae are separated by sparse amounts of tissue fluid. Pacinian corpuscles respond to vibrations and pressure that distort the lamellae. Meissner corpuscles, responsible for touch, are tapered terminals located at the tips of dermal papillae. Ruffini endings, which are sensitive to mechanical stresses, possess a thin connective tissue capsule that surrounds a fluid-filled space. Cold receptors are naked nerve endings that respond to temperatures below 25°C-30°C.

2-C. Aqueous humor is produced by the ciliary processes and passes into the posterior chamber of the eye, then through the opening of the pupil into the anterior chamber of the eye. The fluid exits the eye by passing into the canal of Schlemm. The vitreous body is a refractile gel that fills the chamber of the globe posterior to the lens and is not related to the aqueous humor.

3-D. The choroid is the vascular tunic of the eye that is loosely adhered to the sclera of the tunica fibrosa. It contains many melanocytes, which impart dark pigment to the eye.

4-C. The scala vestibuli and the scala tympani are actually one perilymphatic space separated by the cochlear duct (scala media). The scala vestibuli and tympani communicate with each other at the helicotrema.

5-D. The bony ossicles of the middle ear cavity articulate in a series from the tympanic membrane to the oval window.

6-A. Neuroepithelial hair cells in the maculae of the saccule and the utricle detect linear movement of the head. These cells are connected to the vestibular portion of the acoustic nerve.

7-B. Rods and cones synapse with bipolar cells and horizontal cells.

8-A. The cornea is the transparent anterior portion of the tunica fibrosa, the outer covering of the eye; thus it forms the anterior wall of the anterior chamber of the eye. It is also rich in sensory nerve endings. The sclera represents the posterior portion of the tunica fibrosa. The tunica vasculosa (middle coat) is composed of the choroid and ciliary body and the iris.

9-C. A patient exhibiting high intraocular pressure is exhibiting the symptom of glaucoma. This condition prohibits the aqueous humor from escaping from the anterior chamber of the eye via the canal of Schlemm. If left untreated, a gradual loss of peripheral sight occurs, and over time blindness results. People with cataract exhibit opacity of the lens that causes vision to be blurred. Conjunctivitis is an inflammation of the conjunctiva of the eye with severe reddening of the sclera and the conjunctival surface of the lids, perhaps with a discharge. This condition may be contagious, and if left untreated, blindness may occur.

Comprehensive Examination

1. Which one of the following statements concerning ribonucleic acid (RNA) synthesis is true?

(A) RNA synthesis does not require deoxyribonucleic acid (DNA) to act as a template.
(B) Syntheses of ribosomal RNA (rRNA), messenger RNA (mRNA), and transfer RNA (tRNA) are all catalyzed by the same RNA polymerase.
(C) To yield messenger ribonucleoproteins (mRNPs), introns are excised whereas exons are spliced together.
(D) Protein moieties are removed from mRNPs within the nucleolus, yielding functional mRNAs to exit via the nuclear pores.

2. Which one of the following factors is primarily responsible for causing osteoporosis in older women?

(A) Decreased bone formation
(B) Lack of physical exercise
(C) Diminished estrogen secretion
(D) Calcium deficiency

3. Which one of the following bases that make up deoxyribonucleic acid (DNA) and ribonucleic acid (RNA) is unique to RNA?

(A) Thymine
(B) Adenine
(C) Cytosine
(D) Guanine
(E) Uracil

4. The centroacinar cells of the pancreas secrete

(A) an alkaline, enzyme-poor fluid
(B) pancreatic digestive enzymes
(C) secretin
(D) cholecystokinin
(E) glucagon

5. The zona fasciculata of the adrenal cortex synthesizes and secretes

(A) mineralocorticoids
(B) glucagon
(C) epinephrine
(D) aldosterone
(E) glucocorticoids

6. Which one of the following statements about bony joints is true?

(A) Long bones are generally united by synarthroses.

(B) Diarthroses are classified as synovial joints.
(C) Type A cells of the synovial membrane secrete synovial fluid.
(D) Type B cells of the synovial membrane are phagocytic.
(E) Synarthroses are usually surrounded by a two-layered capsule.

7. All of the following characteristics can be used to distinguish neutrophils and basophils histologically except one. Which one is the exception?

(A) Size of specific granules
(B) Shape of the nucleus
(C) Number of azurophilic granules
(D) The presence or absence of peroxidase
(E) The presence of mitochondria

8. A long-time user of chewing tobacco noticed several whitish, thickened, painless patches on the lining of his cheeks. The most probable diagnosis is

(A) aphthous ulcers
(B) adenocarcinoma
(C) keloids
(D) oral leukoplakia
(E) epidermolysis bullosa

9. Primordial follicles of the ovary possess

(A) a layer of cuboidal follicular cells
(B) an oocyte arrested in prophase of meiosis I
(C) an oocyte arrested in metaphase of meiosis II
(D) well-defined thecae interna and externa
(E) a thick zona pellucida

10. All of the following statements regarding the membranous labyrinth of the inner ear are true except for one. Which is that exception?

(A) It contains the saccule and utricle.
(B) Maculae contain neuroepithelial cells, which possess numerous stereocilia and a single kinocilium.
(C) Cristae ampullares in the semicircular canals detect angular acceleration of the head.
(D) The otolithic membrane contains small calcified particles.
(E) It contains the vestibular membrane.

11. Which of the following statements concerning euchromatin is true?

(A) It represents about 90% of the chromatin.
(B) It appears as basophilic clumps of nucleoprotein when seen under the light microscope.
(C) It is concentrated near the periphery of the nucleus.
(D) It is transcriptionally active.
(E) It is transcriptionally inactive.

12. Intercalated disks function in which one of the following?

(A) End-to-end attachments of smooth muscle cells
(B) Intercellular movement of large proteins
(C) Ionic coupling of cardiac muscle cells
(D) Storage of Ca^{2+} ions
(E) Release of neurotransmitters

13. Release of thyroid hormones from the follicular cells of the thyroid gland depends on thyroid-stimulating hormone (TSH). TSH stimulation involves

(A) binding of TSH to receptors on the apical plasma membrane
(B) formation of apical microvilli
(C) exocytosis of colloid droplets
(D) change in cell shape from flattened to columnar
(E) secretion of lysosomes from the basal cell surface

14. Which one of the following statements about the development of the tooth crown is true?

(A) The enamel organ is derived from ectomesenchyme (neural crest).
(B) The dental papilla is derived from the epithelium.
(C) The four-layer enamel organ appears during the cap stage.
(D) Dentin and enamel are formed during the appositional stage.
(E) Cementum is formed at the same time as enamel.

15. Which one of the following statements concerning liver sinusoids is true?

(A) Their lining includes Ito cells (fat storing cells).
(B) They receive bile from the hepatocytes.
(C) They are lined by nonfenestrated endothelial cells.
(D) The space of Disse is located between sinusoidal cells and hepatocytes.
(E) Sinusoids convey blood from the central vein to the portal vein.

16. A young college student experienced nausea, vomiting, visual disorders, and muscular paralysis after eating canned tuna fish. The probable diagnosis is botulism, caused by ingestion of the *Clostridium botulinum* toxin. The physiologic effect of this toxin is to

(A) inactivate acetylcholinesterase
(B) bind to, and thus inactivate, acetylcholine receptors at myoneural junctions
(C) prevent release of calcium ions from the sarcoplasmic reticulum, thus inhibiting muscle contraction
(D) inhibit release of acetylcholine from presynaptic membranes
(E) inhibit hydrolysis of adenosine triphosphate (ATP) during the contraction cycle

17. Which of the following statements about nucleosomes is true?

(A) Histones form the nucleosome core around which the double-helix deoxyribonucleic acid (DNA) is wound.
(B) Nucleosomes without histones form the structural unit of the chromosome.
(C) Nucleosomes are composed of ribonucleic acid (RNA) molecules and two copies each of four different histones.
(D) Histone H1 forms the core of the nucleosome.

18. A high-school student complains of fatigue and a sore throat. She has swollen, tender lymph nodes and a fever. Blood test results show an increased white blood cell count with many atypical lymphocytes; the number and appearance of the erythrocytes are normal. This student is likely to be suffering from

(A) AIDS
(B) pernicious anemia
(C) infectious mononucleosis
(D) Hodgkin disease
(E) factor VIII deficiency

19. Which one of the following statements about the gallbladder is true?

(A) The gallbladder dilutes bile.
(B) Bile enters the gallbladder via the common bile duct.
(C) Bile leaves the gallbladder via the cystic duct.
(D) The gallbladder is lined by a simple squamous epithelium.
(E) Secretin stimulates the wall of the gallbladder to contract, forcing bile from its lumen.

20. Which one of the following statements concerning mitochondria is true?

(A) They change from the orthodox to the condensed form in response to the uncoupling of oxidation from phosphorylation.
(B) They are unable to divide.
(C) They possess the enzymes of the Krebs cycle in their cristae.
(D) They contain elementary particles in their matrix.
(E) They do not contain deoxyribonucleic acid (DNA).

21. Which one of the following components is present in muscular arteries but absent from elastic arteries?

(A) Fenestrated membranes
(B) Vasa vasorum
(C) Factor VIII
(D) A thick, complete internal elastic lamina
(E) Smooth muscle cells

22. Which one of the following statements concerning the pancreas is true?

(A) Islets of Langerhans secrete enzymes.
(B) It possesses mucous acinar cells.
(C) The endocrine pancreas has more beta cells than delta cells.
(D) Its alpha cells secrete insulin.
(E) Its delta cells secrete amylase.

23. Which one of the following stimulates the production of hydrochloric acid in the stomach?

(A) Somatostatin
(B) Gastrin
(C) Secretin
(D) Cholecystokinin
(E) Urogastrone

24. A deficiency or an excess of which of the following vitamins results in short stature?

(A) Vitamin A
(B) Vitamin C
(C) Vitamin D
(D) Vitamin K

25. The intercellular spaces in the stratum spinosum of the epidermis contain lipid-containing sheets that are impermeable to water. This material is released from

(A) keratohyalin granules
(B) Langerhans cells
(C) membrane-coating granules
(D) sebaceous glands
(E) melanosomes

26. Which one of the following statements concerning basophilic erythroblasts is true?

(A) The nucleus has a fine chromatin network.
(B) The nucleus is in the process of being extruded.
(C) The cytoplasm contains specific granules.
(D) The cytoplasm is pink and reveals a dense reticulum.
(E) The nucleus is bilobed.

27. The ileum includes which of the following structures?

(A) Rugae
(B) Peyer patches
(C) Brunner glands
(D) Parietal cells
(E) Chief cells

28. A 25-year-old woman complains about a frequently recurring painful lesion on her upper lip that exudes a clear fluid. She is probably suffering from

(A) oral leukoplakia
(B) herpetic stomatitis
(C) aphthous ulcer
(D) bullous pemphigoid

29. Which one of the following statements concerning the functions of the skin is true?

(A) Infrared radiation, necessary for synthesis of vitamin D, is absorbed by the skin.
(B) The skin provides no protection against desiccation.
(C) The skin contains temperature receptors and plays a role in regulating body temperature.
(D) Melanin is synthesized by melanocytes, which are located in the dermis.

30. A person with glomerulonephritis will have which of the following signs or symptoms?

(A) Hypotonic urine
(B) Polyuria
(C) Proteinuria
(D) Dehydration
(E) Polydipsia

31. Stratified squamous keratinized epithelium is always present in the

(A) rectum
(B) esophagus
(C) pyloric stomach
(D) jejunum
(E) anus

32. Which of the following properties is exhibited in all three types of cartilage?

(A) Is involved in bone formation
(B) Possesses type II collagen
(C) Possesses type I collagen
(D) Grows interstitially and appositionally
(E) Has an identifiable perichondrium

33. Which of the following is a receptor for fine touch?

(A) Pacinian corpuscle
(B) Crista ampullaris
(C) Ruffini ending
(D) Krause end-bulb
(E) Meissner corpuscle

34. A premature infant presents with labored breathing, which is eventually alleviated by administration of glucocorticoids. The most probable diagnosis is

(A) immotile cilia syndrome
(B) emphysema
(C) hyaline membrane disease
(D) asthma
(E) chronic bronchitis

35. Which of the following statements concerning the cribriform plate is true?

(A) It is the inner layer of the alveolar bone.
(B) It lacks Sharpey fibers.
(C) It is also known as the spongiosa.
(D) It is composed of cancellous bone.

36. Which one of the following substances is synthesized in the pituitary gland?

(A) Oxytocin
(B) Antidiuretic hormone
(C) Somatotropin
(D) Neurophysin

37. Which one of the following is true in breast cancer cells that involve the BRCA1 gene?

(A) Mutated cells are unable to divide because of increased expression of the BRCA1 gene.
(B) Mutated cells fail to reach the G1-S checkpoint.
(C) Mutated cells have an abnormal number of endosomes.
(D) Mutated cells lose their G2-M cell cycle checkpoint.
(E) Mutated cells become haploid due to the interaction of the BRCA1 and p53 genes.

38-40. A 42-year-old male patient arrives in the emergency room with a rash over much of his face, hands, and arms. He states that he was gardening and pulled out a number of weeds. On questioning the patient, the physi-

cian realized that the patient inadvertently came in contact with poison ivy.

38. Which of the following cells is responsible for the release of the pharmacologic agents that caused the rash?

(A) Diffuse neuroendocrine system (DNES) cells
(B) Myofibroblasts
(C) Mast cells
(D) Pericytes
(E) Plasma cells

39. Which one of the following is a secondary mediator produced by the cell in Question 38?

(A) Histamine
(B) Chondroitin sulfate
(C) Neutral protease
(D) Bradykinin
(E) Aryl sulfatase

40. Which one of the following white blood cells can also participate in the reaction of this patient to poison ivy?

(A) T cells
(B) B cells
(C) Neutrophils
(D) Basophils
(E) Eosinophils

41. A 26-year-old female patient goes to the dentist because of a toothache. On examination the dentist notes that the patient has a carious lesion that involves not only the enamel but also the dentin and the cementum of the tooth. Which of these substances cannot repair itself?

(A) Dentin
(B) Enamel
(C) Cementum
(D) Dentin and cementum
(E) Dentin, cementum, and enamel

42. Lisa was small and thin in stature. She spent her days indoors and rarely ate dairy products. When she became pregnant at age 25, she experienced severe lower back and leg pain and tenderness when pressure was applied over bony regions of her body. Radiographs revealed excessive amounts of poorly mineralized osteoid in both femurs. Beneficial treatment of her condition involved high doses of vitamin D, calcium, and phosphorus, a regimen that continued after successful delivery of her baby. Which of the following describes Lisa's disease?

(A) Osteogenesis imperfecta
(B) Osteoporosis
(C) Osteomalacia
(D) Osteopetrosis
(E) Osteopenia

43. Michael experienced visual problems within 1 day after being hit in the head with a ball during a soccer game. He saw large floaters and noticed that a dark film was blocking part of the vision in his right eye. An examination by an ophthalmologist revealed that he had a detached retina, and emergency surgery was done to save the vision in that eye. Which of the following retinal layers were separated from one another to cause Michael's condition?

(A) Layer 1 from layer 2
(B) Layer 2 from layer 3
(C) Layer 3 from layer 4
(D) Layer 5 from layer 6
(E) Layer 6 from layer 7

44. Blood coagulation involves a cascade of reactions that occur in two interrelated pathways, the extrinsic and intrinsic. All of the following are associated with the intrinsic pathway of blood coagulation EXCEPT

(A) conversion of fibrinogen to fibrin
(B) platelet aggregation
(C) release of tissue thromboplastin
(D) von Willebrand factor
(E) calcium

45. A scientist spent a summer in a remote region of Africa, where he studied exotic plants. On returning to the United States, he developed a cough and fever that would not go away. Laboratory tests revealed that he had contracted a roundworm, *Ascaris lumbricoides*. Which of the following cells would be expected to be significantly elevated in a differential count of his blood?

(A) Erythrocytes
(B) Lymphocytes
(C) Monocytes
(D) Eosinophils
(E) Neutrophils

46. A 45-year-old musician who played guitar for 26 years in a rock band noticed he was having difficulty hearing. An otolaryngologist confirmed his loss of hearing and associated it with prolonged exposure to loud sounds. Which of the following structures would show degenerative changes that would account for this man's hearing loss?

(A) The epithelium lining the inner portion of the tympanic membrane
(B) Hair cells in the ampullae of the semicircular ducts
(C) The auditory (eustachian) tube extending to the middle ear cavity
(D) Hair cells of the organ of Corti
(E) Ossicles in the middle ear

Questions 47-49

The response options for the next three items are the same. Each item will state the number of options to select. Choose exactly this number.

(A) Chondronectin
(B) Plasma fibronectin
(C) Osteonectin
(D) Matrix fibronectin

For each of the following descriptions, select the appropriate glycoprotein.

47. A circulating protein in the blood that functions in wound healing (**SELECT 1 GLYCOPROTEIN**)

48. An adhesive glycoprotein that forms fibrils in the extracellular matrix (**SELECT 1 GLYCOPROTEIN**)

49. A multifunctional protein that attaches chondrocytes to type II collagen (**SELECT 1 GLYCOPROTEIN**)

Questions 50-54

The response options for the next five items are the same. Each item will state the number of options to select. Choose exactly this number.

(A) Pericytes
(B) Macrophages
(C) T lymphocytes
(D) Plasma cells
(E) B lymphocytes
(F) Foreign-body giant cells
(G) Mast cells
(H) Eosinophils

Although fibroblasts are the predominant cells present in connective tissue, many other cell types also are found in connective tissue. For each of the following descriptions, select the appropriate type of cell.

50. Antibody-manufacturing cells (**SELECT 1 TYPE OF CELL**)

51. Principal phagocytes of connective tissue (**SELECT 1 TYPE OF CELL**)

52. Cells that can bind immunoglobulin E antibodies and mediate immediate hypersensitivity reactions (**SELECT 1 TYPE OF CELL**)

53. Cells responsible for initiating cell-mediated immune responses (**SELECT 1 TYPE OF CELL**)

54. Pluripotential cells located primarily along capillaries (**SELECT 1 TYPE OF CELL**)

Questions 55-59

The response options for the next five items are the same. Each item will state the number of options to select. Choose exactly this number.

(A) Phospholipid
(B) Glycocalyx
(C) Carrier protein
(D) Band 3 protein
(E) G protein
(F) K+ leak channel

The plasma membrane is a complex structure that functions in membrane transport processes and cell-to-cell communication. For each of the following descriptions, select the appropriate component of the plasma membrane.

55. Functions in activation of secondary messenger systems (**SELECT 1 COMPONENT**)

56. Is primarily responsible for establishing the potential difference across the plasma membrane (**SELECT 1 COMPONENT**)

57. Is an amphipathic molecule (**SELECT 1 COMPONENT**)

58. Is a carbohydrate-containing covering associated with the outer leaflet of the plasma membrane (**SELECT 1 COMPONENT**)

59. Functions in antiport transport (**SELECT 1 COMPONENT**)

Questions 60-64

The response options for the next five items are the same. Each item will state the number of options to select. Choose exactly this number.

(A) CD4
(B) CD8
(C) Interleukin 1
(D) Interleukin 2
(E) Perforin
(F) Interferon γ

For each description, select the most appropriate molecule associated with the immune response.

60. Is a surface marker on T suppressor cells (**SELECT 1 MOLECULE**)

61. Is a surface marker on T helper cells (**SELECT 1 MOLECULE**)

62. Mediates lysis of tumor cells (**SELECT 1 MOLECULE**)

63. Is released by macrophages and stimulates activated T helper cells (**SELECT 1 MOLECULE**)

64. Stimulates activation of natural killer (NK) cells (**SELECT 1 MOLECULE**)

Questions 65-69

The response options for the next five items are the same. Each item will state the number of options to select. Choose exactly this number.

(A) Sertoli cells
(B) Primary spermatocytes
(C) Secondary spermatocytes
(D) Interstitial cells of Leydig
(E) Spermatids

Match each description with the most appropriate type of cell.

65. Replicate their deoxyribonucleic acid (DNA) during the S phase of the cell cycle and undergo meiosis (**SELECT 1 TYPE OF CELL**)

66. Form a temporary cylinder of microtubules called the manchette (**SELECT 1 TYPE OF CELL**)

67. Possess receptors for luteinizing hormone (LH) and produce testosterone in response to binding of LH (**SELECT 1 TYPE OF CELL**)

68. Are responsible for formation of the blood-testis barrier (**SELECT 1 TYPE OF CELL**)

69. Synthesize androgen-binding protein when stimulated by follicle-stimulating hormone (**SELECT 1 TYPE OF CELL**)

Questions 70-73

The response options for the next four items are the same. Each item will state the number of options to select. Choose exactly this number.

(A) Rough endoplasmic reticulum
(B) Smooth endoplasmic reticulum
(C) Mitochondrion
(D) Annulate lamellae
(E) Lysosomes
(F) Polysomes

Match each description with the most appropriate cytoplasmic organelle.

70. Possesses mixed-function oxidases that detoxify phenobarbital and other drugs (**SELECT 1 CYTOPLASMIC ORGANELLE**)

71. Contains ribophorins (**SELECT 1 CYTOPLASMIC ORGANELLE**)

72. Are parallel stacks of membranes that resemble the nuclear envelope (**SELECT 1 CYTOPLASMIC ORGANELLE**)

73. Are the sites where hemoglobin and other cytosolic proteins are synthesized (**SELECT 1 CYTOPLASMIC ORGANELLE**)

Questions 74-78

The response options for the next five items are the same. Each item will state the number of options to select. Choose exactly this number.

(A) Gastrin
(B) Somatostatin
(C) Motilin
(D) Pepsinogen
(E) Lysozyme
(F) Urogastrone

Various enzymes and hormones are secreted by the cells lining the digestive tract. Match each physiologic activity with the most appropriate hormone or enzyme.

74. Stimulates secretion of pepsinogen (**SELECT 1 HORMONE OR ENZYME**)

75. Is produced by Brunner glands and inhibits HCl secretion by parietal cells (**SELECT 1 HORMONE OR ENZYME**)

76. Functions as an antibacterial agent (**SELECT 1 HORMONE OR ENZYME**)

77. Stimulates contraction of smooth muscle in the wall of the digestive tract (**SELECT 1 HORMONE OR ENZYME**)

78. Inhibits secretion by nearby enteroendocrine cells (**SELECT 1 HORMONE OR ENZYME**)

Questions 79-80

The response options for the next two items are the same. Each item will state the number of options to select. Choose exactly this number.

(A) Lysine
(B) Desmosine
(C) Hydroxyproline
(D) Arginine

Match each description with the appropriate amino acid.

79. Is partly responsible for the elasticity of elastin (**SELECT 1 AMINO ACID**)

80. Cross-links elastin molecules to form an extensive network (**SELECT 1 AMINO ACID**)

Questions 81-82

The response options for the next two items are the same. Each item will state the number of options to select. Choose exactly this number.

(A) Simple goiter
(B) Exophthalmic goiter
(C) Graves disease
(D) Addison disease

Match each characteristic with the condition that it most often causes.

81. Inadequate amounts of iodine in the diet (**SELECT 1 CONDITION**)

82. Destruction of the adrenal cortex (**SELECT 1 CONDITION**)

Questions 83-85

The response options for the next three items are the same. Each item will state the number of options to select. Choose exactly this number.

(A) Troponin C
(B) Globular head (Sl fragment) of myosin
(C) Myoglobin
(D) Actin

For each function, select the muscle protein that performs it.

83. Hydrolyzes adenosine triphosphate (ATP) (**SELECT 1 MUSCLE PROTEIN**)

84. Binds oxygen (**SELECT 1 MUSCLE PROTEIN**)

85. Binds calcium ions (**SELECT 1 MUSCLE PROTEIN**)

Questions 86-87

The response options for the next two items are the same. Each item will state the number of options to select. Choose exactly this number.

(A) Epidermolysis bullosa
(B) Basal cell carcinoma
(C) Malignant melanoma

Match each description with the disease to which it applies.

86. A rare form of skin cancer that may be fatal (**SELECT 1 DISEASE**)

87. A hereditary skin disease characterized by blister formation after minor trauma (**SELECT 1 DISEASE**)

Questions 88-91

The response options for the next three items are the same. Each item will state the number of options to select. Choose exactly this number.

(A) Trachea
(B) Nasopharynx
(C) Terminal bronchiole
(D) Alveolar duct
(E) Intrapulmonary bronchi

Match each description with the most appropriate structure of the respiratory system.

88. Is the region of concern in patients with adenoids (**SELECT 1 STRUCTURE**)

89. Possesses C-shaped rings of hyaline cartilage (**SELECT 1 STRUCTURE**)

90. Contains smooth muscle at openings into alveoli (**SELECT 1 STRUCTURE**)

91. Is lined by an epithelium containing ciliated cells and Clara cells (**SELECT 1 STRUCTURE**)

Questions 92-95

The response options for the next four items are the same. Each item will state the number of options to select. Choose exactly this number.

(A) Endometriosis
(B) Cervical cancer
(C) Ectopic tubal pregnancy
(D) Breast cancer

For each characteristic, select the condition most closely associated with it.

92. Is associated with rupture of the oviduct and hemorrhaging into the peritoneal cavity (**SELECT 1 CONDITION**)

93. May be classified as lobular carcinoma (**SELECT 1 CONDITION**)

94. May be detected by a Papanicolaou smear (**SELECT 1 CONDITION**)

95. Commonly results in hemorrhaging into the peritoneal cavity dependent on the stage of the menstrual cycle (**SELECT 1 CONDITION**)

Questions 96-100

The response options for the next five items are the same. Each item will state the number of options to select. Choose exactly this number.

(A) Pepsin
(B) Enzyme associated with the glycocalyx of the intestinal striated border
(C) Lipase
(D) Chylomicron
(E) Gastric intrinsic factor

Match each process with the substance associated with it.

96. Absorption of vitamin B_{12} (**SELECT 1 SUBSTANCE**)

97. Digestion of carbohydrates (**SELECT 1 SUBSTANCE**)

98. Digestion of proteins (**SELECT 1 SUBSTANCE**)

99. Transport of triglycerides into lacteals (**SELECT 1 SUBSTANCE**)

100. Manufactured and released by parietal cells (**SELECT 1 SUBSTANCE**)

Answers and Explanations

1-C. To yield messenger ribonucleoproteins (mRNPs), introns are excised, whereas exons are spliced together. Deoxyribonucleic acid (DNA) does act as the template for the synthesis of ribonucleic acid (RNA). Three RNA polymerases (I, II, and III) are needed to synthesize ribosomal RNA (rRNA), messenger RNA (mRNA), and transfer RNA (tRNA), respectively. Protein moieties are removed from the mRNPs as they leave the nucleus to yield functional mRNAs outside the nucleus.

2-C. The most common cause of osteoporosis in older women is diminished estrogen secretion.

3-E. Although adenine, cytosine, and guanine are found in both deoxyribonucleic acid (DNA) and ribonucleic acid (RNA), uracil is found only in RNA. Uracil substitutes for the base thymine in DNA.

4-A. Pancreatic centroacinar cells form the initial segment of the intercalated duct and are part of the exocrine pancreas. They secrete an alkaline, enzyme-poor fluid when stimulated by secretin. Pancreatic digestive enzymes are synthesized by the acinar cells of the exocrine pancreas; their release is stimulated by cholecystokinin. Glucagon is produced in the endocrine pancreas (islets of Langerhans).

5-E. The zona fasciculata, the largest region of the adrenal cortex, produces glucocorticoids (cortisol and corticosterone). The zona glomerulosa produces mineralocorticoids, primarily aldosterone. Epinephrine is produced in the adrenal medulla. Glucagon is produced in the pancreas, not in the adrenal gland.

6-B. Diarthroses, the type of joint connecting two long bones, are classified as synovial joints, which are surrounded by a two-layered capsule housing a synovial membrane. Type A cells of the synovial membrane are phagocytic, whereas type B cells secrete the synovial fluid. Synarthroses joints are those found joining the bones of the skull, which are immovable.

7-E. Neutrophils have a nucleus with three or four lobes, many azurophilic granules, and small specific granules that lack peroxidase. In contrast, basophils have an S-shaped nucleus, few azurophilic granules, and large specific granules that contain peroxidase. Both neutrophils and basophils possess mitochondria.

8-D. Oral leukoplakia, which results from epithelial hyperkeratosis, is usually of unknown etiology but often is associated with the use of chewing tobacco. Although the characteristic painless lesions are benign, they may transform into squamous cell carcinoma. Aphthous ulcers are painful lesions of the oral mucosa that are surrounded by a red border. Adenocarcinoma is a form of cancer arising in glandular tissue. Keloids are swellings in the skin that arise from increased collagen formation in hyperplastic scar tissue. Epidermolysis bullosa is a group of hereditary skin diseases characterized by blister formation after minor trauma.

9-B. A primordial follicle is composed of a flattened layer of follicular cells surrounding a primary oocyte, which is arrested in prophase of meiosis I. Well-defined thecal layers and a thick zona pellucida are found in growing follicles. A graafian (mature) follicle possesses a secondary oocyte, which becomes arrested in metaphase of meiosis II just before ovulation.

10-E. The first four statements are true. Linear acceleration of the head is detected by the neuroepithelial hair cells of the maculae, which are specialized regions of the saccule and utricle. However, the vestibular membrane is a part of the organ of Corti.

11-D. Euchromatin, the transcriptionally active form of chromatin, represents only about 10% of the chromatin. In the light microscope, it appears as a light-staining, dispersed region of the nucleus.

12-C. Large protein molecules cannot move across intercalated disks (the step-like junctional complexes present in cardiac, not smooth, muscle). These junctional structures possess three specializations: desmosomes, which provide end-to-end attachment of cardiac muscle cells; fascia adherentes, to which the thin myofilaments attach; and gap junctions, which permit intercellular movement of small molecules and ions (ionic coupling).

13-D. Flattened, squamous cells are characteristic of an unstimulated, inactive thyroid gland. Thyroid-stimulating hormone (TSH) binds to G-protein-linked receptors on the basal surface of follicular cells. Under TSH stimulation, thyroid follicular cells become columnar and form pseudopods, which engulf colloid. Lysosomal enzymes split thyroxine (T_4) and triiodothyronine (T_3) from thyroglobulin; the hormones are then released basally.

14-D. The enamel organ is epithelially derived, whereas the dental papilla comes from ectomesenchyme. The bell, not the cap, stage of odontogenesis is characterized by possessing a fourth layer in its enamel organ. Formation of dentin and enamel occurs during the appositional stage of tooth development. Cementum is located on the root and is formed only after the crown is completed and enamel ceases to be elaborated.

15-D. Liver sinusoids convey blood to the central vein. Their endothelial cells are fenestrated, and material from the sinusoids may enter the space of Disse through the fenestrae, where it may be endocytosed by hepatocytes. The space of Disse houses Ito cells (fat-storing cells). Because bile is the exocrine secretion of hepatocytes, it does not enter the sinusoids.

16-D. The toxin from *Clostridium botulinum* inhibits the release of acetylcholine, the neurotransmitter at myoneural junctions. As a result, motor nerve impulses cannot be transmitted across the junction and muscle cells are not stimulated to contract.

17-A. The nucleosome, the structural unit of chromatin packing, does not contain ribonucleic acid (RNA). In extended chromatin, two copies each of histones H2A, H2B, H3, and H4 form the nucleosome core around which a deoxyribonucleic acid (DNA) molecule is wound. Condensed chromatin contains additional histones (H1), which bind to nucleosomes, forming the condensed 30-nm chromatin fiber.

18-C. Only infectious mononucleosis is characterized by all of the signs and symptoms indicated. AIDS is associated with a decreased lymphocyte count (particularly of T helper cells). Pernicious anemia is associated with a decreased red blood cell count. Hodgkin disease is associated with fatigue and enlarged lymph nodes, but the nodes are not painful and the presence of Reed-Sternberg cells is diagnostic of this disease. Factor VIII deficiency, a coagulation disorder, is not associated with any of these signs and symptoms.

19-C. The gallbladder, which concentrates and stores bile, is lined by a simple columnar epithelium. Cholecystokinin stimulates contraction of the gallbladder wall, forcing bile from the lumen into the cystic duct; this joins the common hepatic duct to form the common bile duct, which delivers bile to the duodenum.

20-A. Uncoupling of oxidation from phosphorylation induces mitochondria to change from the orthodox to the condensed form. Condensed mitochondria are often present in brown fat cells, which produce heat rather than adenosine triphosphate. Mitochondria possess circular deoxyribonucleic acid (DNA), and they reproduce (divide) by fission.

21-D. Muscular (distributing) arteries have a thick, complete internal elastic lamina in the tunica intima, whereas elastic (conducting) arteries have an incomplete internal elastic lamina. Both types of arteries have vasa vasorum, factor VIII, and smooth muscle cells in their walls. Muscular arteries possess numerous layers of muscle cells in the tunica media, but elastic arteries do not. Only elastic arteries possess fenestrated (elastic) membranes in the tunica media, in which smooth muscle cells are dispersed.

22-C. In the endocrine pancreas, beta cells account for about 70% of the secretory cells; alpha cells, about 20%; and delta cells, less than 5%. Polypeptide hormones are synthesized by and released from the islets of Langerhans (endocrine pancreas). The exocrine pancreas possesses serous (not mucous) acinar cells. Insulin is produced by beta cells.

23-B. Somatostatin and urogastrone both inhibit the production of HCl, whereas gastrin enhances it. Secretin and cholecystokinin act on the pancreas to facilitate secretion of buffer and pancreatic enzymes, respectively.

24-A. A deficiency of vitamin A inhibits bone formation and growth, whereas an excess stimulates ossification of the epiphyseal plates, thus leading to premature closure of the plates. Both conditions result in short stature. A deficiency of vitamin D reduces calcium absorption from the small intestine and results in soft bones, whereas an excess of vitamin D stimulates bone resorption. A deficiency of vitamin C results in poor bone growth and fracture repair. Vitamin K plays no role in bone formation.

25-C. Membrane-coating granules are present in keratinocytes in the stratum spinosum (and stratum granulosum). The contents of these granules are released into the intercellular spaces to help "waterproof" the skin. Keratinocytes in the stratum granulosum also possess keratohyalin granules; these contain proteins that bind keratin filaments together.

26-A. The nucleus of erythroblasts is not in the process of being extruded and is round with a very fine chromatin network. The cytoplasm is blue and possesses no granules.

27-B. The ileum includes Peyer patches. Rugae, parietal cells, and chief cells are located in the stomach. Brunner glands are present in the submucosa of the duodenum.

28-B. Herpetic stomatitis is characterized by painful fever blisters on the lips or near the nostrils. These blisters exude a clear fluid. Aphthous ulcers (canker sores) do not exude fluid. Bullous pemphigoid is an autoimmune disease marked by chronic, generalized blisters in the skin.

29-C. The skin, which consists of the epidermis and dermis, is important in the regulation of body temperature and contains temperature receptors in the dermis. Ultraviolet (not infrared) radiation absorbed by the skin is necessary for the synthesis of vitamin D. Protection against desiccation is provided by the contents of the membrane-coating granules of the epidermis. Melanocytes are located in the deepest layer of the epidermis (stratum basale).

30-C. Patients who suffer from glomerulonephritis excrete protein in their urine. All of the other symptoms are characteristic of patients with diabetes insipidus, who are incapable of producing adequate amounts of antidiuretic hormone and therefore suffer from polyuria (large volume of hypotonic urine production), polydipsia, and dehydration.

31-E. The anus is lined by stratified squamous keratinized epithelium. The rectum, jejunum, and pyloric stomach are lined by simple columnar epithelium. The esophagus is lined by stratified squamous (nonkeratinized) epithelium.

32-D. Hyaline cartilage, elastic cartilage, and fibrocartilage all exhibit both interstitial and appositional growth. Hyaline and elastic cartilage have type II collagen in their matrix and are surrounded by a perichondrium, whereas fibrocartilage has type I collagen and lacks an identifiable perichondrium. Only hyaline cartilage is involved in endochondral bone formation.

33-E. Meissner corpuscles, located in the papillary layer of the dermis, are fine-touch receptors. Pacinian corpuscles perceive pressure, touch, and vibration; they are located in the dermis, hypodermis, and connective tissue of mesenteries and joints. Cristae ampullares are special regions of the semicircular canals that detect circular movements of the head. Ruffini endings, located in the dermis and joints, function in pressure and touch perception. Krause end-bulbs are cold and pressure receptors located in the dermis.

34-C. Hyaline membrane disease, which results from inadequate amounts of pulmonary surfactant, is characterized by labored breathing and typically is observed in premature infants. Glucocorticoids stimulate synthesis of surfactant and can correct the condition.

35-A. The cribriform plate, the inner layer of the alveolar bone, is composed of compact bone. It is attached to the principal fiber groups of the periodontal ligament via Sharpey fibers. The outer layer of the alveolar bone is the cortical plate. The spongiosa is the region of cancellous (spongy) bone enclosed between the cortical and cribriform plates.

36-C. Somatotropin (growth hormone) is synthesized by cells called somatotrophs, which are acidophils located in the pars distalis of the anterior lobe of the pituitary gland. Oxytocin and antidiuretic hormone (also called vasopressin) are produced in the hypothalamus and transported to the pars nervosa of the pituitary. Neurophysin, a binding protein, aids in this transport.

37-D. Mutations in the BRCA1 gene, a breast tumor suppressor gene, are the major cause of breast cancer. In most breast cancers involving this gene deoxyribonucleic acid (DNA) synthesis proceeds, indicating that the G1-S checkpoint is not affected; however, the mutated cells cannot control the transition between G2 and M. Moreover, the mutated cells had abnormal centrosome numbers and their nuclear division did not proceed normally, leading to aneuploidy and genetic instability of the daughter cells.

38-C. The patient is suffering from an immediate (type I) hypersensitivity reaction. The cells responsible for releasing the primary and secondary mediators are the mast cells.

39-D. Bradykinins are the only pharmacologic agents listed in this question that are produced via the arachidonic acid pathway. All of the others are primary mediators because they are stored in the storage granules of mast cells.

40-D. Basophils are very similar in function to mast cells. They also participate in the immediate (type I) hypersensitivity reaction.

41-B. The only hard tissue of the tooth that cannot be repaired by the body is enamel because ameloblasts, the cells that manufacture enamel, are eliminated as the tooth emerges into the oral cavity.

42-C. Osteomalacia, or adult rickets, is Lisa's disorder. It is characterized by a failure of newly formed osteoid to calcify because there is a lack of calcium, vitamin D, and phosphorus. The bones gradually soften and bend, and pain accompanies the condition, which often becomes severe during pregnancy as the fetus removes calcium from the mother's body. Osteogenesis imperfecta is a genetic condition affecting the synthesis of type I collagen in the bone matrix and results in extreme bone fragility and breakage. Osteoporosis is a disease characterized by low bone mineral density and structural deterioration, making bone more susceptible to fracture, and osteopenia is reduced bone mass caused by inadequate osteoid synthesis. Osteopetrosis is the excessive formation of bone, which obliterates the marrow cavities and thus impairs the formation of blood cells.

43-B. The pigmented epithelium (layer 2) was separated from the layer of rods and cones (layer 3), which make up the light-sensitive part of the neural retina.

44-C. The extrinsic pathway is initiated by the release of tissue thromboplastin after trauma to extravascular tissue. Platelet aggregation is promoted by von Willebrand factor, which is associated with the intrinsic pathway only. Calcium is required in both pathways, and the final reaction—the conversion of fibrinogen to fibrin—is the same in both.

45-D. Eosinophils are increased in parasitic infections and allergic reactions. Both eosinophils and basophils have receptors for immunoglobulin E, which seems to be important in the destruction of parasites. Both neutrophils and monocytes lack immunoglobulin E receptors.

46-D. The inner ear is where sound waves are transduced into nerve impulses that convey auditory information to the brain, and the hair cells of the organ of Corti play a key role in this process. Sound waves are initially received by the outer ear and transmitted via the tympanic membrane (eardrum) to the middle ear, where ossicles transmit the vibrations to the inner ear. The inner ear has an auditory system for hearing (the organ of Corti) and a vestibular system of semicircular ducts that control equilibrium and spatial orientation.

47-B. Plasma fibronectin functions in wound healing, blood clotting, and phagocytosis of material from the blood.

48-D. Matrix fibronectin mediates cell adhesion to the extracellular matrix by binding to fibronectin receptors on the plasma membrane.

49-A. Chondronectin has binding sites for collagen, proteoglycans, and cell-surface receptors.

50-D. Plasma cells, which arise from antigen-activated B lymphocytes, produce antibodies and thus are directly responsible for humoral-mediated immunity.

51-B. Macrophages, the principal phagocytes of connective tissue, remove large particulate matter and assist in the immune response by acting as antigen-presenting cells.

52-G. Mast cells (and basophils) have receptors for immunoglobulin E antibodies on their surface. These cells release histamine, heparin, leukotriene C (slow-reacting substance of anaphylaxis), and eosinophil chemotactic factor, which has effects that constitute immediate hypersensitivity reactions.

53-C. T lymphocytes initiate cell-mediated immune responses.

54-A. Pericytes are smaller than fibroblasts and are located along capillaries. When necessary, they assume the pluripotential role of embryonic mesenchymal cells.

55-E. G proteins are membrane proteins that are linked to certain cell-surface receptors. On binding of a signaling molecule to the receptor, the G protein functions as a signal transducer by activating a secondary messenger system that leads to a cellular response.

56-F. K^+ leak channels are ion channels that are responsible for establishing a potential difference across the plasma membrane.

57-A. The term amphipathic refers to molecules, such as phospholipids, that possess both hydrophobic (nonpolar) and hydrophilic (polar) properties. The plasma membrane contains two phospholipid layers (leaflets) with the hydrophobic tails of the molecules projecting into the interior of the membrane and the hydrophilic heads facing outward.

58-B. The glycocalyx (cell coat) is associated with the outer leaflet of the plasma membrane. It is composed primarily of proteoglycans, which possess polysaccharide side chains.

59-C. Membrane carrier proteins are highly folded transmembrane proteins that undergo reversible conformational alterations, resulting in transport of specific molecules across the membrane. The Na^+-K^+ pump is a carrier protein that mediates antiport transport, which is the transport of two molecules concurrently in opposite directions.

60-B. T suppressor cells and T cytotoxic cells have CD8 marker molecules on their surfaces.

61-A. T helper cells have CD4 marker molecules on their surfaces.

62-E. Perforin, which is released by cytotoxic T cells, mediates lysis of tumor cells and virus-infected cells.

63-C. Interleukin 1, which is produced by macrophages, stimulates activated T helper cells. In turn, activated T helper cells produce interleukin 2 and other cytokines involved in the immune response.

64-F. Interferon γ (macrophage-activating factor) stimulates activation of natural killer (NK) cells and macrophages, thereby increasing their cytotoxic and/or phagocytic activity.

65-B. Primary spermatocytes undergo the first meiotic division following deoxyribonucleic acid (DNA) replication in the S phase. The resulting secondary spermatocytes undergo the second meiotic division, without an intervening S phase, forming spermatids.

66-E. During spermiogenesis, the manchette is formed. This temporary structure aids in elongation of the spermatid.

67-D. Interstitial cells of Leydig produce testosterone when they are stimulated by luteinizing hormone.

68-A. Sertoli cells are columnar cells that extend from the basal lamina to the lumen of the seminiferous tubules. Adjacent Sertoli cells form basal tight junctions, which are responsible for the blood-testis barrier, thus protecting the developing sperm cells from autoimmune reactions.

69-A. Sertoli cells also produce androgen-binding protein, which binds testosterone and maintains it at a high level in the seminiferous tubules.

70-B. Smooth endoplasmic reticulum possesses mixed-function oxidases that detoxify phenobarbital and certain other drugs.

71-A. The membrane of the rough endoplasmic reticulum contains ribophorins, which are receptors that bind the large ribosome subunit.

72-D. Annulate lamellae are parallel stacks of membranes that resemble the nuclear envelope. They are present in the cytoplasm of rapidly growing cells, but their function remains obscure.

73-F. Cytosolic proteins are synthesized on polysomes (polyribosomes). In contrast, secretory proteins are synthesized at the rough endoplasmic reticulum.

74-A. Gastrin, a paracrine hormone secreted in the pylorus and duodenum, stimulates pepsinogen secretion by chief cells in the gastric glands.

75-F. Urogastrone, produced by Brunner glands in the duodenum, inhibits gastric HCl secretion and enhances division of epithelial cells.

76-E. Lysozyme, manufactured by Paneth cells in the crypts of Lieberkühn, is an enzyme that has antibacterial activity.

77-C. Motilin, a paracrine hormone secreted by cells in the small intestine, increases gut motility by stimulating smooth muscle contraction.

78-B. Somatostatin, produced by enteroendocrine cells in the pylorus and duodenum, inhibits secretion by nearby enteroendocrine cells.

79-B. Four lysine molecules in different chains form desmosine links, which are responsible for elastin's elasticity.

80-A. Lysine cross-links elastin molecules, forming a network. Fibrillin is a glycoprotein that organizes elastin into fibers.

81-A. Simple goiter is an enlargement of the thyroid gland resulting from inadequate dietary iodine (less than 10 µg/d). It is common where the food supply is low in iodine.

82-D. Addison disease is most commonly caused by an autoimmunity that destroys the adrenal cortex. As a result, inadequate amounts of glucocorticoids and mineralocorticoids are produced. Unless these are replaced by steroid therapy, the disease is fatal.

83-B. The globular head of the myosin molecule has adenosine triphosphatase (ATPase) activity, but interaction with actin is required for the noncovalently bound reaction products [adenosine diphosphate (ADP) and P_i] to be released. This ATPase activity is retained by the S1 fragment resulting from digestion of myosin with proteases.

84-C. Myoglobin is a sarcoplasmic protein that, like hemoglobin, can bind and store oxygen. The myoglobin content of red (slow) muscle fibers is higher than that of white (fast) muscle fibers.

85-A. Troponin C is one of the three subunits of troponin, which along with tropomyosin binds to actin (thin) filaments in skeletal muscle. Binding of calcium ions by troponin C results in unmasking of the myosin-binding sites on thin filaments.

86-C. Malignant melanoma, a relatively rare form of skin cancer, arises from melanocytes. It is aggressive and invasive. Surgery and chemotherapy usually are necessary for successful treatment of this cancer.

87-A. Epidermolysis bullosa is a group of hereditary skin diseases characterized by the separation of the layers in skin with consequent blister formation.

88-B. The nasopharynx is the site of the pharyngeal tonsil; when enlarged and infected, this tonsil is known as adenoids.

89-A. The trachea and extrapulmonary (primary) bronchi have walls supported by C-shaped hyaline cartilages (C-rings), whose open ends face posteriorly.

90-D. The alveolar duct has alveoli with openings that are rimmed by sphincters of smooth muscle. Alveoli more distal than these have only elastic and reticular fibers in their walls.

91-C. Terminal bronchioles are lined by a simple cuboidal epithelium containing ciliated cells and Clara cells. Clara cells can divide and regenerate both cell types.

92-C. An ectopic tubal pregnancy occurs when the embryo implants in the wall of the oviduct (rather than in the uterus). Because the oviduct cannot support the developing embryo, the duct eventually bursts, causing hemorrhaging into the peritoneal cavity.

93-D. Breast cancer that originates from the epithelium lining the terminal ductules of the mammary gland is classified as lobular carcinoma.

94-B. Abnormal cells associated with cervical cancer are revealed in a Papanicolaou smear, providing a simple method for the early detection of this cancer.

95-A. Endometriosis is a condition in which uterine endometrial tissue is located in the pelvic peritoneal cavity. The misplaced endometrial tissue undergoes cyclic hormone-induced changes, including menstrual breakdown and bleeding.

96-E. Gastric intrinsic factor, which is produced by parietal cells in the gastric glands, is necessary for absorption of vitamin B_{12} in the ileum.

97-B. Disaccharidases located in the glycocalyx of the striated border hydrolyze disaccharides to monosaccharides.

98-A. Digestion of proteins begins with the action of pepsin in the stomach, forming a mixture of polypeptides. Activation of pepsinogen to pepsin only occurs at a low pH.

99-D. After free fatty acids and monoglycerides, in micelles, enter the surface absorptive cells of the small intestine, they are re-esterified to form triglycerides. These are complexed with proteins, forming chylomicrons, which are released from the lateral cell membrane and enter lacteals in the lamina propria.

100-E. Parietal cells are responsible for establishing the low pH of the stomach by manufacturing hydrochloric acid. Another function of parietal cells is the synthesis and release of gastric intrinsic factor, necessary for the absorption of vitamin B_{12} in the ileum.

Index

Page numbers in *italics* denote figures; those followed by a *t* denote tables.